DEVELOPMENTS IN POLYMER STABILISATION—6

CONTENTS OF VOLUMES 3–5

DEVELOPMENTS IN POLYMER STABILISATION—6

Edited by

GERALD SCOTT

M.A., M.Sc., F.R.S.C., F.P.R.I.

Professor of Polymer Science and Technology
University of Aston in Birmingham, Birmingham, UK

APPLIED SCIENCE PUBLISHERS
LONDON and NEW YORK

APPLIED SCIENCE PUBLISHERS LTD
Ripple Road, Barking, Essex, England

Sole Distributor in the USA and Canada
ELSEVIER SCIENCE PUBLISHING CO., INC.
52 Vanderbilt Avenue, New York, NY 10017, USA

British Library Cataloguing in Publication Data
Developments in polymer stabilisation.—6—
 (The Developments series)
 1. Polymers and polymerization—Deterioration
 —Periodicals 2. Stabilizing agents—
 Periodicals
 I. Series
 668.9   TP1087

ISBN 0-85334-168-0

WITH 57 TABLES AND 150 ILLUSTRATIONS

© APPLIED SCIENCE PUBLISHERS LTD 1983

Filmset and printed in Northern Ireland at The Universities Press (Belfast) Ltd.

PREFACE

The purpose of the present series of publications is two-fold. In the first place it is intended to review progress in the development of practical stabilising systems for a wide range of polymers and applications. A complementary and ultimately more important objective is to accommodate these practical developments within the framework of antioxidant theory, since there can be little question that further major advances in the practice of stabilisation technology will only be possible on the basis of a firm mechanistic foundation.

The discovery of a new theory in the well-established field of antioxidant mechanisms is rare but the authors of Chapter 1, Berger, Bolsman and Brouwer, must be credited with such a discovery. Although it has been recognised in principle that under the appropriate conditions chain-breaking electron donor and chain-breaking electron acceptor antioxidants may function together in a complementary way, Bolsman and his co-workers indicate the relevance of this mechanism in real systems and demonstrate the reality of antioxidants acting catalytically. Chapters 2 and 3 outline the increasing importance in polymer stabilisation of sulphur antioxidants acting by a peroxidolytic mechanism. The chemistry of their complex transformation to give sulphur acids is summarised and the reasons for the difference in behaviour of the alkyl sulphides which are not UV stabilisers (Scott, Chapter 2) and the metal thiolate antioxidants which generally are (Al-Malaika, Chakraborty and Scott, Chapter 3) are clarified. The mechanisms of UV stabilisation are treated systematically for a range of UV stabilisers acting by different mechanisms in Chapter 4 (Tudos, Bálint and Kelen). Although some of the data are in conflict with

v

earlier UV stabilisation mechanisms, the explanations proposed should provide a starting point for further definitive studies.

The last three chapters provide a valuable insight into the extensive investigations in progress in the USSR into polymer stabilisation mechanisms. Minsker and co-authors (Chapter 5) review the work which has led to the identification of defects in PVC which are the subsequent cause of further degradation and the role of PVC stabilisation is reviewed in this context. The equally important topic of the attack of ozone on polymers and the protective role of antioxidants is discussed by Razumovskii and Zaikov in Chapter 6. Finally, the chemistry of polymer oxidation at very high temperatures which has previously been rather neglected is examined from a fundamental point of view by Gladyshev and Vasnetsova in Chapter 7. The interesting conclusion that emerges is that under these very severe conditions, the scavenging of oxygen itself is frequently more important than inhibition of the radical chain reaction. This represents a return to one of the earliest (pre-1920) theories of antioxidant action, which was popular before the radical chain theory of autoxidation was developed by Bäckström. Significantly the next theory to be developed in the 1920s, that of peroxide decomposition, was also discarded with the establishment of the radical chain theory and was not given its rightful place until Denison and his co-workers revived it in 1944. These examples should be a salutory reminder that no theory should be lightly discarded because it does not accord with current fashion.

GERALD SCOTT

CONTENTS

LIST OF CONTRIBUTORS

M. I. ABDULLIN
  Bashkir State University, Ufa, USSR.

SAHAR AL-MALAIKA
  Department of Chemistry, University of Aston in Birmingham, Gosta Green, Birmingham B4 7ET, UK.

G. BÁLINT
  Central Research Institute for Chemistry of the Hungarian Academy of Sciences, H-1025 Budapest, Pusztaszeri ut 59–67, Hungary.

H. BERGER
  Koninklijke/Shell-Laboratorium, Shell Research BV, Postbus 3003, 1003 AA Amsterdam, The Netherlands.

T. A. B. M. BOLSMAN
  Koninklijke/Shell-Laboratorium, Shell Research BV, Postbus 3003, 1003 AA Amsterdam, The Netherlands

D. M. BROUWER
  Koninklijke/Shell-Laboratorium, Shell Research BV, Postbus 3003, 1003 AA Amsterdam, The Netherlands.

KHIRUD B. CHAKRABORTY
  Department of Chemistry, University of Aston in Birmingham, Gosta Green, Birmingham B4 7ET, UK.

G. P. GLADYSHEV
Institute of Chemical Physics, Academy of Sciences of the USSR, Vorobyevskoye Chausee 2b, 117334 Moscow, USSR.

T. KELEN
Central Research Institute for Chemistry of the Hungarian Academy of Sciences, H-1025 Budapest, Pusztaszeri ut 59–67, Hungary.

S. V. KOLESOV
Bashkir State University, Ufa, USSR.

K. S. MINSKER
Bashkir State University, Ufa, USSR.

S. D. RAZUMOVSKII
Institute of Chemical Physics, Academy of Sciences of the USSR, Vorobyevskoye Chausee 2b, 117334 Moscow, USSR.

GERALD SCOTT
Department of Chemistry, University of Aston in Birmingham, Gosta Green, Birmingham B4 7ET, UK.

F. TUDOS
Central Research Institute for Chemistry of the Hungarian Academy of Sciences, H-1025 Budapest, Pusztaszeri ut 59–67, Hungary.

O. A. VASNETSOVA
Institute of Chemical Physics, Academy of Sciences of the USSR, Vorobyevskoye Chausee 2b, 117334 Moscow, USSR.

G. E. ZAIKOV
Institute of Chemical Physics, Academy of Sciences of the USSR, Vorobyevskoye Chausee 2b, 117334 Moscow, USSR.

Chapter 1

CATALYTIC INHIBITION OF HYDROCARBON AUTOXIDATION BY SECONDARY AMINES AND NITROXYLS

H. BERGER, T. A. B. M. BOLSMAN and D. M. BROUWER

Koninklijke/Shell-Laboratorium (Shell Research BV), Amsterdam, The Netherlands

SUMMARY

Secondary amines and derivatives thereof (hydroxylamines and nitroxyls) act as catalytic radical scavengers in autoxidising hydrocarbons at 130°C. Certain dialkylamines are much more efficient than diarylamines due to their better resistance towards side-reactions. Some of the dialkylamines are capable of scavenging several hundreds of propagating radicals per molecule of inhibitor in autoxidising alkanes and a few thousand in autoxidising n-hexylbenzene.

A reaction cycle is discussed in which alkyl and alkylperoxyl radicals are scavenged alternately. In this cycle, alkyl radicals are scavenged by a nitroxyl with formation of an O-alkylhydroxylamine, which undergoes thermolysis to give the corresponding hydroxylamine and olefin. The former scavenges the alkylperoxyl radicals with simultaneous regeneration of a nitroxyl.

Kinetic data on the individual steps of this cycle show that it accounts for the observed inhibiting effect for tertiary, but not for secondary alkyl(peroxyl) radicals.

This chapter describes in detail the kinetics of partially inhibited autoxidation at high temperature, from which a method for the determination of the stoichiometric inhibition coefficient (f) can be derived. It also includes a brief description of the experimental procedure for the determination of f.

1

1. INTRODUCTION

The inhibition of autoxidation has been studied thoroughly for large numbers of inhibitors and substrates including high molecular weight materials.[1] Detailed mechanisms have been established for the inhibition process, particularly in the case of hindered phenols.[2] These inhibitors react stoichiometrically and intercept a maximum of two peroxyl radicals for each phenolic functional group by the reactions shown in Schemes 1 and 2.

SCHEME 1

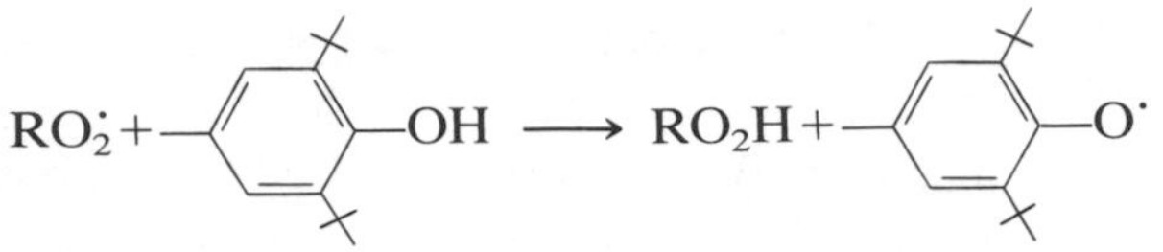

SCHEME 2

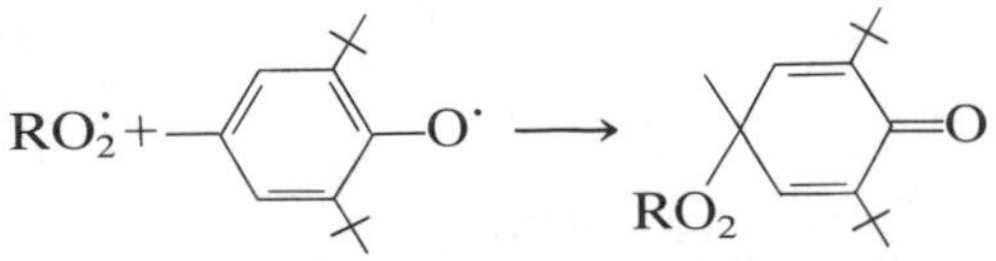

Recently, inhibitors have been reported that act in a catalytic fashion and are thus capable of disarming a larger number of chain-propagating radicals.[3] Arylamine-type compounds such as 1-aminonaphthalene are capable of catalytic radical scavenging in autoxidising alcohols,[1,4] aliphatic amines[5] and 1,3-cyclohexadiene.[1] This catalytic inhibition is due to the reducing capacity of the chain-propagating radicals derived from the substrates mentioned, i.e. the 1-hydroxyalkyl, 1-aminoalkyl and cyclohexadienyl radicals, respectively. After oxidation by a peroxyl radical, the original inhibitor can be regenerated by reduction by these particular alkyl radicals. This results in a reaction cycle in which the antioxidant alternately oxidises an alkyl radical and reduces an alkyl peroxyl radical (Scheme 3) and which leads to the removal of two chain-propagating radicals in each cycle. (See *Developments in Polymer Stabilisation*—4, p. 5 *et seq.*)

For saturated and alkylaromatic hydrocarbons, which lack sufficiently reactive hydrogens, the substrate oxidation step is too slow at temperatures below 100°C to give rise to catalytic inhibition. However, we found that N,N'-diphenyl-p-benzoquinone diimine is capable of scavenging a large number of radicals (70) in an autoxidising

SCHEME 3

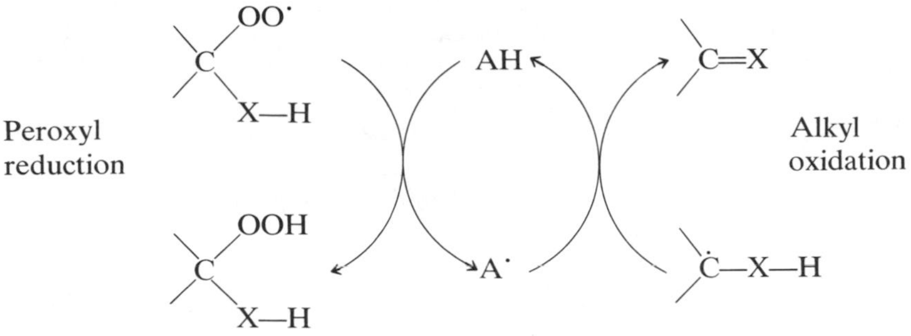

Peroxyl
reduction

$OO^{\cdot}$

$X—H$

OOH

$X—H$

AH

$A^{\cdot}$

$C=X$

Alkyl
oxidation

$\dot{C}—X—H$

$C—X—H$

X = O, N or cyclohexadienyl-C

mixture of alkanes at 130°C.[6] Subsequent work showed that certain simple diarylamines, such as diphenylamine, have similar radical scavenging efficiencies.[16]

Multiple radical scavenging has also been reported by other workers in this field. Denisov et al.[7] reported the catalytic inhibition of autoxidation of pentadecane at 120°C by an O-silylated 4-hydroxydiphenylamine, the number of radicals scavenged being eight. Rozantsev et al.,[8] observing very long inhibition periods of the radiation-induced autoxidation of n-octane at 100°C in the presence of some bisnitroxyls, attributed this finding to catalytic radical scavenging. Regeneration of nitroxyl radicals during autoxidation of polypropylene at a temperature of 114°C has been reported by Denisov et al.[9,10] Also the protective action of secondary amines and the corresponding nitroxyls in polypropylene, which was earlier attributed to non-radical processes,[11,12] has recently been discussed in terms of radical processes.[13-15]

The present contribution deals with our work on the catalytic inhibition by a number of secondary amines and related compounds, such as nitroxyls and substituted hydroxylamines. The mechanism, the kinetics of the catalytic cycle and the method by which the number of radicals scavenged per molecule of inhibitor (the stoichiometric inhibition coefficient, f) has been determined, will be described.

2. INFLUENCE OF THE STRUCTURE OF THE INHIBITOR ON THE STOICHIOMETRIC INHIBITION COEFFICIENT

Table 1 summarises the stoichiometric inhibition coefficients (f) of some diarylamines and derivatives thereof for the autoxidation of a

TABLE 1

STOICHIOMETRIC INHIBITION COEFFICIENTS (f) OF SOME SUBSTITUTED DIARYLAMINES AND DERIVATIVES IN A PARAFFINIC OIL AT $130°C^a$

	Compound[b]	f
IA		41
IB		36
IC		53
ID		52
IE		0
IF		17
IG		0
IH		0
IJ		26
IK		15
IL		35

[a] See Ref. 16 for experimental details.
[b] Concentration of inhibitor at the time of addition, 2×10^{-5} mol litre^{-1}.

paraffinic oil (Shell Ondina 33, a mixture of high-boiling cycloalkanes and isoalkanes) at 130°C. Up to 50 radicals are scavenged per inhibitor molecule, underlining the true catalytic nature of the inhibition. The data in Table 1 suggest that the number of radicals scavenged increases when the para-hydrogen atoms of the aromatic rings are replaced by tertiary alkyl groups. Electron withdrawing groups diminish or even completely destroy the catalytic activity; an ethoxy group has a minor influence.

Compounds **IB**, **IJ** and **IL** of the table show that the amine, hydroxylamine and nitroxyl of the 4-ethoxydiaryl compound are essentially equally active. However, in the case of the 4,4'-dinitrodiaryl system (**IH** and **IK**) the nitroxyl has a moderate activity whereas the parent amine has no activity whatsoever under the same conditions. The implications of these observations will be discussed later.

The efficiency of a catalytic inhibitor as expressed by the f value is determined not only by its activity in the cyclic reaction but also by its resistance towards side-reactions that cause its decomposition. The most probable depletion reaction is that starting from the intermediate nitroxyl. Electron spin resonance spectroscopic studies have shown[17] that the unpaired electron density in diarylnitroxyls is strongly delocalised. As a result, radical reactions will occur not only at the nitroxyl function but also at the aromatic rings, preferably at the para positions.[18] This means that under conditions of autoxidation the intermediate nitroxyl will be attacked by peroxyl radicals on the aromatic ring, eventually resulting in destruction of the inhibitor.

We have oxidised diphenylamine with tert-BuOO· (generated from tert-butyl hydroperoxide and di-tert-butyl peroxide at 130°C). Nitrobenzene and p-benzoquinone were identified among the products, which supports the idea that oxidation at the para position causes depletion. A possible mechanism is given in Scheme 4. This depletion path suggests that compounds lacking one or both of the susceptible phenyl rings would display a better catalytic efficiency.

Table 2 summarises the f values of some tert-alkylaryl- and di-tert-alkylamine-derived compounds. In spite of the fact that the intermediate tert-alkylaryl nitroxyls and di-tert-alkyl nitroxyls are thermally less stable than diaryl nitroxyls,[19,20] the tert-alkylaryl- and di-tert-alkylamine-type compounds display a much better catalytic efficiency than the diarylamines; f values of several hundreds are observed.

The cyclic compounds are markedly more efficient than the open-chain compounds (compare **IIG** and **IIK**). This difference can be

SCHEME 4

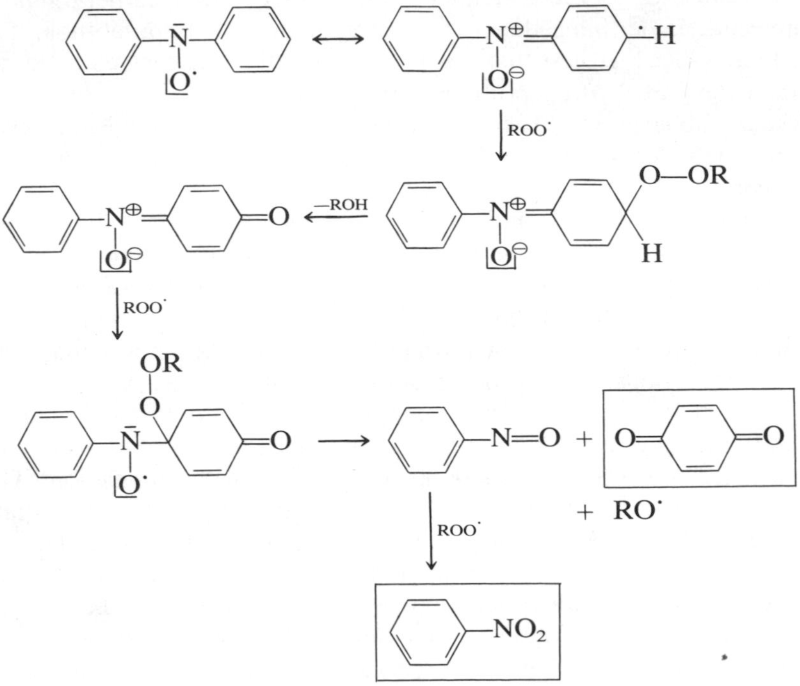

attributed to the fact that open-chain aliphatic nitroxyls are thermally less stable than cyclic aliphatic nitroxyls. The most likely reason for this is that the reactive sites formed after α-scission, which is the primary thermal reaction,[21] have a much higher chance of collapsing back to the nitroxyl in the case of the cyclic compounds where they cannot diffuse apart (Scheme 5). Di-*tert*-butyl nitroxyl has a half-life of only 13·4 minutes ($k = 8·6 \times 10^{-4}\,s^{-1}$ in *tert*-butylbenzene, 130°C, both in oxygen and in an argon atmosphere). After 50 min, 82% of 2-methyl-2-nitrosopropane was found as one of the decomposition products of di-*tert*-butyl nitroxyl. The cyclic nitroxyls, by contrast, show no detectable decomposition for at least 200 min. Yet, in autoxidising

SCHEME 5

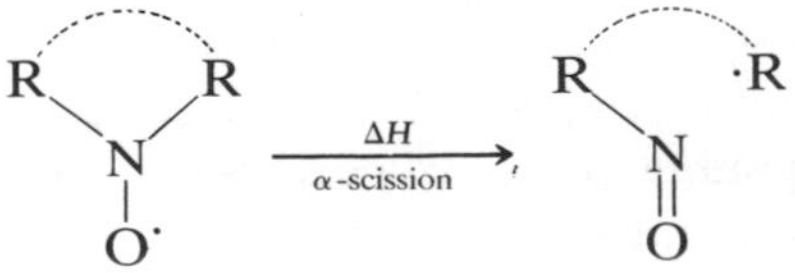

TABLE 2

STOICHIOMETRIC INHIBITION COEFFICIENTS (f) OF SOME *tert*-ALKYLARYL- AND DI-*tert*-ALKYLAMINE-BASED COMPOUNDS IN A PARAFFINIC OIL AT 130°C

Compound[a]	f
IIA	95
IIB	250
IIC	110
IID	420
IIE	400
IIF	275
IIG	510

TABLE 2—*contd.*

Compound[a]	f
IIH	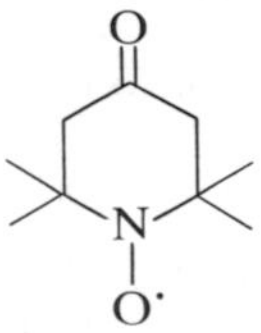 410
IIJ	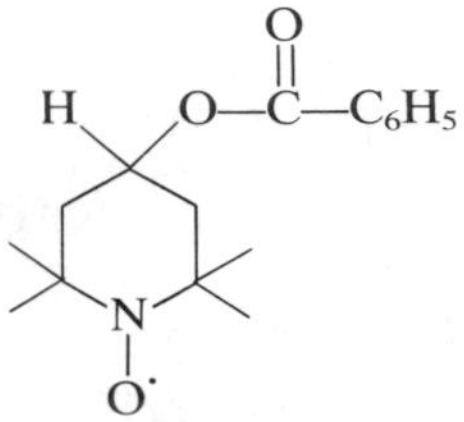 630
IIK $(H_3C)_3C\!-\!N\!-\!C(CH_3)_3$ $\overset{\mid}{O^{\cdot}}$	225

[a] Concentrations of inhibitor at the time of addition, $1\cdot0\times10^{-6}$–$4\cdot0\times10^{-6}$ mol litre^{-1}.

paraffinic oil these inhibitors display catalytic activity only for about 50 min.[16] This implies that their depletion must be effected by a different process, most likely by autoxidation.[22]

3. MECHANISMS OF THE CATALYTIC INHIBITION

Recently, various mechanisms have been proposed in the literature for the catalytic scavenging of radicals in autoxidising hydrocarbons. Denisov *et al.*[3,7] proposed the following cyclic mechanisms (Schemes 6 and 7) for the catalytic inhibition of autoxidising pentadecane by *O*-silylated 4-hydroxydiphenylamine:

SCHEME 6

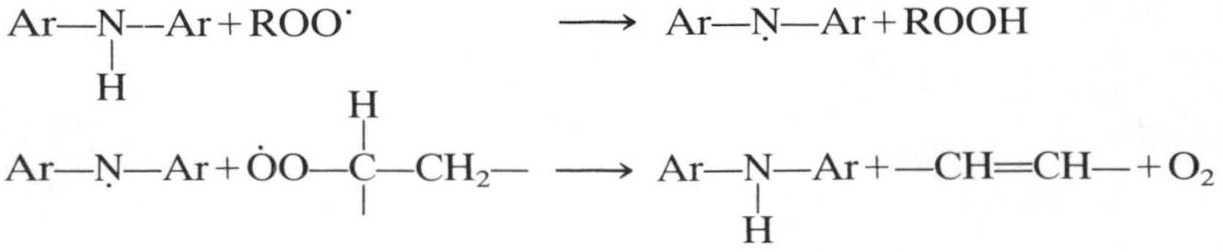

$$Ar\!-\!\underset{\underset{H}{\mid}}{N}\!-\!Ar + ROO^{\cdot} \longrightarrow Ar\!-\!\underset{\cdot}{N}\!-\!Ar + ROOH$$

$$Ar\!-\!\underset{\cdot}{N}\!-\!Ar + \overset{\cdot}{O}O\!-\!\underset{\mid}{\overset{\overset{H}{\mid}}{C}}\!-\!CH_2\!- \longrightarrow Ar\!-\!\underset{\underset{H}{\mid}}{N}\!-\!Ar + -CH\!=\!CH\!- + O_2$$

or, alternatively,

SCHEME 7

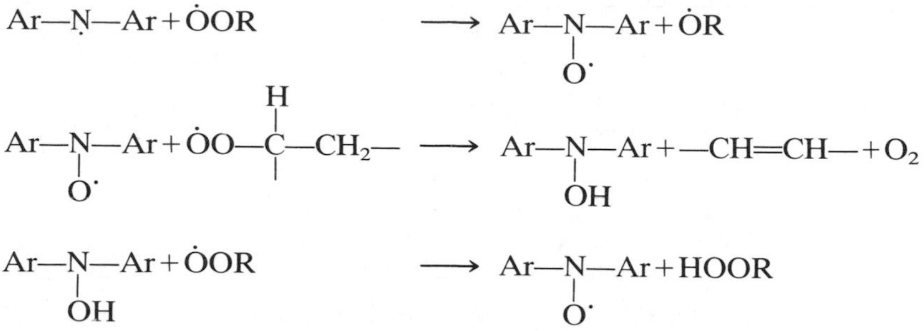

$$Ar—\underset{\cdot}{N}—Ar + \dot{O}OR \longrightarrow Ar—\underset{\underset{O^{\cdot}}{|}}{N}—Ar + \dot{O}R$$

$$Ar—\underset{\underset{O^{\cdot}}{|}}{N}—Ar + \dot{O}O—\overset{\overset{H}{|}}{\underset{|}{C}}—CH_2— \longrightarrow Ar—\underset{\underset{OH}{|}}{N}—Ar + —CH{=}CH— + O_2$$

$$Ar—\underset{\underset{OH}{|}}{N}—Ar + \dot{O}OR \longrightarrow Ar—\underset{\underset{O^{\cdot}}{|}}{N}—Ar + HOOR$$

Rozantsev et al.,[8] in order to explain the very long inhibition periods of the radiation-induced autoxidation of n-octane at 100°C in the presence of some bisnitroxyls, suggested the mechanism shown in Scheme 8.

SCHEME 8

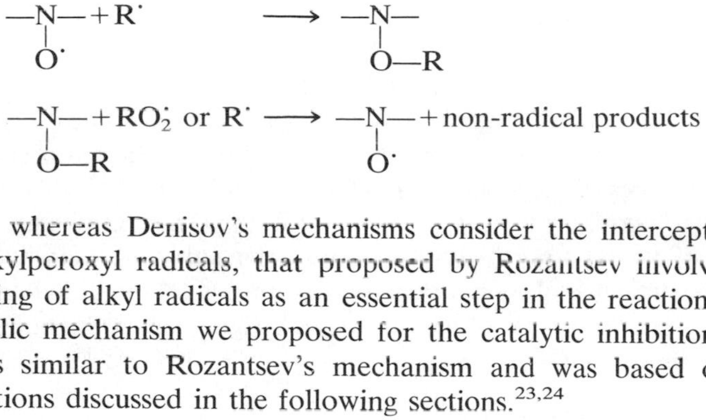

$$—\underset{\underset{O^{\cdot}}{|}}{N}— + R^{\cdot} \longrightarrow —\underset{\underset{O—R}{|}}{N}—$$

$$—\underset{\underset{O—R}{|}}{N}— + RO_2^{\cdot} \text{ or } R^{\cdot} \longrightarrow —\underset{\underset{O^{\cdot}}{|}}{N}— + \text{non-radical products}$$

Thus, whereas Denisov's mechanisms consider the interception of only alkylperoxyl radicals, that proposed by Rozantsev involves the scavenging of alkyl radicals as an essential step in the reaction cycle. The cyclic mechanism we proposed for the catalytic inhibition (vide infra) is similar to Rozantsev's mechanism and was based on the observations discussed in the following sections.[23,24]

3.1. Reaction of Nitroxyl Radicals with Alkyl and Alkylperoxyl Radicals at 130°C

According to Kovtun et al.,[25] dialkyl nitroxyls react rapidly with alkyl radicals but not with peroxyl radicals at 60–75°C. In order to find out whether this is also the case at 130°C (the temperature at which we observe catalytic inhibition) we have investigated the reactions of the 1-cyano-1-methylethyl radical and the corresponding peroxyl radical

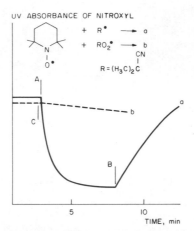

FIG. 1. Changes in concentration of 2,2,6,6-tetramethylpiperidine-N-oxyl upon introduction of AIBN in argon atmosphere (A), followed by oxygen purging (B), and upon introduction of AIBN in oxygen atmosphere (C), at 130°C in *tert*-butylbenzene.

with 2,2,6,6-tetramethylpiperidine-N-oxyl (Fig. 1). The nitroxyl appears to react rapidly with the alkyl radical but remains virtually unchanged in the presence of peroxyl radicals.

3.2. Effect of Hydrocarbon Structure on f

If the catalytic cycle includes the scavenging of alkyl radicals as well as alkylperoxyl radicals, one expects the structure of the hydrocarbon to have a significant influence on f. The data in Table 3 show a particularly large increase of the f value of inhibitor **IIJ** upon changing from the paraffinic oil to n-hexylbenzene as the substrate. On the other hand, the presence of diphenylmethane, but not 1,1-diphenylethane, as a co-substrate in the paraffinic oil causes a significant reduction of the f value of inhibitor **IIG**.

3.3. Effect of Temperature on f

In contrast to the very efficient catalytic inhibition with n-hexylbenzene at 130°C, inhibitors **IIG** and **IIK** were found to give only stoichiometric inhibition (f not exceeding 2) in autoxidising tetralin and cumene at 60°C. The inference is that at least one step in the catalytic cycle must have a fairly high energy of activation, which in its turn would be indicative of the occurrence in the cycle of a unimolecular step.

TABLE 3
EFFECT OF STRUCTURE OF THE HYDROCARBON ON THE STOICHIOMETRIC INHIBITION COEFFICIENT (f) AT 130°C

Hydrocarbon	Inhibitor[a]	f
Paraffinic oil	IIJ	630
n-Hexylbenzene	IIJ	3900
β,β-Dideuterio-n-hexylbenzene	IIJ	1100
Paraffinic oil	IIH	410
n-Hexylbenzene	IIH	720
Paraffinic oil	ID	52
n-Hexylbenzene	ID	65
β,β-Dideuterio-n-hexylbenzene	ID	25
Hexadecane	ID	12
Paraffinic oil	IIG	510
Paraffinic oil + diphenylmethane[b]	IIG	190
Paraffinic oil + 1,1-diphenylethane[b]	IIG	580

[a] Concentrations of inhibitor at the time of addition: IIJ and IIH, 1×10^{-6} mol litre^{-1}; ID and IIG, 2×10^{-5} mol litre^{-1}.
[b] 10 wt %.

3.4. Product of the Reaction of Alkyl Radicals with Nitroxyls
Aliphatic nitroxyls are known to react very efficiently at 60°C with alkyl radicals (1-cyano-1-methylethyl) to give the corresponding N,N,O-trisubstituted hydroxylamine[17,26] (Scheme 9(a)). No disproportionation (Scheme 9(b)) was observed at this temperature, but this does not rule out the possible occurrence of such a reaction at 130°C. We have therefore measured k_d/k_c for the reaction between 1,1-diphenylethyl radicals and 2,2,6,6-tetramethylpiperidine-N-oxyl (IIG) at various temperatures. As the source of the alkyl radicals we used 2,2,3,3-tetraphenylbutane, which decomposes at 60°C with a rate constant

SCHEME 9

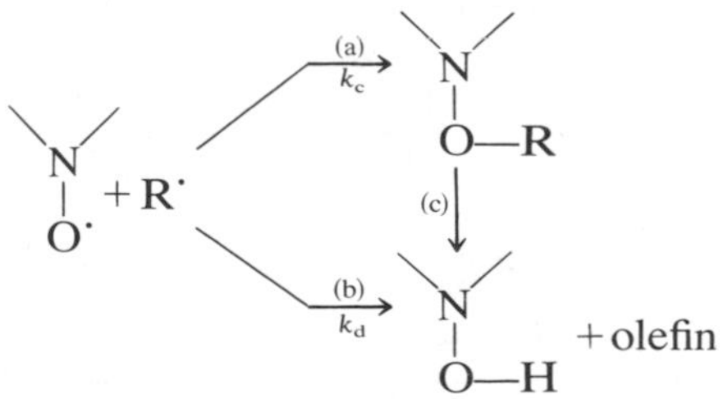

combination

disproportionation

$2 \times 10^{-4}\,\text{s}^{-1}$.[27] At each temperature a comparison was made between product yields obtained with and without nitroxyl. No change in k_d/k_c was observed upon raising the temperature from 60 to 80°C ($k_d/k_c =$ 0·21); it is not possible to determine this ratio at substantially higher temperatures because the thermolysis of the combination product, which affords the same products as are formed in the disproportionation reaction (Scheme 9(b)), becomes too fast. This result is in agreement with the observation that for the reaction between two carbon-centred radicals the ratio is hardly influenced by an increase in temperature.[28]

TABLE 4

VALUES OF f FOR SOME POSSIBLE INTERMEDIATES FROM AMINES IN A PARAFFINIC OIL AT 130°C

	Compound[a]	f
IIIA		0
IIIB		0
IIIC		205

IIID	$(H_3C)_3C$—N—$C(CH_3)_3$
	          $\mid$
	         O—$C(CH_3)_3$
		170

IIA		95

[a] Concentrations, $1·0 \times 10^{-6}$–$4·0 \times 10^{-6}$ mol litre^{-1}.

Thus, even at 130°C, the reaction between nitroxyl and alkyl radicals will occur predominantly by way of combination.

3.5. Catalytic Inhibition by Hydroxylamines

If the catalytic cycle proceeds via an intermediate compound such as a hydroxylamine or an O-alkylated hydroxylamine, these compounds should be as efficient inhibitors as the corresponding nitroxyls. Table 4 shows that the catalytic efficiencies of some hydroxylamines and their O-sec-alkyl and O-tert-alkyl derivatives are comparable with those of the nitroxyls. By contrast, O-n-alkylhydroxylamines (**IIIA** and **IIIB**) do not show any catalytic inhibition.

3.6. Regeneration of Nitroxyls from O-tert-Alkylhydroxylamines

There are two possible mechanisms for the regeneration of nitroxyls from O-tert-alkylhydroxylamines: (a) abstraction of β-hydrogen atoms by peroxyl radicals (Scheme 10), and (b) thermolysis (Scheme 11). The

SCHEME 10

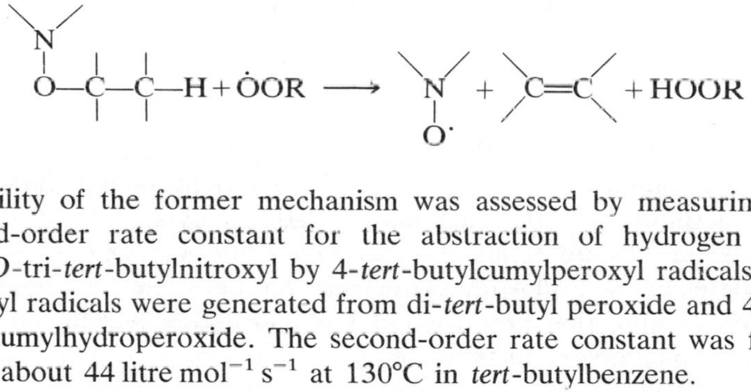

$$\overset{\diagup}{\underset{\diagdown}{N}} - O - \overset{|}{\underset{|}{C}} - \overset{|}{\underset{|}{C}} - H + \dot{O}OR \longrightarrow \overset{\diagup}{\underset{\diagdown}{N}}_{\overset{|}{O^{\cdot}}} + \overset{\diagup}{\underset{\diagdown}{C}} = \overset{\diagup}{\underset{\diagdown}{C}} + HOOR$$

feasibility of the former mechanism was assessed by measuring the second-order rate constant for the abstraction of hydrogen from N,N,O-tri-tert-butylnitroxyl by 4-tert-butylcumylperoxyl radicals. The peroxyl radicals were generated from di-tert-butyl peroxide and 4-tert-butylcumylhydroperoxide. The second-order rate constant was found to be about 44 litre mol^{-1} s^{-1} at 130°C in tert-butylbenzene.

The thermolytic mechanism involves decomposition of the O-tert-alkylhdroxylamine to N,N-disubstituted hydroxylamine and olefin (Scheme 11, reactions (a) and (b)), followed by oxidation of the N,N-disubstituted hydroxylamine to the nitroxyl (11(c)).

The decomposition of N,N,O-tri-tert-butylhydroxylamine (**IIID**) at 130°C in an atmosphere of argon (C$_6$D$_5$Br or tert-butylbenzene as the solvent) affords exclusively equimolar amounts of N,N-di-tert-butylhydroxylamine and isobutene. The rates of decomposition in C$_6$D$_5$Br at 100, 110 and 130°C were found to be $k = 0.58$, 1.2 and

$5.7 \times 10^{-4} \, s^{-1}$, respectively ($E_A = 96.2 \, kJ \, mol^{-1}$). As the decomposition in chlorine-donating solvents such as hexachloroacetone resulted in incorporation of chlorine into the products it was concluded that the thermolysis of the N,N,O-tri-*tert*-butylhydroxylamine proceeds via homolysis of the O-alkyl bond (Scheme 11) rather than via a concerted reaction (Scheme 9(c)). Recent extensive studies by Howard *et al.*[29,30] of the thermal decomposition of some O-alkyl derivatives of 1-hydroxy-2,2,6,6-tetramethyl-4-oxopiperidine have established well the homolytic mechanism of this reaction.

The homolytic fission of the O-alkyl bond is followed by transfer of a β-hydrogen atom to produce the decomposition products, for example

SCHEME 11

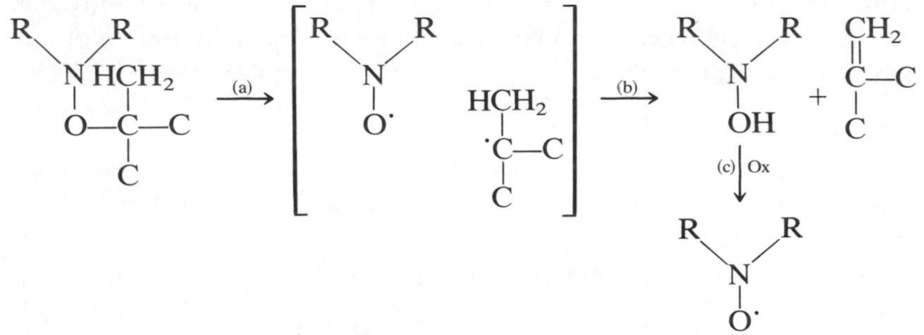

Thus, the combination of alkyl and nitroxyl radicals and the subsequent thermolysis of the combination product eventually afford the same products as are obtained directly by disproportionation. Scheme 12, where $k_c' > k_d'$, illustrates the relation between the kinetically favoured combination product and the thermodynamically favoured disproportionation products.

SCHEME 12

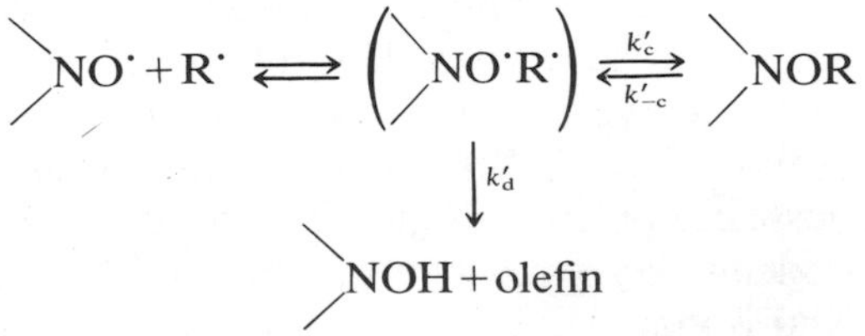

3.7. Regeneration of Nitroxyls from N,N-Disubstituted Hydroxylamines

SCHEME 13

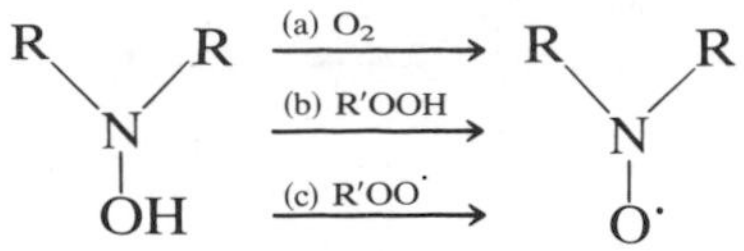

Based on the above evidence, N,N-dialkylhydroxylamines can be formed by thermolysis of trialkylhydroxylamines as well as by disproportionation of nitroxyls and alkyl radicals. The regeneration of a nitroxyl from the corresponding hydroxylamine involves an oxidation. This oxidation may be effected by molecular oxygen (Scheme 13(a)), hydroperoxide (13(b)) and/or peroxyl radicals (13(c)). We have examined each of these possible reactions. The second-order rate constant of the oxidation of 1-hydroxy-2,2,6,6-tetramethylpiperidine by molecular oxygen in *tert*-butylbenzene at 130°C was measured and found to be $1 \cdot 3 \times 10^{-2}$ litre mol^{-1} s^{-1}, and that for the oxidation by 4-*tert*-butylcumyl hydroperoxide $0 \cdot 46$ litre mol^{-1} s^{-1}.

The oxidation of N,N-dialkylhydroxylamines by peroxyl radicals is known to be an extremely fast reaction.[30] The second-order rate constant for the abstraction of hydrogen from 1-hydroxy-2,2,6,6-tetramethyl-4-oxopiperidine by polyperoxystyrylperoxyl radicals has been reported to be $k = 5 \times 10^5$ litre mol^{-1} s^{-1} at 65°C; therefore, since we may assume the pre-exponential factor in the Arrhenius equation for this bimolecular reaction to be of the order of 10^8, the rate constant at 130°C must be $1 \times 10^6 - 2 \times 10^6$ litre mol^{-1} s^{-1}.

3.8. The Proposed Reaction Cycle

The reaction cycle proposed[24] for the catalytic radical scavenging in autoxidations, proceeding via tertiary alkyl and alkylperoxyl radicals, is shown in Scheme 14. It assumes that if secondary amines are used as the inhibitors, these are first oxidised to the corresponding nitroxyls before entering the catalytic cycle. The proposed cycle involves the alternating scavenging of alkyl and alkylperoxyl radicals. The alkyl radicals are scavenged by the nitroxyl to give an O-alkylhydroxylamine, which undergoes thermolysis to give the corresponding hydroxylamine and olefin. The former reduces an alkylperoxyl radical with simultaneous regeneration of nitroxyl.

SCHEME 14

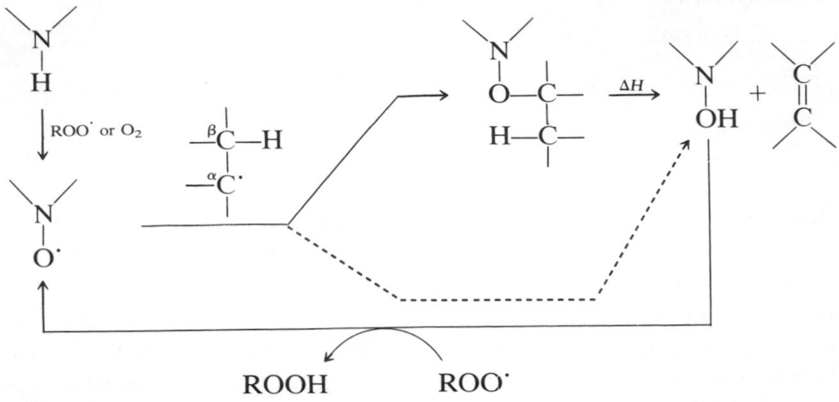

In the following section we shall discuss the various steps in more detail. It goes without saying that for any reaction cycle all the steps must be fast enough to account for the observed reduction in rate of oxidation. This kinetic aspect has been evaluated for the partially inhibited oxidation of a paraffinic oil at 130°C in the presence of inhibitor **IIG** for which the required rates of the various steps in the reaction cycle have been calculated (Section 5).

4. DISCUSSION OF THE MECHANISM OF THE CATALYTIC INHIBITION

4.1. Conversion of Secondary Amines to Nitroxyls

On the basis of literature data we have assumed that secondary amines enter the catalytic cycle as a result of their oxidation to the corresponding nitroxyls. It is known that abstraction of hydrogen from an aromatic amine by a peroxyl radical (Scheme 15) is a facile process ($k \simeq 4 \times 10^4$ litre mol^{-1} s^{-1} at 65°C) and leads to the formation of an aminyl radical.[31,32] This reaction is fast enough to convert all the amine within a few minutes. Oxidation of aminyl radicals to nitroxyls by

SCHEME 15

$$\text{Ar}{-}\text{N}{-}\text{Ar} + \text{ROO}^{\cdot} \longrightarrow \text{Ar}{-}\dot{\text{N}}{-}\text{Ar} + \text{ROOH}$$
$$\underset{\text{H}}{|}$$

SCHEME 16

$$Ar—\overset{\cdot}{N}—Ar + ROO^{\cdot} \longrightarrow \underset{\overset{|}{O^{\cdot}}}{Ar—N—Ar} + RO^{\cdot}$$

peroxyl radicals (Scheme 16) has been reported to be very fast, although no accurate data are known.

There is ample evidence for the occurrence of nitroxyls in amine-inhibited autoxidations.[1,33,34] Our assumption that the parent amines themselves are not involved in the catalytic cycle is supported by the observation that 4,4'-dinitrodiphenylamine (**IH**), which is difficult to oxidise, shows no inhibiting effect, whereas the corresponding nitroxyl (**IK**) is a catalytic inhibitor.

The abstraction of hydrogen from a secondary aliphatic amine was recently reported[22] to be much slower than that from an aromatic amine. Yet, we have observed that with the piperidine-type inhibitors, too, the maximum reduction in oxidation rate is obtained in a few minutes after their addition. It is possible that oxidation by hydroperoxides may also be involved in this case.

4.2. Reaction of Nitroxyls with Alkyl Radicals

As dialkyl nitroxyls do not react with peroxyl radicals but react very rapidly with alkyl radicals we conclude that the scavenging of alkyl radicals by nitroxyls is one of the steps in the catalytic cycle. The rate constant of the reaction between alkyl radicals and nitroxyls can be estimated to be 10^8–10^9 litre mol^{-1} s^{-1},[25,35] which makes this reaction fast enough to meet the rate required (cf. Section 5) for its participation in the catalytic cycle.

As outlined above, disproportionation takes place to a minor extent only at lower temperatures. Thus, an N,N,O-trisubstituted hydroxylamine will be formed as an intermediate in the catalytic cycle.

4.3. Regeneration of Nitroxyls from O-tert-Alkylhydroxylamines

The next question concerns the way in which an O-tert-alkylhydroxylamine reacts further with eventual regeneration of the nitroxyl.

If abstraction of a β-hydrogen from an O-tert-alkylhydroxylamine by peroxyl radicals were responsible for the regeneration of nitroxyl, the rate constant of this reaction should be at least $k = 10^5$ litre mol^{-1} s^{-1} (see Section 5). As we have found the rate constant to be only $k = 44$ litre mol^{-1} s^{-1} for tri-tert-butylhydroxylamine, the possibility that this reaction acts as the regenerative step can be ruled out.

In the preceding section we have seen that N,N,O-tri-*tert*-butyl-hydroxylamine undergoes a rapid thermolysis in which homolysis of the C–O bond and subsequent transfer of a β-hydrogen atom produces the corresponding hydroxylamine and olefin (Scheme 11).

The first-order rate constant of this reaction at 130°C is $k = 4.7 \times 10^4 \, \text{s}^{-1}$. In the paraffinic oil the rate of thermolysis of the O-*tert*-alkylhydroxylamines will be higher because the oxygen-bound tertiary alkyl group is far bulkier than the *tert*-butyl group. The combined effects of repulsion between the N-bound alkyl groups and the O-bound *tert*-alkyl group and of the increased back strain in the latter group[36] is believed to be sufficient to increase the thermolysis rate to the required $k \geqslant 4 \times 10^{-2} \, \text{s}^{-1}$ (Section 5).

4.4. Oxidation of N,N-Disubstituted Hydroxylamines to Nitroxyls

The last step in the catalytic cycle is the regeneration of the nitroxyl radical by oxidation of the corresponding hydroxylamine. The possibility of oxidation by molecular oxygen or hydroperoxide can be ruled out on kinetic grounds. For both reactions the experimental rate constants (see Section 3.7) ($k = 0.013$ litre $\text{mol}^{-1} \text{s}^{-1}$ and $k = 0.46$ litre $\text{mol}^{-1} \text{s}^{-1}$, respectively, are much lower than those required ($k \geqslant 40$ litre $\text{mol}^{-1} \text{s}^{-1}$ and $k \geqslant 100$ litre $\text{mol}^{-1} \text{s}^{-1}$, respectively; Section 5). The oxidation by peroxyl radicals is fast enough to account for the regeneration of the nitroxyl. The estimated rate constant for the abstraction of the hydroxyl hydrogen by peroxyl radicals is 10^6–10^7 litre $\text{mol}^{-1} \text{s}^{-1}$, which is well above the value required (10^5 litre $\text{mol}^{-1} \text{s}^{-1}$).

4.5. Scavenging of Secondary Radicals

In our original paper[24] it had been tacitly assumed that the scavenging of the secondary benzyl and benzylperoxyl radicals in the inhibited autoxidation of n-hexylbenzene proceeded by precisely the same mechanism as that for the tertiary alkyl radicals. The assumed thermolysis of O-(1-phenylhexyl)hydroxylamines via homolysis of the O–C bond and subsequent transfer of a β-hydrogen atom was thought to be responsible for the strong deuterium isotope effect observed with β,β-dideuterio-n-hexylbenzene ($f(^1\text{H})/f(^2\text{H}) = 2$–3.5; Table 3); the relatively high energy of activation of this kind of thermolysis would have explained the change of catalytic inhibition at 130°C into stoichiometric inhibition at 60°C. Recent work by Howard et al.,[29] however, has shown that 1-cumyloxy-2,2,6,6-tetramethyl-4-oxopiperidine is fairly stable at 100°C. From their data it must be concluded that the decomposition at 130°C of an O-(1-phenylhexyl)hydroxylamine into the hydroxylamine and olefin is much too slow to be part of the radical-scavenging cycle that causes the observed reduction in rate of oxidation of n-hexylbenzene. In this respect the situation is the same as that for alkanes that autoxidise via secondary alkyl and alkylperoxyl radicals. For these, the occurrence of an efficient radical-scavenging cycle is well demonstrated by the inhibited autoxidation of hexadecane (Table 3) and also by the fact that the oxidation of the paraffinic oil proceeds via secondary as well as tertiary radicals. Yet, the O-sec-butylhydroxylamine **IIIC** is thermally stable at 130°C.[24]

Clearly, some mechanism other than that depicted in Scheme 13 is responsible for the catalytic scavenging of secondary alkyl(peroxyl) radicals. Further detailed, quantitative studies will be needed to find out in what respect it differs from that shown in Scheme 13.

5. KINETICS OF PARTIALLY INHIBITED AUTOXIDATIONS AT HIGH TEMPERATURE AND THE DETERMINATION OF STOICHIOMETRIC INHIBITION COEFFICIENTS

The number of radicals scavenged by one inhibitor molecule (SIC; hereafter designated by f) can be simply determined by the addition of an inhibitor (AH) to an initiated autoxidation if the oxidation takes place at low temperature (say, not above 80°C) and the addition of the inhibitor leads to complete, or at least a very large, reduction of the rate of oxidation. In that case (Fig. 2a) the number f can be calculated from the equation

$$f[\text{AH}]_0 = R_i t_{\text{inh}} \tag{i}$$

where $[\text{AH}]_0$ is the concentration of added inhibitor, R_i is the rate of initiation and t_{inh} is the inhibition time (or the induction period). The same procedure can be used to determine R_i if one uses an inhibitor for which f is known, e.g. a hindered phenol ($f = 2$).

If the oxidation takes place at high temperature (say, above 120°C) or if the addition of the inhibitor effects only a moderate reduction of the rate of oxidation ('partial inhibition') eqn. (i) cannot be used.

The important difference between low-temperature and high-temperature oxidations is that in the latter case the hydroperoxides formed undergo homolysis, which produces chain-initiating radicals. Consequently, the rate of initiation, R_i, is no longer constant but increases with time.

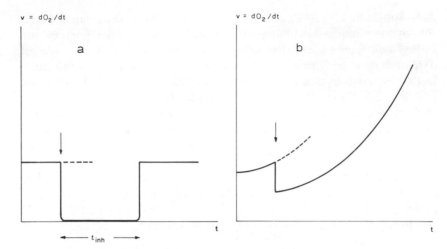

FIG. 2. Effect of inhibitor on the rate of oxidation for 'complete' inhibition at low temperature (a) and for partial inhibition at high temperature (b). Arrow marks the moment of addition of the inhibitor.

In the case of partial inhibition, eqn. (i) is no longer valid because a large fraction of the radicals disappears by self-termination of the peroxyl radicals instead of by reaction with the inhibitor. Moreover, if the partially inhibited oxidation takes place at high temperature, a well-defined end point of the inhibition period is no longer observed (Fig. 2b).

The method we have used[16] to determine f for non-stoichiometric radical scavengers under these conditions of partial inhibition at high temperature is essentially the same as that used earlier by Berger[37] for the determination of initiation rates under such conditions. It consists in using a modified form of eqn. (i), namely

$$f[AH]_0 = (R_i)_{t=t^*} . T \qquad \text{(ii)}$$

where

$$T = \int \{(v_0)_{t'}^2 - (v_A)_t^2\} \, dt/(v)_{t=t^*}^2 \qquad \text{(iii)}$$

in which $(v_0)_{t'}$ and $(v_A)_t$ are the rates of the uninhibited and inhibited oxidations, respectively, after the same amounts of oxygen have been consumed, and $(R_i)_{t=t^*}$ and $(v)_{t=t^*}$ are the rates of initiation and oxygen consumption, respectively, immediately before the addition of the inhibitor.

5.1. Kinetics

In an initiated autoxidation in the presence of an inhibitor that scavenges alkyl and/or alkylperoxyl radicals the following reactions take place:

$$\text{Initiator} + O_2 + RH \xrightarrow{R_i} R^{\cdot}$$

$$R^{\cdot} + O_2 \xrightarrow{k_{ox}} RO_2^{\cdot}$$

$$RO_2^{\cdot} + RH \xrightarrow{k_p} RO_2H + R^{\cdot}$$

$$2RO_2^{\cdot} \xrightarrow{2k_t} \text{non-radical products}$$

$$R^{\cdot} + X \xrightarrow{k_x} \text{non-radical products}$$

$$RO_2^{\cdot} + Y \xrightarrow{k_y} \text{non-radical products}$$

These give rise to the steady-state equations

$$d[R^{\cdot}]/dt = 0 = R_i + k_p[RO_2^{\cdot}][RH] - k_{ox}[R^{\cdot}][O_2] - k_x[R^{\cdot}][X] \quad \text{(iv)}$$

$$d[RO_2^{\cdot}]dt = 0 = k_{ox}[R^{\cdot}][O_2] - k_p[RO_2^{\cdot}][RH] - 2k_t[RO_2^{\cdot}]^2 - k_y[RO_2^{\cdot}][Y] \quad \text{(v)}$$

also

$$R_i = 2k_t[RO_2^{\cdot}]^2 + k_x[R^{\cdot}][X] + k_y[RO_2^{\cdot}][Y] \quad \text{(via)}$$

or

$$2k_t[RO_2^{\cdot}]^2 = R_i(1 - \alpha) \quad \text{(vib)}$$

where

$$\alpha = \{k_x[R^{\cdot}][X] + k_y[RO_2^{\cdot}][Y]\}/R_i \quad \text{(vii)}$$

Provided that in the partially inhibited reaction the chain length is sufficiently long, $k_{ox}[R^{\cdot}][O_2]$ and $k_p[RO_2^{\cdot}][RH]$ are large compared with the other terms in eqns. (iv) and (v) and the oxygen consumption by the initiation reaction can be neglected; the rate of oxygen uptake is then given by

$$v = dO_2/dt = k_{ox}[R^{\cdot}][O_2] = k_p[RO_2^{\cdot}][RH] \quad \text{(viii)}$$

Combination of eqns. (viii) and (vib) gives the rate of the inhibited oxidation

$$(v_A)_t = \sqrt{\frac{(R_i)_t}{2k_t}}\, k_p[RH]\sqrt{(1-\alpha)} \qquad \text{(ix)}$$

For the corresponding uninhibited oxidation ($k_x = k_y = 0$; $\alpha = 0$)

$$(v_0)_t = \sqrt{\frac{(R_i)_t}{2k_t}}\, k_p[RH] \qquad \text{(x)}$$

In eqns. (ix) and (x) we have added the subscript t to R_i, v_0 and v_A to emphasise that for an autoxidation at high temperature R_i is time-dependent through its increase with increasing concentration of ROOH. As long as only a small fraction of the hydroperoxide produced is used for the additional initiation, the amount of ROOH present is equal to the amount of oxygen consumed, so that

$$(R_i)_t = (R_i)_0 + \text{constant} \times [\text{ROOH}]_t = (R_i)_0 + \text{constant} \times [O_2]_{\text{cons}} \qquad \text{(xi)}$$

where $(R_i)_0$ is the rate of initiation by the initiator and $[O_2]_{\text{cons}}$ is the number of moles of oxygen consumed per litre of the solution at time t. Thus, for the uninhibited reaction a plot of $(v)^2$ against $[O_2]_{\text{cons}}$ gives a straight line (Fig. 3a).

An important feature of the plot of $(v)^2$ against $[O_2]_{\text{cons}}$ is the

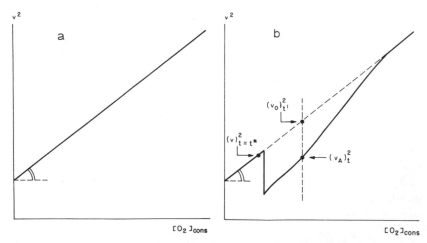

FIG. 3. Plots of v^2 versus $[O_2]_{\text{cons}}$ for uninhibited (a) and inhibited (b) autoxidations at high temperature.

following: if the oxidation is interrupted by the addition of an inhibitor the plot shows a dip but *it returns to the original straight line of the uninhibited oxidation once all of the inhibitor is consumed* (Fig. 3b). The catalytic inhibitors discussed in this chapter, for which the inhibition reactions are relatively slow and for which f is quite large, produce a rather shallow, very broad dip (Fig. 3b), whereas a hindered phenol, when used in sufficiently high concentrations, results in a sharp dip.

From eqns. (ix) and (x) it is seen that at a given rate of initiation the inhibitor reduces the rate of oxidation by a factor given by eqn. (xii).

$$(v_A)/(v_0) = \sqrt{(1-\alpha)} \qquad \text{(xii)}$$

For the special case of a catalytic inhibitor that alternately scavenges $R^{\bullet}$ and $RO_2^{\bullet}$ radicals in a cyclical process, the steady state of the inhibitor species ($[X]+[Y]=$ constant) requires that the ratio $[X]/[Y]$ adjusts itself so that

$$k_x[R^{\bullet}][X] = k_y[RO_2^{\bullet}][Y] \qquad \text{(xiii)}$$

Equation (xiii) implies that the scavenging of $R^{\bullet}$ and of $RO_2^{\bullet}$ make equal contributions to α (eqn. (vii)). That is, they make an equal contribution to the reduction of the rate of oxidation.

Equations (xii) and (vii), or alternatively the relationship $(v_A)/(v_0) = \sqrt{1/(1+\beta)}$ where $\beta = \{k_x[R^{\bullet}][X]+k_y[RO_2^{\bullet}][Y]\}/2k_t[RO_2^{\bullet}]^2$, can be used to calculate the values of the rates or rate constants demanded of the inhibition reactions in order to effect the observed reduction of the rate of oxidation. We note that, since eqn. (xii) applies to the rates v_0 and v_A at the same rate of initiation, we have to compare the rates of the inhibited and uninhibited reactions not at the same time but after the same amount of oxygen has been consumed (cf. eqn. (xi)).

For the autoxidation of the paraffinic oil at 130°C, $2k_t = 0.7 \times 10^7$ litre mol^{-1} s^{-1}, $k_p[RH] = 90$ s^{-1} and $[O_2] = 10^{-3}$ mol litre^{-1}. In a typical experiment, in which $(R_i)_t$ was 1.2×10^{-6} mol litre^{-1} s^{-1} and 4×10^{-6} mol litre^{-1} of inhibitor **IIG** had been added, $(v_A)_t/(v_0)_{t'}$ was 0.86 at the beginning of the inhibition period. Hence, for this case

$$k_x[X][R^{\bullet}] = k_y[Y](RO_2^{\bullet}) = 1.5 \times 10^{-7} \text{ mol litre}^{-1} \text{ s}^{-1}$$

where, according to eqn. (vib), $[RO_2^{\bullet}] = 3.6 \times 10^{-7}$ mol litre^{-1} so that $k_y[Y] = 0.4$ s^{-1}, which requires k_y to be at least 10^5 litre mol^{-1} s^{-1}.

The concentration of alkyl radicals can be estimated with the aid of eqn. (viii). Assuming that k_{ox} is not much smaller than that for a diffusion-limited reaction, say $k_{ox} = 10^8 – 10^9$ litre $mol^{-1} s^{-1}$, we obtain $[R^{\cdot}] = 3 \times 10^{-11} – 3 \times 10^{-10}$ mol litre^{-1}, so that $k_x[X] = 5 \times 10^2 – 5 \times 10^3 s^{-1}$, which requires k_x to be at least $1 \cdot 3 \times 10^8 – 1 \cdot 3 \times 10^9$ litre $mol^{-1} s^{-1}$.

If the catalytic cycle includes a step in which an intermediate species (Z) undergoes unimolecular decomposition

$$X \xrightarrow[k_x]{R^{\cdot}} Z \xrightarrow[k_z]{} Y \xrightarrow[k_y]{RO_2^{\cdot}} X$$

the steady state for the inhibitor species requires that

$$k_z[Z] = k_x[X][R^{\cdot}] = k_y[Y][RO_2^{\cdot}]$$

so that k_z must be at least $0 \cdot 04 \, s^{-1}$.

If the conversion of any intermediate (U) in the cycle were to be effected by a bimolecular reaction with oxygen or hydroperoxide (under the above conditions $[ROOH] = 3 \times 10^{-4} – 4 \times 10^{-4}$ mol litre^{-1}), then the products $k_u[U][O_2]$ or $k_u[U][ROOH]$, respectively, must be equal to $k_x[X][R^{\cdot}]$, i.e. k_u must be at least 40 litre $mol^{-1} s^{-1}$ for the reaction with oxygen and at least 10^2 litre $mol^{-1} s^{-1}$ for the reaction with hydroperoxide.

5.2. Determination of Stoichiometric Inhibition Coefficient (f)

The total number of radicals ($R^{\cdot}$ and/or $RO_2^{\cdot}$) scavenged by a given amount of an inhibitor during the inhibition period is given by

$$N = \int \{k_x[X]_t[R^{\cdot}]_t + k_y[Y]_t[RO_2^{\cdot}]_t\} \, dt = f . [AH]_0 \qquad \text{(xiv)}$$

The subscripts t have been added to $[X]$ and $[Y]$ in order to emphasise that, although in a cyclical process $[X]$ and $[Y]$ remain related via eqn. (xiii), they both decrease with time as the inhibitor is consumed by side-reactions. In order to evaluate the integral in terms of measurable quantities, the steady-state equation for the formation and disappearance of radical species (eqn. (via)) may be written in the form

$$\{k_x[X]_t[R^{\cdot}]_t + k_y[Y]_t[RO_2^{\cdot}]_t\} = (R_i)_t - 2k_t[RO_2^{\cdot}]_t^2 \qquad \text{(vic)}$$

The last term of the right-hand side of this equation is related to the rate of the inhibited reaction, v_A, by means of eqns. (vii) and (viii)

$$2k_t[RO_2^{\cdot}]_t^2 = (v_A)_t^2 . (2k_t/k_p^2[RH]^2) \qquad \text{(xv)}$$

The first term of the right-hand side of eqn. (vic) can be related to the rate of the uninhibited reaction, v_0. If the time is t' at which $[O_2]_{cons}$ in the uninhibited reaction is equal to that at the time t in the inhibited reaction, then (eqn. (xi)) $(R_i)_{t'} = (R_i)_t$. Then, according to eqn. (x),

$$(R_i)_t = (R_i)_{t'} = (v_0)_{t'}^2 \cdot (2k_t/k_p^2[RH]^2) \qquad (xvi)$$

Similarly, for the inhibited reaction just before the inhibitor is added $(t = t^*)$,

$$(R_i)_{t=t^*} = v_{t=t^*}^2 \cdot (2k_t/k_p^2[RH]^2) \qquad (xvii)$$

Substitution of eqn. (xvii) into eqns. (xv) and (xvi) and of the latter two into eqn. (vic) affords eqn. (xviii),

$$\{k_x[X]_t[R^{\cdot}]_t + k_y[Y]_t[RO_2^{\cdot}]_t\} = \frac{(v_0)_{t'}^2 - (v_A)_t^2}{v_{t=t^*}^2} \cdot (R_i)_{t=t^*} \qquad (xviii)$$

which upon substitution in eqn. (xiv) affords eqns. (ii) and (iii).

5.3 Experimental Procedure for the Determination of f

By means of eqns. (ii) and (iii) f may be obtained by the following experimental procedure.

The oxidation is started by the addition of a suitable initiator and both the rate of oxygen uptake and the total amount of oxygen consumed are recorded as a function of time. After the oxidation has proceeded for some time, the inhibitor is added and the recordings are continued until all of the inhibitor is consumed. From plots of v^2 against $[O_2]_{cons}$ for the inhibited reaction and for a corresponding uninhibited reaction, $(v_0)_{t'}^2 - (v_A)_t^2$ is calculated for pairs of rates at various values of $[O_2]_{cons}$ during the inhibition period (cf. Fig. 3b). Subsequently, the values of $(v_0)_{t'}^2 - (v_A)_t^2$ are plotted against the time t (of the inhibited reaction). The area below the curve thus obtained gives the integral of eqn. (iii), division of which by $(v)_{t=t^*}^2$ affords T.

Calculation of f still requires the appropriate value of $(R_i)_{t=t^*}$. For the hydrocarbon–initiator combination in question, this value is obtained by precisely the same procedure as outlined above, using a not too large amount of 2,6-di-*tert*-butyl-4-methylphenol (for which $f = 2$) as the inhibitor.[37]

REFERENCES

1. DENISOV, E. T., *Uspekhi Khimii*, **42** (1973) 361.
2. INGOLD, K. U., *Free Radicals*, Vol. I (1973) pp. 72–80. John Wiley, New York.

3. DENISOV, E. T., *Developments in Polymer Stabilisation*—3, Ed. G. Scott (1980) p. 1. Applied Science Publishers, London.
4. VARDANYAN, R. L., KHARITONOV, V. V. and DENISOV, E. T., *Izv. Akad. Nauk. SSSR, Ser. Khim.* (1970) 1536.
5. KOVTUN, G. A. and ALEKSANDROV, A. L., *Izv. Akad. Nauk. SSSR, Ser. Khim.* (1973) 2208.
6. BERGER, H., unpublished results.
7. VARLAMOV, V. T., KHARITONOV, V. V. and DENISOV, E. T., *Dokl. Akad. Nauk. SSSR*, **220** (1975) 620.
8. SUDNIK, M. V., ROMANTSEV, M. F., SHAPIRO, A. B. and ROZANTSEV, E. G., *Izv. Akad. Nauk. SSSR, Ser. Khim.* (1975) 2813.
9. SHILOV, YU. B., BATTALOVA, R. M. and DENISOV, E. T., *Dokl. Akad. Nauk. SSSR*, **207** (1972) 388.
10. SHILOV, YU. B. and DENISOV, E. T., *Vysokomolek. Soed.*, **A16** (1974) 2313.
11. GRATTAN, D. W., REDDOCH, A. H., CARLSSON, D. J. and WILES, D. M., *J. Polym. Sci.*, **16** (1978) 143.
12. ALLEN, N. S., MCKELLAR, J. F. and WILSON, D., *Chem. Ind.*, (1978) 887.
13. CHAKRABORTY, K. B. and SCOTT, G., *Chem. Ind.* (1978) 237.
14. CARLSSON, D. J., GARTON, A. and WILES, D. H., *Developments in Polymer Stabilisation*—1, Ed. G. Scott (1979) p. 219. Applied Science Publishers, London.
15. CHAKRABORTY, K. B. and SCOTT, G., *Polymer*, **21** (1980) 252.
16. BOLSMAN, T. A. B. M., BLOK, A. P. and FRIJNS, J. H. G., *Rec. Trav. Chim. Pays-Bas*, **97** (1978) 310.
17. FORRESTER, A. R., HAY, J. M. and THOMSON, R. H., *Organic Chemistry of Stable Free Radicals* (1968). Academic Press, New York.
18. CALDER, A. and FORRESTER, A. R., *J. Chem. Soc.* (C) (1969) 1459.
19. AURICH, H. G. and WEISS, W., *Topics in Current Chemistry*, **59** (1975) 65.
20. KAYEN, A. H. M., BOLSMAN, T. A. B. M. and DE BOER, TH. J., *Rec. Trav. Chim. Pays-Bas*, **95** (1976) 14.
21. ROZANTSEV, E. G. and SHOLLE, V. D., *Synthesis* (1971) 401.
22. GRATTAN, D. W., CARLSSON, D. J. and WILES, D. M., *J. Polym. Deg. Stab*, **1** (1979) 69.
23. BOLSMAN, T. A. B. M., BLOK, A. P. and FRIJNS, J. H. G., *Rec. Trav. Chim. Pays-Bas*, **97** (1978) 313.
24. BOLSMAN, T. A. B. M. and BROUWER, D. M., *Rec. Trav. Chim. Pays-Bas*, **97** (1978) 320.
25. KOVTUN, G. A., ALEKSANDROV, A. L. and GOLUBEV, V. A., *Izv. Akad. Nauk. SSSR, Ser. Khim.* (1974) 2197.
26. MURAYAMA, K., MORIMURA, S. and YOSHIOKA, T., *Bull. Chem. Soc. Jap.*, **42** (1969) 1640.
27. MAHONEY, L. R., *J. Am. Chem. Soc.*, **88** (1966) 3035.
28. GRIBIAN, M. J. and CORLEY, R. C., *Chem. Rev.*, **73** (1973) 441.
29. HOWARD, J. A. and TAIT, J. C., *J. Org. Chem.*, **43** (1978) 4279.
30. GRATTAN, D. W., CARLSSON, D. J., HOWARD, J. A. and WILES, D. M., *Can. J. Chem.*, **57** (1979) 2834.
31. BOWMAN, D. F., MIDDLETON, B. S. and INGOLD, K. U., *J. Org. Chem.*, **34** (1969) 3456.

32. BROWNLIE, I. T. and INGOLD, K. U., *Can. J. Chem.*, **45** (1967) 2419.
33. ADAMIC, K., DUNN, M. and INGOLD, K. U., *Can. J. Chem.*, **47** (1969) 287.
34. THOMAS, J. R. and TOLMAN, C. A., *J. Am. Chem. Soc.*, **84** (1962) 2930.
35. KHLOPTYANKINA, M. S., BUCHACHENKO, A. L., NEIMAN, M. B. and VASIL'EVA, A. G., *Kinetika i Kataliz* (1965) 394.
36. DUISMANN, W., HERTEL, R., MEISTER, J. and RÜCHARDT, Chr., *Ann. Chem.*, **109** (1976) 1820.
37. BERGER, H., *Abstracts Annual Meeting, Chemical Society No. 2*, 4A, Manchester (1972).

Chapter 2

PEROXIDOLYTIC ANTIOXIDANTS: SULPHUR ANTIOXIDANTS AND AUTOSYNERGISTIC STABILISERS BASED ON ALKYL AND ARYL SULPHIDES

GERALD SCOTT

Department of Chemistry, University of Aston in Birmingham, UK

SUMMARY

The common mechanisms involved in the antioxidant activity of compounds containing one or more sulphur atoms attached to carbon are discussed. The ability to destroy hydroperoxides catalytically is a general feature of the sulphur antioxidants but the parallel occurrence of pro-oxidant reactions limits the use of simple alkyl or aryl sulphides alone. However, in combination with an antioxidant acting by the chain-breaking (CB–D) mechanism, the peroxidolytic (PD–C) sulphides show powerful synergism under a variety of conditions. Recent developments in which a PD–C function is combined with a different antioxidant function in the same molecule are also described.

1. INTRODUCTION

1.1. Conceptual Basis of Peroxidolytic (PD–C) Antioxidants

In view of the critical role of hydroperoxides in autoxidation[1] two broadly complementary antioxidant mechanisms have been recognised.[2] The first, the kinetic chain-breaking mechanism, involves the removal of chain-propagating radicals to give inactive products. The second, the preventive mechanism, provides an alternative means of

29

inhibiting the chain reaction by removing or deactivating radical-generating species responsible for its initiation. By far the most important class of radical generators is the hydroperoxides since they are the universal products of the reaction of oxygen of the atmosphere (including singlet oxygen and ozone) with polymers.

A variety of peroxidolytic agents functioning by a catalytic mechanism (PD–C) have been recognised following the pioneering studies of Denison[3] and later of Kennerley and Patterson.[4] The latter showed that the products formed in the catalysed breakdown of hydroperoxides (e.g. α,α-dimethylbenzyl hydroperoxide, cumyl hydroperoxide, CHP; **I**) by transition metal dithiolates were different from those formed by transition metal ions. The early theories propounded to explain the action of peroxidolytic antioxidants are now largely of historical interest[5] since the fundamental studies of Kharasch and his co-workers[6] showed the characteristically different decomposition behaviour of aralkyl hydroperoxides in the presence of acidic catalysts (Scheme 1(a)) and transition metal ions (Scheme 1(b)). Analysis of the

SCHEME 1

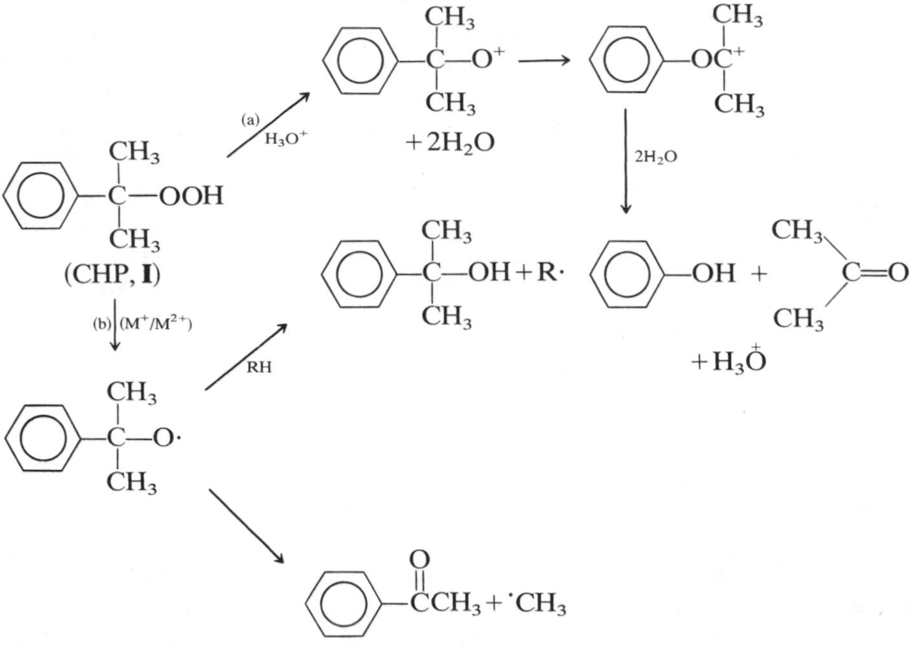

products formed in the catalytic decomposition of CHP (**I**) has proved to be an invaluable diagnostic technique in the study of the mechanism of peroxidolytic reactions. As will be seen in later sections, not only does it distinguish between the two extreme types of catalytic activity, but when, as frequently happens, the contribution of the two alternative processes varies depending on the conditions, it allows an estimate to be made of the relative contribution of each.

1.2. The Scope of the Peroxidolytic Antioxidant Mechanism in Polymer Stabilisation

The study of the peroxidolytic mechanism began independently in a number of different technologies. Early studies of antioxidant-active naturally occurring sulphur impurities in lubricating oils have been reviewed[5] and will not be considered in detail here except to note that the Group II dithiophosphate (**II**) complexes which were discovered about that time still constitute one of the most important classes of antioxidants/anticorrosive agents for lubricating oils. Their use in polymers will be discussed in Chapter 3.

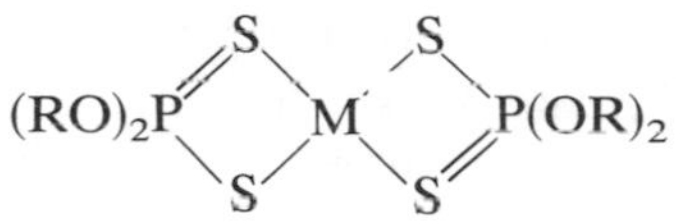

$$(RO)_2P \overset{S}{\underset{S}{\diagdown}} \quad \underset{S}{\overset{S}{M}} \quad P(OR)_2 \qquad\qquad (ROCOCH_2CH_2)_2S$$

(II)   (III)

About the same time dialkyl monosulphides were found to be effective antioxidants in foodstuffs[7] and this was followed by the acceptance of a number of esters of thiodipropionic acid (**III**, R = alkyl) as additives for foodstuffs by the US Government in 1949.[8]

In the field of polymers, the most important discovery was that rubbers cured with thiuram disulphides (**IV**) in the absence of sulphur but in the presence of zinc oxide gave vulcanisates with a high level of resistance to thermal oxidation.[9] This was shown[10,11] to be associated

SCHEME 2

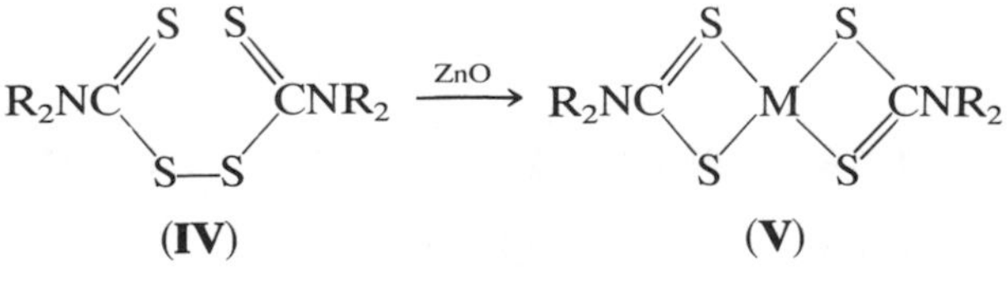

$$R_2NC \overset{S}{\diagup} \quad \overset{S}{\diagdown} CNR_2 \quad \overset{ZnO}{\longrightarrow} \quad R_2NC \overset{S}{\diagup} \quad M \quad \overset{S}{\diagdown} CNR_2$$

(IV)   (V)

with the formation of the corresponding zinc dialkyl dithiocarbamates (**V, M = Zn**) (Scheme 2). Like the metal dithiophosphates (**II**), the dithiocarbamates (**V**) had previously been used as antioxidants in lubricating oils[12] and similar activity was found for the former in rubber.[13]

With the later development of the polyolefins, the high activity of the metal dithiocarbamates (**V**) and metal dithiophosphates (**II**) as oxidation stabilisers in their own right was quickly recognised.[14] The simple monosulphides (**III**) on the other hand were found not to be very effective when used alone in polyolefins, although in combination with chain-breaking antioxidants they became the basis of most modern heat stabilising systems.[15,16] The chemistry involved in the antioxidant function of monosulphides proved to be very complex; although it was recognised that sulphur acids formed from them were catalysts for the decomposition of hydroperoxides,[17,18] the reaction sequence leading to their formation has only recently been elucidated.

1.3. Common Mechanisms of Sulphur Compounds as PD–C Antioxidants

From the brief historical survey presented above, it will be evident that the sulphur antioxidants discussed share a common mechanism or mechanisms. Subsequent investigations have shown that they have the following features in common.

(a) The effective antioxidant is not the parent sulphur compound but an oxidation product formed from it in the autoxidising medium.[17,18]

(b) In all cases, the effective antioxidant is a catalyst for the ionic decomposition of hydroperoxides (PD–C).[17]

(c) The antioxidant stage is frequently preceded by a pro-oxidant stage which varies in intensity. Hence the oxidations are generally autoretarding.[17]

In this chapter, the antioxidant mechanisms of sulphide antioxidants and their synergistic combinations with other types of antioxidant will be discussed and in Chapter 3 the related mechanisms of the metal thiolates will be considered.

2. THE BEHAVIOUR OF SULPHIDES IN AUTOXIDATION

The characteristic behaviour of sulphides in autoxidation has become apparent from studies in three quite independent technologies. The earliest was the discovery[3] that small amounts of naturally occurring sulphur compounds were responsible for the long term oxidative stability of lubricating oils. Ten years later, Stafford drew attention to the characteristic autoretarding characteristics of sulphur-vulcanised rubbers during autoxidation.[19] One of the most puzzling features of these and other studies was the observation that increasing sulphur content in the substrate increased the initial oxidation rate and resulted in a more rapid loss of mechanical properties in the vulcanisate.[20] The third key to the mechanism of action of sulphur antioxidants was the discovery of the extraordinary synergism between the thiodipropionate esters and chain-breaking antioxidants in polypropylene.[21] When used alone, these antioxidants were relatively ineffective, particularly at high temperatures during the processing of polypropylene[22] where they showed the characteristic autoretarding behaviour referred to above. An important function of the chain-breaking antioxidants (e.g. the hindered phenols) appeared to be to reduce or eliminate an initial pro-oxidant process involved in the oxidation of the sulphide to an effective antioxidant product.[22]

2.1. The Sulphur Cross-link in Rubber

In an attempt to rationalise the complex behaviour of sulphur vulcani-sates in rubber ageing, Bateman and his co-workers[23-26] studied the behaviour of a wide range of model sulphides related to the sulphur cross-link. Their conclusions can be summarised as follows.

(a) Model sulphides like sulphur vulcanisates oxidise in an auto-retarding mode,[23] suggesting the formation of an inhibitor during oxidation.

(b) Chain-branching on the carbon atom to sulphur increases the antioxidant effectiveness of the sulphide.[24]

(c) Dialkyl disulphides are more effective than dialkyl mono-sulphides.[24]

(d) Derived sulphoxides and thiolsulphinates are more effective than the parent sulphides and disulphides.[24]

(e) The effectiveness of sulphoxides and thiolsulphinates as antiox-idants appears to be related to their thermal instability.[25]

(f) All the active sulphur compounds are capable of destroying hydroperoxides.[26]

(g) Sulphoxides form stoichiometric hydrogen-bonded complexes with hydroperoxides and the antioxidant activity of the sul-phides is associated with the 'deactivation' of hydroperoxides by this mechanism.[25]

Subsequent work, notably by Shelton[16,27] on simple alkyl sulphides and by Scott and co-workers[28-30] on the thiodipropionate esters, has confirmed the main experimental conclusions of the MRPRA group. However, although there is no doubt that sulphoxides do form hydrogen-bonded complexes with hydroperoxides, this process appears to have no relevance to their antioxidant function since under conditions where complexation of hydroperoxides might be expected to be maximal (i.e. molar excess of sulphoxide), the sulphoxide (or to be more precise its decomposition products) catalyses the homolytic decomposition of hydroperoxides,[28,29] leading to pro-oxidant rather than antioxidant effects.

The salient conclusions of Shelton on the simple alkyl sulphides[27] are in accord with the parallel studies of the thiodipropionate esters and will be referred to as appropriate in the following section.

2.2. Esters of Thiodipropionic Acid

The dialkyl thiodipropionates (**III**) are not by themselves very effective antioxidants in polyolefins. Their potential is not realised unless they

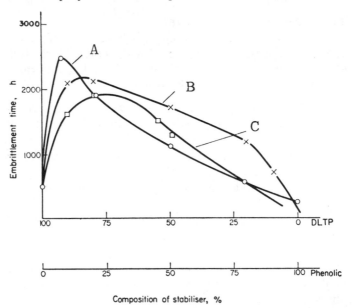

FIG. 1.   Synergism between DLTP and hindered phenols at constant total weight concentration at 150°C. A, Topanol CA; B, Tetrabis(4-hydroxy-3-*tert*-2-methylphenyl)butane; C, Nonox WSP.

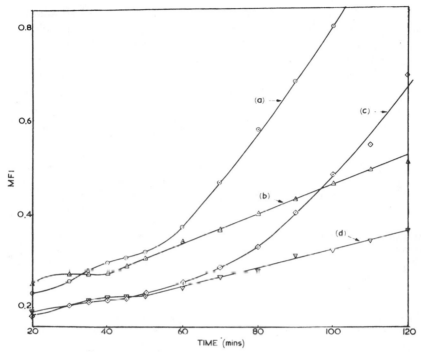

FIG. 2.   Change in melt flow index (MFI) of polypropylene. (a) DLTP, 2×10^{-3} mol/100 g; (b) DLTP, 4×10^{-3} mol/100 g; (c) as (a) with TBC, 2×10^{-3} mol/100 g; (d) as (b) with TBC, 4×10^{-3} mol/100 g. (After ref. 22 with permission.)

are used in admixture with an electron donor chain-breaking (CB–D) antioxidant.[22,31] A relatively small percentage of a high molecular weight phenolic antioxidant appears to be all that is required to give optimum stabilising effectiveness during the thermal oxidation of polypropylene at 150°C (Fig. 1).

The melt-stabilising effectiveness of the thiodipropionate esters was found to pass through a maximum with increasing concentration and, during the initial stages of processing, autoretardation was more pronounced the higher the concentration of the sulphide.[22] The addition of a phenolic antioxidant (TBC; **XXXII**, R = CH$_3$) not only increased the long-term effectiveness of dilauryl thiodipropionate (DLTP; **III**, R = C$_{12}$H$_{25}$) as a melt stabiliser but it also reduced the initial oxidative degradation produced by the sulphide alone (Fig. 2). A similar autoretardation was observed during oven ageing of polypropylene films at

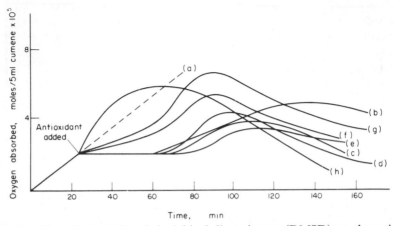

FIG. 3. The effect of dimethylsulphinyl dipropionate (DMSD) on the oxidation of cumene initiated by cumene hydroperoxide ([CHP] = 0·05 M) at 75°C at the following molar ratios [CHP]/[DMSD]: (a) no DMSD; (b) 18·5; (c) 3·0; (d) 2·0; (e) 1·0; (f) 0·85; (g) 0·59; (h) 0·23. (After ref. 29 with permission.)

140°C[29] and studies in model systems using cumene hydroperoxide (CHP) as initiator showed a parallel behaviour.[30] The extent of the pro-oxidant behaviour was found to be strongly dependent upon the ratio of hydroperoxide to sulphide[32] or to the derived sulphoxide[29] (Fig. 3). At low (0·33) molar [ROOH]/[S] ratios, an immediate pro-oxidant effect was evident. With increasing [ROOH]/[S] an induction period became evident followed by a delayed pro-oxidant effect. The change in mechanism is also reflected in a change in the products of decomposition of cumene hydroperoxide (CHP) with variation of molar [CHP]/[S] ratio. When the above equals 1, phenol and acetone are the main products by reaction (a) in Scheme 1. At [CHP]/[S] < 1, reaction (b) becomes increasingly important.[29]

As indicated earlier, sulphides are thermally stable but the derived sulphoxides undergo a facile breakdown to give the corresponding sulphenic acid.[27,33,34] In the case of the sulphinyl dipropionate esters (VI), this process (Scheme 3) was found to be reversible and became

SCHEME 3

$$(ROCOCH_2CH_2)_2S=O \rightleftharpoons ROCOCH_2CH_2SOH + ROCOCH=CH_2$$

(VI)　　　　　　　　　　　(VII)

first order in the presence of an oxidising agent which removed the corresponding sulphenic acid (**VII**) as it was formed.[28] Two such reagents which have been shown to facilitate the attainment of first-order thermolysis throughout its decomposition are the 'stable' radical, galvinoxyl (G·; **VIII**) and CHP.[34] The sulphinyl radical (**IX**) was found

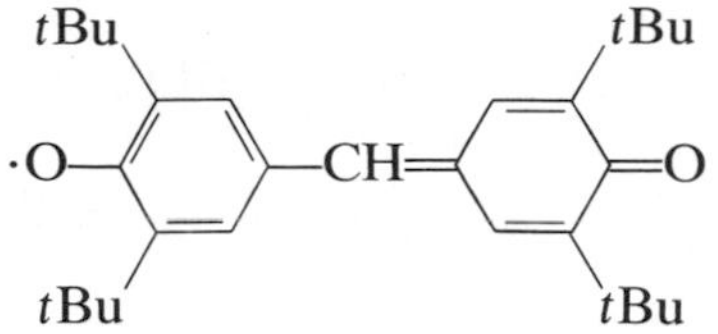

(**VIII**, G·)

to be inert under these conditions and dimerised to the thiolsulphonate.[28] However, the combination of sulphoxide and hydroperoxide initiated both oxidation and polymerisation.[28] From this it was concluded that the alkoxy radical formed in (b) is the most probable cause of the observed pro-oxidant effects at low [ROOH]/[S] (Scheme 4).

SCHEME 4

$$\underset{(\mathbf{VII})}{ROCOCH_2CH_2SOH} \xrightarrow[G\cdot]{(a)} GH + ROCOCH_2CH_2S\dot{O}\ (\mathbf{IX})$$

$$\underset{(\mathbf{VII})}{ROCOCH_2CH_2SOH} \xrightarrow[ROOH]{(b)} RO\cdot + H_2O + ROCOCH_2CH_2S\dot{O}\ (\mathbf{IX})$$

In the absence of hydroperoxide, the products formed from dimethyl sulphinyl dipropionate (**VI**, R = Me) were the corresponding thiolsulphinate (**X**) and its stable disproportionation products (**XI** and **XII**)[28,30] (Scheme 5). However, in the presence of a 10 : 1 molar excess of *tert*-butyl hydroperoxide (TBH), the corresponding sulphinic acid (**XIII**) was a major reaction product.[30]

Methyl β-sulphopropionate (MSP; **XIII**, R = Me) was found to be relatively stable at room temperature but it rapidly disproportionated at 60°C to give the corresponding thiolsulphinate (**XIV**) and sulphonic acid, MSPO (**XV**)[30] (Scheme 6).

At high temperatures, MSP was found to pyrolyse to give sulphur dioxide and dimethyl adipate as major products.[30] Homolysis (Scheme 7) was proposed to account for these products and the minor by-products, methyl propionate and methyl acrylate. The formation of

SCHEME 5

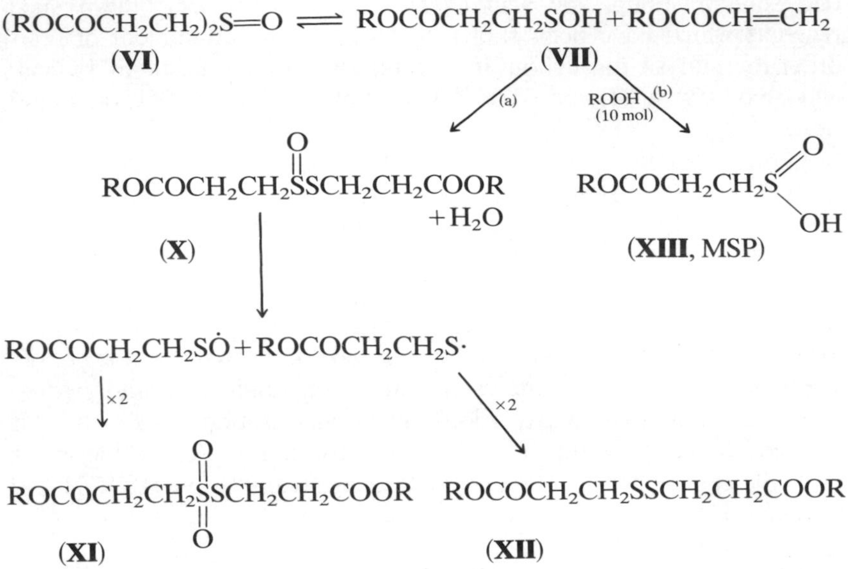

$$(ROCOCH_2CH_2)_2S{=}O \rightleftharpoons ROCOCH_2CH_2SOH + ROCOCH{=}CH_2$$

(VI)   **(VII)**

(a)   ROOH (b)
(10 mol)

$$ROCOCH_2CH_2\overset{O}{\overset{\|}{S}}SCH_2CH_2COOR + H_2O$$

(X)

$$ROCOCH_2CH_2\overset{O}{\overset{\|}{S}}{-}OH$$

(XIII, MSP)

$$ROCOCH_2CH_2S\dot{O} + ROCOCH_2CH_2S\cdot$$

$\times 2$ $\times 2$

$$ROCOCH_2CH_2\overset{O}{\underset{O}{\overset{\|}{\underset{\|}{S}}}}SCH_2CH_2COOR \qquad ROCOCH_2CH_2SSCH_2CH_2COOR$$

(XI)   **(XII)**

SCHEME 6

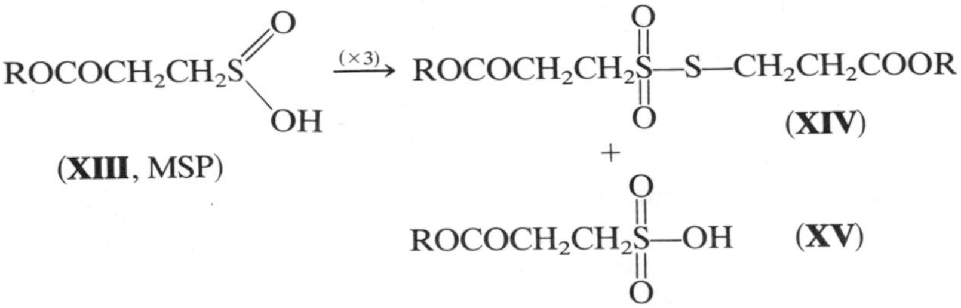

$$ROCOCH_2CH_2\overset{O}{\overset{\|}{S}}{-}OH \xrightarrow{(\times 3)} ROCOCH_2CH_2\overset{O}{\underset{O}{\overset{\|}{\underset{\|}{S}}}}{-}S{-}CH_2CH_2COOR$$

(XIII, MSP)   **(XIV)**

$+$

$$ROCOCH_2CH_2\overset{O}{\underset{O}{\overset{\|}{\underset{\|}{S}}}}{-}OH \qquad \textbf{(XV)}$$

SCHEME 7

$$MeOCOCH_2CH_2SO_2H \longrightarrow MeOCOCH_2CH_2\cdot + H\dot{S}O_2$$

(XIII, MSP)

$\times 2$ $\times 2$

$$(MeOCOCH_2CH_2)_2 \qquad\qquad MeOCOCH_2CH_3 + SO_2$$

$$MeOCOCH{=}CH_2$$
$$+$$
$$MeOCOCH_2CH_3$$

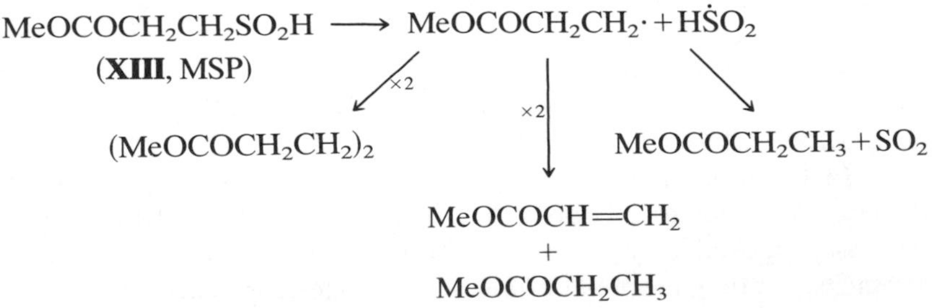

methyl propionate is consistent with the observation[35] that a propionate ester can be identified as a transformation product of DLTP (**III**, $R = C_{12}H_{25}$) in polypropylene.

Methyl β-sulphinopropionate (MSP), methyl-β-sulphopropionate (MSPO) and sulphur dioxide (see Section 3) were all found to be powerful antioxidants in oxidations initiated by both an alkyl radical generator (azobisisobutyronitrile, AIBN) and by a hydroperoxide (CHP).[30] The presence of the acidic hydrogen is necessary for antioxidant activity in both the protonic acids and under both initiating conditions. A simple CB–D activity (Scheme 8) has been proposed[30,36] to account for the activity of both compounds in the presence of an alkyl radical generator. This proposal was confirmed when MSP was

SCHEME 8

$$\text{MeOCOCH}_2\text{CH}_2\text{SO}_x\text{H} \xrightarrow[\text{CB–D}]{\text{ROO·}} \text{MeOCOCH}_2\text{CH}_2\dot{\text{S}}\text{O}_x + \text{ROOH}$$

(**XVI**)

found to react rapidly with a variety of free radicals including the relatively stable cyanoisopropyl and galvinoxyl (**VIII**),[30] just as the corresponding sulphenic acid (**VII**) does.[31] It seems likely then that all three sulphur acids (**XVI**, $x = 1, 2, 3$) are involved in the CB–D activity of the thiodipropionates (Scheme 8). In each case, the radical produced appears to be incapable of continuing the chain reaction.

A more complicated process is observed in the autoxidation of cumene initiated by CHP. At high [CHP]/[S] ratios, immediate inhibition (with gas evolution) was observed (see Fig. 4). At [CHP]/[S] ratios less than 1, an initial pro-oxidant effect was observed which was not masked by gas evolution (Fig. 4).[30] Complete cessation of oxidation was ultimately observed in all cases. The similarity of this behaviour to that of the thiodipropionates is evident (cf. Fig. 3). No pro-oxidant effect was observed, however, with MSPO at any ratio. Again it seems likely that the reducing sulphur acids undergo redox reactions with hydroperoxides (Scheme 9, $x = 0, 1, 2$) whereas MSPO, which is not a

SCHEME 9

$$\text{MeOCOCH}_2\text{CH}_2\text{SO}_x\text{H} + \text{ROOH}$$
$$\rightarrow \text{MeOCOCH}_2\text{CH}_2\dot{\text{S}}\text{O}_x + \text{H}_2\text{O} + \text{RO·}$$

(**XVI**)

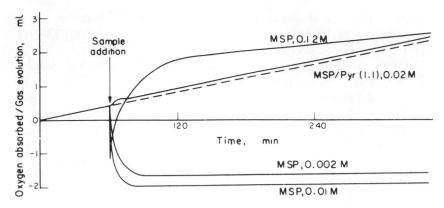

FIG. 4.   Oxidation of cumene initiated by CHP (0·1 M) in the presence of MSP and its pyridine salt. Numbers on curves are additive concentrations. (After ref. 30 with permission.)

reducing agent, does not. The evidence discussed suggests that both alkyl sulphinyl and alkyl sulphonyl propionates are involved in the peroxidolytic antioxidant action of the thiodipropionate esters at low temperatures, and sulphur dioxide and its further oxidation products at high temperatures. As will be seen below, this is almost certainly an oversimplification. All three sulphur compounds are powerful catalysts for the ionic decomposition of hydroperoxides. Comparison of the rates at approximately equivalent molar concentrations (Fig. 5) shows that SO_2 and MSP are considerably more active than the sulphonic acid, MSPO. The first two tend to the same limiting rate with time and satisfy the requirement that the necessary concentration of the catalyst is produced during the induction period. The sulphonic acid does not.[30]

Dimethyl sulphinyl dipropionate (VI, R = Me) showed an induction period at 75°C before it became an effective catalyst for the ionic decomposition of cumene hydroperoxide. The same compound, after heating at 75°C for 20 h before being used as a catalyst, showed no induction period and produced the maximum observed yield ($\simeq$72%) of phenol in about one-third of the time. As indicated earlier, the corresponding thiolsulphinate (X) is the first soluble sulphur-containing reaction product under these conditions (see Scheme 5). This compound was found to be a powerful peroxidolytic agent whereas its further disproportionation products (XI) and (XII) were not. The formation of an acrylate ester in a solution of thiolsulphinate (X)

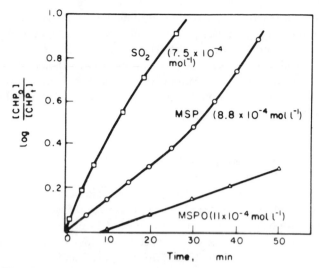

FIG. 5. Kinetics of decomposition of CHP in the presence of sulphur compounds at the concentrations indicated ([CHP] = 0·15 M). (After ref. 30 with permission.)

heated in the absence of air indicated the formation of thiolsulphoxylic acid (**XVII**), a reaction previously reported by Block[37] for the decomposition of *tert*-butyl thiolsulphinate (Scheme 10(a)).

SCHEME 10

$$ROCOCH_2CH_2\overset{\overset{\displaystyle O}{\|}}{S}SCH_2CH_2COOR \xrightarrow{(a)} ROCOCH=CH_2$$

$$+$$

$$ROCOCH_2CH_2SSOH$$

$$(\textbf{XVII})$$

$$\xrightarrow[(b)]{\diagup 2\,ROOH}$$

$$ROCOCH_2CH_2SOH + SO_2$$

(X)

A similar reaction has been reported by Shelton[27] in the decomposition of di-*tert*-butyl thiolsulphinate and this author attributes at least part of the ionic catalytic activity to this species. However, it seems likely that thiolsulphoxylic acids will be readily oxidised further to sulphur acids with the elimination of sulphur dioxide (Scheme 10(b)).

The direct formation of sulphur dioxide has been reported during the pyrolysis of diaryl thiolsulphinates and sulphonates at 120°C.[38] No evidence was found for the direct formation of SO_2 from **X** at 75°C and the thiolsulphonate **XI** was quite stable under these conditions.[30] However, the possibility that direct SO_2 formation from these species by pyrolysis occurs under high temperature processing conditions in polypropylene cannot be ruled out.

3. THE ROLE OF SULPHUR DIOXIDE

It is a frequent assumption that sulphur dioxide is the essential peroxidolytic catalyst produced from a variety of sulphur antioxidants.[38] The proposal was originally made independently by Hawkins[18] and by Scott[17] on the basis of circumstantial evidence. Although sulphur dioxide is more effective than sulphuric acid as a catalyst for peroxide decomposition, it seems unlikely that SO_2 can survive in the strongly oxidising environment of a typical autoxidation and it is therefore probable that the effective catalyst(s) are either sulphur trioxide or sulphuric acid. Experimental studies with SO_2 in a model system supports this view.[39] Like most of the sulphur compounds discussed in the previous section, SO_2 was found to cause a pro-oxidant effect at $[CHP]/[SO_2]$ molar ratios in the region of 40 (Fig. 6). At higher ratios, no pro-oxidant effect was observed and at molar

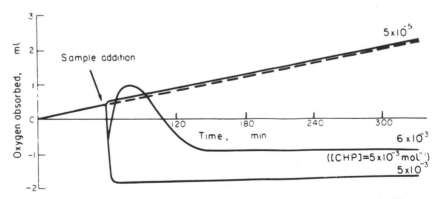

FIG. 6. Effect of sulphur dioxide on the oxidation of cumene initiated by CHP (0·1 M). Numbers on curves are SO_2 concentrations, mol litre^{-1}. (After ref. 39 with permission.)

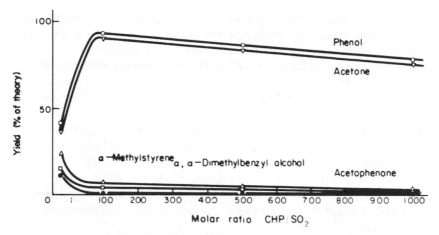

FIG. 7.   Effect of molar [CHP]/[SO$_2$] ratio on the product distribution in the catalytic decomposition of CHP by SO$_2$ at 75°C. (After ref. 39 with permission.)

[CHP]/[SO$_2$] = 100, the decomposition products of CHP were found to be almost entirely the expected heterolysis products phenol and acetone (Fig. 7) whereas at [CHP]/[SO$_2$] < 1, they were predominantly homolytic.[39] These findings are consistent with the formation of hydroxyl and sulphinyl radicals observed by ESR during the reaction of hydroperoxides with SO$_2$[40] (reactions (a) and (b) in Scheme 11).

SCHEME 11

$$\text{ROOH} + \text{SO}_2 \xrightarrow{\text{(a)}} \left[\begin{array}{c} \text{OH} \\ | \\ \text{ROOS}{=}\text{O} \end{array} \right]$$

(b) ↙   ↘ (c)

$$\text{RO}\cdot + \cdot\text{OS} \diagup^{\text{OH}}_{\diagdown\text{O}}$$ $$\text{ROH} + \text{SO}_3$$

(Pro-oxidants)   (Antioxidants)

At high (>45) molar [CHP]/[SO$_2$] ratios, the decomposition of CHP is pseudo first order over most of its decomposition. At lower [CHP]/[SO$_2$] ratios (17·5), the rate is initially non-integral but tends to zero order with separation of an organic insoluble phase which was found to be essentially H_2SO_4. At high ratios, it seems likely that a soluble strong acid is responsible for the catalysis. The same catalyst appears to be formed from MSP (**XIII**) under the same conditions and is almost certainly SO_3 formed by the alternative ionic breakdown of the unstable peracid (Scheme 11(c)). Evidence for the conversion of SO_2 to a catalyst for hydroperoxide decomposition was found by incremental addition of TBH to an excess of SO_2 in organic solution (Fig. 8). The reaction was initially stoichiometric but became catalytic well before a stoichiometric equivalent of hydroperoxide had been added. It is clear from this that Lewis acid catalysis can operate even in the presence of an excess of the radical-generating redox system: a conclusion of some importance for the behaviour of sulphur-containing antioxidants in polymers.

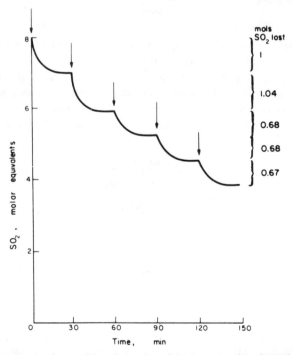

FIG. 8. Incremental reaction of *tert*-butyl hydroperoxide (TBH) with excess sulphur dioxide in chlorobenzene at 75°C. 1 mol equivalent of TBH was added at points indicated. (After ref. 39 with permission.)

4. THE BEHAVIOUR OF THIODIPROPIONATE ESTERS IN POLYMERS

Scheme 12 summarises the known chemistry involved in the antioxidant mechanism of the thiodipropionate esters. This shows a complex series of redox reactions associated with pro-oxidant effects, donor chain-breaking antioxidant effects (CB–D) and catalytic ionic peroxidolytic effects (PD–C) through a broad programme from the first two to the last with time. Although some details remain to be clarified, this scheme explains many of the technological phenomena associated with the thiodipropionate esters and with the autosynergistic antioxidants based on sulphides to be discussed in later sections.

4.1. During Processing

The situation existing during the early stages of processing when a thiodipropionate ester is mixed with a polymer at higher temperatures favours the formation of free radicals, due to the presence of high [S] and low [ROOH]. This leads to an initial pro-oxidant effect (see Fig. 2) which autoretards to give a substantial induction period. Conventional CB–D antioxidants effectively inhibit the pro-oxidant effect and synergise with the derived PD–C antioxidant. Thiodipropionates are therefore always used in combination with radical scavengers.

Somewhat surprisingly, dilauryl sulphinyl (dipropionate (**VI**, R = $C_{12}H_{25}$) has been found to be a highly effective melt stabiliser for polypropylene under high temperature processing conditions (270°C).[41] Under these conditions, the sulphenic acid (**VII**) is formed extremely rapidly. This agent is known to be a powerful trap for both alkyl[28] and alkylperoxyl[29,36] radicals, both of which are important in mechano-chemically initiated oxidation.[42] Sulphoxides which cannot break down to give sulphenic acids (e.g. dimethyl or diphenyl sulphoxide) are not effective. There is evidence to suggest that cyclical regenerative processes may be involved in this activity.[42]

4.2. During Oven Ageing

Thiodipropionate esters are quite effective thermal antioxidants in fabricated polymers even in the absence of added synergists. This can be seen from Fig. 1, which shows that at 150°C, DLTP is more effective when used alone than typical high molecular weight hindered phenols. However, the two types of antioxidant powerfully augment

SCHEME 12. Main reactions of thiodipropionate esters in autoxidation. Abbreviations: ex, excess; defic, deficiency.

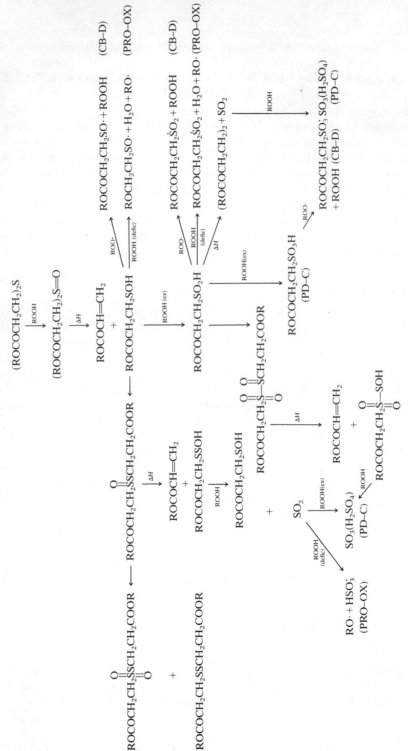

each other by virtue of their different mechanisms. The greater effectiveness of DLTP in the fabricated polymer than in the melt is due to the fact that it has been partially converted during the processing operation to the PD–C species. The same phenomenon is observed during the later stages of the thermal ageing of rubber vulcanisates containing sulphide cross-links[43] although in this case it comes too late in the oxidation sequence to be technologically useful, since the properties of the vulcanisate are essentially destroyed by then.

4.3. During Fatigue Ageing

Fatigue in polymers resembles the processing operation and differs from normal thermal oxidation in that initiation is primarily due to the formation of macro-alkyl radicals and termination involves both macro-alkyl and macro-alkylperoxyl radicals.[42,44] Sulphur vulcanisates are known to be much more resistant to fatigue than peroxide (carbon–carbon) cross-linked vulcanisates. This appears to be associated with the liability of the partially oxidised polysulphidic cross-links. However, this is not simply a relaxation of stress by cross-link rearrangement, since if sulphur derivatives of thiodipropionic acid (notably DLTP, **III**, $R = C_{12}H_{25}$; DLSP, **VI**, $R = C_{12}H_{25}$) and dilauryl dithiodipropionate (DLTDP; **XVIII** $= (ROCOCH_2CH_2S)_2$ where $R = C_{12}H_{25}$) are incorporated into a peroxide vulcanisate by swelling, the fatigue life increases with the number of sulphur atoms and with decrease in thermal stability of the molecules[45] (see Table 1). A high degree of polysulphide cross-linking favours good fatigue resistance but gives the least satisfactory thermal oxidative resistance. The mechanistic link

TABLE 1

FATIGUE LIFE OF PEROXIDE-VULCANISED RUBBERS CONTAINING SULPHIDES AND DERIVATIVES (2 g/100 g) SWOLLEN INTO THE VULCANISATES[45]

Vulcanised rubber	Fatigue life (hours to failure)
Peroxide vulcanisate (P) (no additive)	19
P + DLTP (**III**, $R = C_{12}H_{25}$)	50
P + DLSP (**VI**, $R = C_{12}H_{25}$)	58
P + DLTDP (**XVIII**, $R = C_{12}H_{25}$)	90
Extracted polysulphide vulcanisate (CBS 0·5 g/100 g; S 2·5 g/100 g)	163

between the opposed phenomena is the formation of sulphenic and sulphinic acids which have been shown to be readily converted to the corresponding sulphur-centred radicals by redox reaction with hydroperoxides (Scheme 12). Once formed, the sulphur-centred radicals are relatively stable and are effective alkyl radical traps. It seems possible that they are involved in regenerative processes similar to those involved with nitroxyl radicals during fatiguing processes[42,46] or with sulphur-centred radicals in the processing of polypropylene, but this remains to be established.

5. DIARYL AND DIARALKYL SULPHIDES

5.1. Diarylsulphides

Diaryl and dialkaryl sulphides, like the dialkyl sulphides, are not effective antioxidants unless they are synergised with chain-breaking antioxidants (see Section 6). Indeed some aryldisulphides, notably **XIX**, are effective pro-oxidants for rubbers[47,48] and polyolefins[31] and

(**XIX**; (**a**) R = H; (**b**) R = NHCOPh)

are used as chemical plasticisers to reduce the viscosity of rubbers before incorporating other compounding ingredients.[47] In common with the dialkyl monosulphides, the diaryl disulphides typified by **XIXb** rapidly autoretard to give a long-term antioxidant effect.[47,48] Evidence that further oxidation products of the disulphides are involved in the antioxidant activity was provided by the work of Hawkins and Sautter[49] in a study of the activity of **XIXa** and its derived oxidation products in the autoxidation of cumene at 120°C (Scheme 13). Whereas the disulphide (**XIXa**) became effective as an antioxidant only after an initial period of inactivity, the derived thiolsulphinate (**XX**) was effective immediately. The end disproportionation product of **XX**, the thiolsulphonate **XXI**, was less effective but became an antioxidant after a longer period of inactivity. Thiolsulphonates undergo thermolysis relatively slowly at temperatures below 130°C[50,51] whereas thiolsulphinates are unstable at much lower temperatures and readily undergo further reaction with hydroperoxides to give sulphur acids.

SCHEME 13

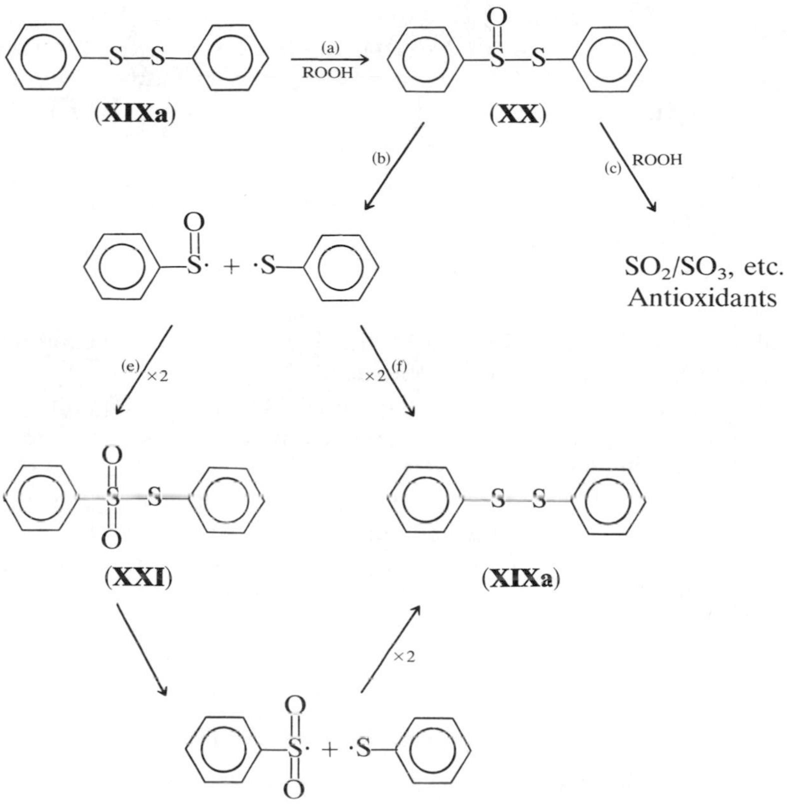

Reaction (b) in Scheme 13 and subsequent homolytic processes tend to predominate during the initial stages of autoxidation, but once hydroperoxides have built up in the system, 13(c) effectively inhibits further oxidation. A practical application of the dual behaviour of the diaryl disulphides is the use of **XIXb** as a pródegradant during the melt spinning of polypropylene fibre and as a long-term stabiliser at lower temperatures.[31]

5.2. Aralkyl Sulphides
Benzyl monosulphides have been found to show the same general behaviour as the diaryl disulphides in polypropylene.[31] Derived benzylic sulphoxides (**XXII**) have been shown to undergo facile reversible

SCHEME 14

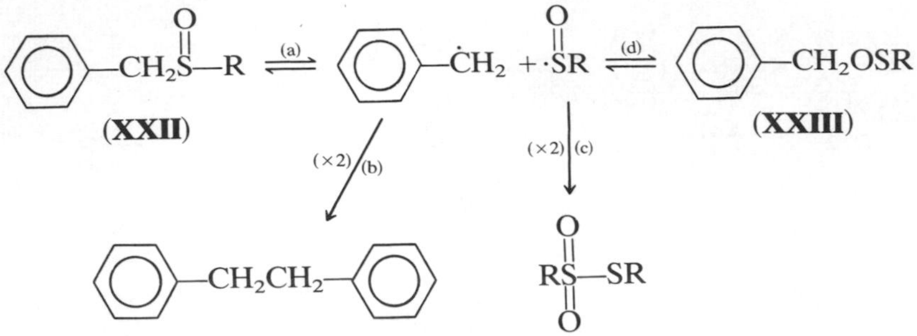

(XXII)   (XXIII)

homolysis[52,53] (Scheme 14) under the high temperature conditions found during the processing of polymers.

Pro-oxidant reactions during the initial stages of oxidation[31] can therefore be explained on this basis. Further oxidation of the rearranged sulphenate esters (XXIII) by hydroperoxides gives rise to an effective ionic catalyst for the decomposition of hydroperoxides[54] (Scheme 15). The same catalyst is obtained irrespective of the nature of the benzyl moiety (see Section 6.1) and the evidence strongly suggests that it is again a low molecular weight sulphur acid.

SCHEME 15

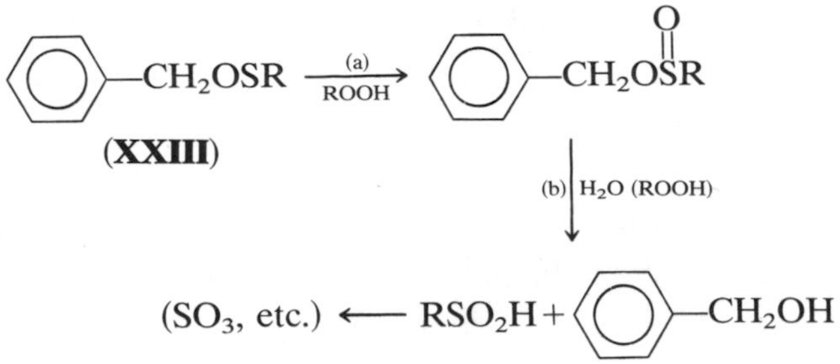

(XXIII)

$(SO_3, etc.) \longleftarrow RSO_2H + \langle O \rangle - CH_2OH$

6. AUTOSYNERGISTIC ANTIOXIDANTS INVOLVING THE PEROXIDOLYTIC MECHANISM

6.1. Phenolic Sulphides as Thermal Antioxidants in Polyolefins

Bisphenolic sulphides (e.g. XXIV, XXV) have been used for many years as stabilisers for polyethylene.[55] They are in general more

(**XXIV**)   (**XXV**)

effective thermal oxidation stabilisers in polyolefins than they are in vulcanised rubbers. Some members of this class were shown by Kennerly and Patterson in their early studies of peroxidolytic antioxidants[4] to be as effective catalysts for the destruction of hydroperoxides as the simple alkyl monosulphides discussed in Section 2. Pospisil and his co-workers have shown[56,57] that **XXV** undergoes sulphur elimination under autoxidative conditions and it has been shown[38] that at 120°C the sulphoxide derived from **XXV** undergoes thermolysis to give SO_2 in an inert solvent. Autosynergism between the two antioxidant functions in the same molecule (hindered phenol, CB–D, and sulphide, PD–C) accounts for the enhanced effectiveness of the bisphenolic sulphides.[58] However, it has been found[54] that aryl monosulphides are much less effective peroxidolytic agents than the 4-hydroxybenzyl monosulphides (**XXVI**) at 70°C (see Table 2).

(**XXVI**; (**a**) R = H(BHBM); (**b**) R = $C_{10}H_{21}$; (**c**) R = CH$_2$— 〈aryl〉 —OH)

The interesting feature to emerge from these studies is the apparently invariant activity on a molar basis of the ionic catalyst produced from the sulphur antioxidants. Not only do the 4-hydroxybenzyl sulphides show a similar first-order rate constant at the same molar concentration, but also do the non-phenolic dibenzyl monosulphide (**XXVII**) whose function was discussed in the last section and the more remotely related thioglycollic acid ester (**XXVIII**).[54,58] However, alkyl sulphides such as **XXVII**, **XXVIII** and **XXIX** are not effective thermal

TABLE 2

INDUCTION PERIODS AND PSEUDO FIRST ORDER RATE CONSTANTS (k) FOR THE
IONIC DESTRUCTION OF CHP $(0.01$ M$)$ BY SULPHUR ANTIOXIDANTS $(0.001$ M$)$ AT
$70°C^{54}$

Antioxidant	Induction period (min)	$10^2\,k(s^{-1})$
XXV	—	0·01
XXVIa	25	3·4
XXVIb	33	3·8
XXVIc	35	3·8
XXVII	28	3·7
XXVIII	31	3·3

(XXVII)

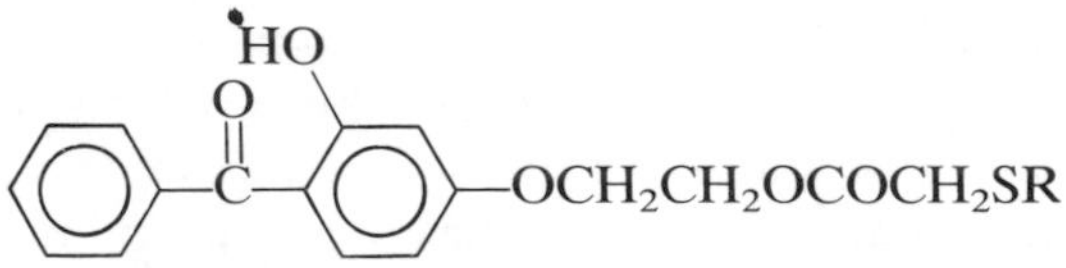

(XXVIII)

antioxidants, unlike **XXX** which also contains a phenolic group in the
same molecule.[60] The latter compound, which is related to the bound
antioxidant structure produced by reacting **XXVIa** with *cis*-poly-
isoprene, is substantially more effective as an antioxidant in tetralin
than a synergistic equimolar combination of **XXIX** and a hindered
phenol, TBC (**XXXII**, R = CH$_3$), which contains the same antioxidant
functional group.[59]

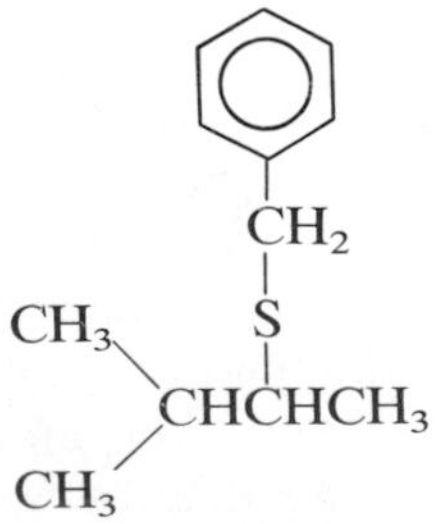

(XXIX)

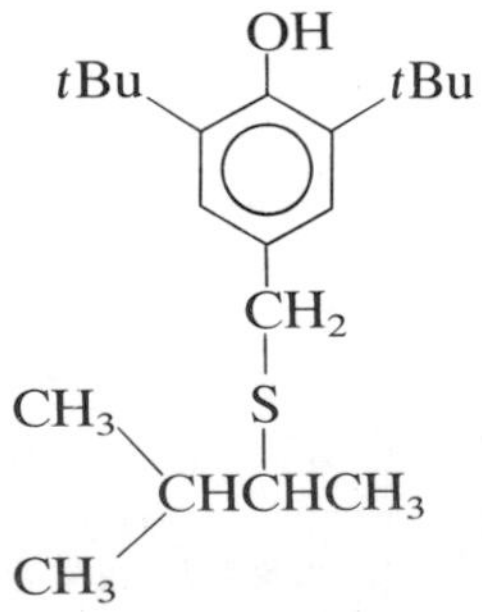

(XXX)

The phenolic benzyl sulphides (**XXVI**) have been found to be extraordinarily effective catalysts for the ionic destruction of hydroperoxides.[54] The stoichiometric coefficient (i.e. the number of mol of cumene hydroperoxide destroyed by 1 mol of sulphur compound), ν, for **XXVI**, $R = C_{12}H_{25}$, was found to be 3×10^5, over two orders of magnitude higher than for an *ortho* analogue, **XXXI**.

$$\text{OH}$$
CH$_2$SC$_7$H$_{15}$

(**XXXI**)

CH$_3$

Table 3 shows that autosynergistic antioxidants with the structure **XXVI** have a much higher intrinsic antioxidant activity on a molar basis (as measured by oxygen absorption in decalin[34]) than the most effective commercial hindered phenols (**XXXII**).[60]

tBu   OH   tBu

(**XXXII**: (a) $R = CH_3$ (TBC);

(b) $R = CH_2CH_2COOC_{18}H_{37}$ (Irganox 1076))

R

In polymers, physical factors other than intrinsic antioxidant activity (as measured by induction period in decalin) become relatively more important.[34,61] In particular, solubility in the substrate correlates broadly with antioxidant performance in a closed system (PP_c). In an open system (air oven), however, volatility factors dominate antioxidant performance; **XXVI**, $R = C_{18}H_{37}$ is less effective than the commercial antioxidant Irganox 1076 (**XXXII**, $R = CH_2CH_2COOC_{18}H_{37}$) under these conditions because it is more volatile (Table 3). Ideally from this point of view, the group R should be a polymer chain. The thiol antioxidant (**XXVIa**) can be made to react with polyolefins in the presence of a UV light generator[65] leading to highly effective nonvolatile autosynergistic antioxidant systems. Table 3 shows the effectiveness of **XXVIa** in polymer-bound form with the same antioxidant as a conventional additive.[65]

TABLE 3

COMPARISON OF ANTIOXIDANT ACTIVITY AT 140°C AND PHYSICAL PROPERTIES OF PEROXIDOLYTIC ANTIOXIDANTS (**XXVI**) WITH CONVENTIONAL HINDERED PHENOLIC ANTIOXIDANTS (**XXXII**) (all antioxidants were at a concentration of 6×10^{-4} mol/100 g)[60]

Antioxidant	IP^a (h) D_c	IP^a (h) PP_c	PP_o	LOR ($cm^3/h/100\,cm^3$) D_c	LOR PP_c	V ($g/h \times 10^3$)	S (g/100 g)
None	—	—	0·2	180·0	280·0	—	—
XXVI, R=H	51·0	38·5	21·0	180·0	5·0	110·0	96
XXVI, R=C$_2$H$_5$	47·5	33·0	5·0	2·4	5·4	100·0	65
XXVI, R=C$_5$H$_{11}$	47·0	38·0	6·0	2·5	5·0	77·5	∞
XXVI, R=C$_8$H$_{17}$	46·5	39·5	7·5	2·4	5·0	43·1	∞
XXVI, R=C$_{12}$H$_{25}$	45·0	44·5	11·0	2·5	5·0	15·0	94
XXVI, R=C$_{18}$H$_{37}$	44·5	27·0	15·0	2·4	5·5	2·0	57
XXVI, R=polypropylene	—	—	83·0	175·0	21·0	—	—
XXXII, R=CH$_3$	10·0	6·0	0·4	175·0	21·0	157·5	98
XXXII, R=CH$_2$CH$_2$COOC$_{18}$H$_{37}$	5·0	18·0	92·0	180·0	25·0	0·7	64

^{a} Abbreviations **IP** = induction period; **LOR** = linear oxidation rate at end of **IP**; D_c = oxygen absorption (in decalin, in a closed system); PP_c = oxygen absorption (in polypropylene in a closed system); PP_o = carbonyl formation (in an air oven); V = volatility of antioxidant (weight loss at 140°C); S = solubility of antioxidant in hexane at 25°C.

6.2. Autosynergistic Antidegradants Containing Chain-breaking and Peroxidolytic Functions in Unsaturated Polymers

6.2.1. Heat Stabilisers for Rubbers and Rubber Modified Plastics

The reaction of a variety of thiol antioxidants with lattices of olefinically unsaturated polymers has been described in earlier volumes in this series.[44,58] The thiol groups in **XXVIa**, **XXVIII**, R = H and **XXXIII**, R = H, provide a convenient method of attaching the antiox-

$$\langle O \rangle \!\!-\!\! NH \!\!-\!\! \langle O \rangle \!\!-\!\! NHCOCH_2SR \qquad (\textbf{XXXIII})$$

idant to the polymer backbone by radical addition to the double bond so that it cannot subsequently be removed by volatilisation or solvent leaching.[62 66] The technological behaviour of such a 'molecularly dispersed' antioxidant in a polymer above its glass transition temperature is a much more reliable measure of its intrinsic antioxidant activity than can be achieved by conventional incorporation as an additive, since subsequent performance in the second case may be dominated by antioxidant loss from the polymer or incompatibility in the polymer matrix. The adduct of **XXVIa** with ABS has been shown to be considerably more active as a thermal antioxidant at 100°C after exhaustive solvent extraction than **XXXIIa** and **XXXIIb** at equivalent molar concentration without extraction.[59,65]

Autosynergistic effects involving sulphur are much less in evidence in vulcanised rubbers due to the presence of a high concentration of mono-, di- and polysulphides in the rubber network. Nevertheless, adducts of **XXXIII** have been found to be substantially more effective under aggressive conditions in natural rubber (NR), styrene–butadiene rubber (SBR) and nitrile–butadiene rubber (NBR) than the best commercially available polymeric amine heat-ageing antioxidants. This is shown typically in a nitrile rubber vulcanisate in Fig. 9.[67] It should be noted that the autosynergist (**XXXIII**, R = NBR) was much superior to BHBM adduct (**XXVI**, R = NBR) even after extraction, although the

$$\left[\begin{array}{c} CH_3 \\ | \\ C \!\!=\!\! CH \\ \langle O \rangle \qquad \quad CH_3 \\ | \\ N \!\!-\!\! C \\ | \qquad \backslash \\ H \qquad CH_3 \end{array} \right]_n \qquad (\textbf{XXXIV})$$

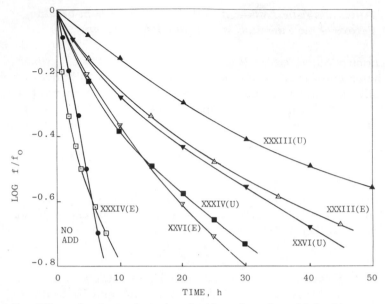

FIG. 9. Effect of polymer-bound antioxidants based on thiol adducts on the continuous stress-relaxation of nitrile–butadiene rubber at 150°C. U, unextracted; E, extracted.

extent of binding was similar (≃75%) in both cases. This is consistent with the known superiority of p-phenylenediamines as chain-breaking antioxidants in vulcanised rubbers. However, BHBM was superior to the commercial stabiliser, Flectol H (**XXXIV**), both before and after extraction.

6.2.2. Antifatigue Agents/Antiozonants

Thioglycollic amide (**XXXIII**) adducts with NR have been found to be very effective thermal antioxidants and antiozonants.[63] In both situations, **XXXIII**, R = NR, appears to be very effective even after extracting the rubber to remove extra-network antiozonants (see Fig. 10). This is not unexpected in the context of heat ageing but it is very surprising that a polymer-bound antioxidant should function as effectively as an antiozonant since it has been suggested that migration of antiozonants to the surface of rubber is necessary for effective protection against ozone. There is little doubt that the sulphide moiety is involved in the antiozonant mechanism since the autosynergist is

$$\langle\bigcirc\rangle\text{—NH—}\langle\bigcirc\rangle\text{—NHCH(CH}_3)_2 \quad (\textbf{XXXV})$$

superior to one of the most effective commercial antiozonants (**XXXV**) containing the same aromatic amine function without extraction.

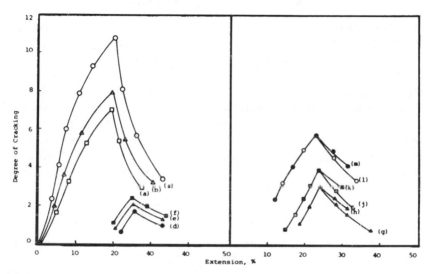

FIG. 10. Relationship between strain and degree of cracking at different ozone exposure times. (a) Control, 4 h; (b) control, 8 h; (c) control, 16 h; (d) MADA-B, 4 h; (e) MADA-B, 8 h; (f) MADA-B, 16 h; (g) IPPD, 4 h; (h) MADA-B, extracted, 4 h; (j) IPPD; (k) MADA-B, extracted, 8 h; (l) IPPD, 16 h; (m) MADA-B, extracted, 16 h. (After ref. 68 with permission.)

6.3. Sulphides as Thermal Stabilisers for PVC

It has been known for many years that sulphur compounds are effective synergistic stabilisers for PVC[69] and although peroxide decomposition has been suggested as one possible method by which they may function[70,72] there was until recently little evidence to suggest that hydroperoxides play a significant part in the thermal degradation of PVC. However, studies in model systems have led to the recognition of the great importance of redox reactions between hydroperoxides and HCl during technological processing operations[71,72] and in this situation, sulphur compounds might be expected to play a part both during processing and during subsequent ageing.[71,73]

6.3.1. Alkyltin Thioglycollate Esters

Derivatives of tetravalent tin have assumed a central role in PVC stabilisation technology in recent years.[73] The most important are

$$R_2Sn(OCOR')_2 \qquad R_2SnS(CH_2COOR')_2$$

$$\textbf{(XXXVI)} \qquad\qquad \textbf{(XXXVII)}$$

XXXVI, where R' contains a maleate group, and the thioglycollates, **XXXVII**, which are particularly effective under conditions of thermal oxidation (oven ageing).[74] Dioctyltin bis(isooctylthioglycollate) (DOTG; **XXXVII**, R = nC_8H_{17}, R' = 2-ethylhexyl) has been shown to be an effective catalyst for the decomposition of cumene hydroperoxide.[74,75] At long reaction times, the products are those expected on the basis of an ionic decomposition (i.e. phenol and acetone). As in the case of other sulphides referred to in the preceding sections, the process is pseudo first order after an initial second-order reaction. Figure 11 shows that cumyl alcohol is the major product formed during the initial second-order reaction in a radical generating process.

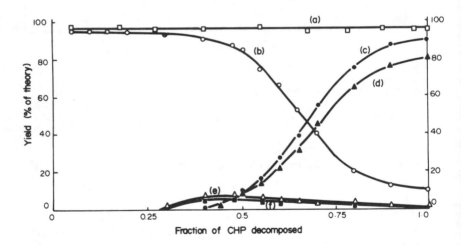

FIG. 11. Decomposition products formed from CHP ($2\cdot6\times10^{-2}$ M) in the presence and absence of DOTG ($2\cdot6\times10^{-3}$ M) in chlorobenzene at 60°C. (a) Cumyl alcohol in the absence of DOTG; (b) cumyl alcohol in the presence of DOTG; (c) phenol in the presence of DOTG; (d) acetone in the presence of DOTG; (e) acetophenone in the presence of DOTG; (f) methanol in the presence of DOTG. (After ref. 75 with permission.)

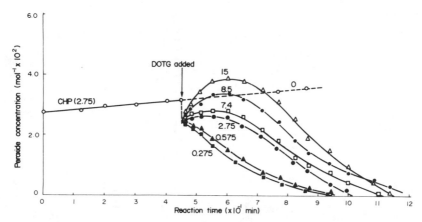

FIG. 12.   The formation of hydroperoxide in cumene initiated by CHP ($2 \cdot 75 \times$ 10^{-2} mol litre^{-1}) at 50°C in the presence of various concentrations of DOTG. Numbers on curves are DOTG concentrations, 10^2 M. (After ref. 75 with permission.)

Evidence for the intervention of free radicals was found in a study of the autoxidation of cumene catalysed by cumene hydroperoxide (Fig. 12). At molar ratios [CHP]/[DOTG] < 1, a pro-oxidant effect is always observed initially, although the peroxide concentration is always reduced to zero eventually. At [CHP]/[DOTG] > 1, pro-oxidant effects are not observed. A further function of DOTG is to neutralise HCl liberated from PVC during processing.[71,72] The effectiveness of DOTG depends on the residual concentration of HCl in the system. Even in the presence of stoichiometric excess of HCl which causes an immediate pro-oxidant reaction, the latter is removed by DOTG to give an antioxidant effect.[71] The additional peroxidolytic antioxidant function of DOTG is responsible for its relatively high thermal antioxidant activity (oven ageing at 140°C) compared with dibutyltin maleate (DBTM; **XXXVI**, R'R' = —CH=C—), under the same conditions.[74]

DOTG reacts rapidly with *tert*-butyl hydroperoxide at equimolar concentrations at 45°C to give the corresponding sulphoxide;[71,74] see Scheme 16. This compound is relatively stable at 45°C but it undergoes thermal breakdown at 65°C, a behaviour analogous to that of the benzyl sulphides discussed in Section 5.2. The formation of sulphur acids from the tin sulphenates (Scheme 16(a)) can be postulated by analogy with the behaviour of the benzyl sulphides. However, this is not the only way in which sulphur acids can be formed, since in the

SCHEME 16

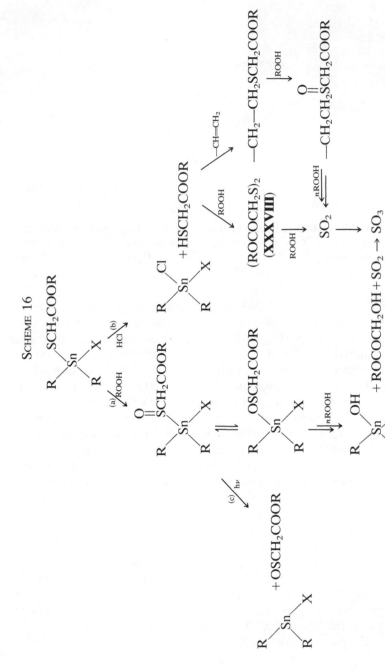

presence of HCl, free thioglycollic esters and further adduct products are also known to be formed.[71,74] These can also lead to the formation of free sulphur acids (under oxidative conditions, Scheme 16(b)). The disulphide (**XXXVIII**) has also been identified as a product formed from **XXXVII** in the presence of hydroperoxides.[76]

6.3.2. Antioxidant Autosynergists as Thermal Stabilisers for PVC

Antioxidant thiols (e.g. **XXVIa**) have some activity as processing stabilisers and as thermal oxidative stabilisers for PVC when used alone. However, they are readily destroyed by HCl liberated during processing and the degradation products cause severe discolouration of the PVC.[77]

In the presence of an effective HCl scavenger they become effective thermal stabilisers by virtue of their ability to react with developing unsaturation in the polymer.[77,78] Even **XXVIII**, which does not contain a recognised chain-breaking function, is an effective synergist for dibutyltin maleate (DBTM). Both BHBM (**XXVIa**) and EBHPT (**XXVIII**) show synergistic optima at 20–25% in combination with DBTM.[78]

In an air oven test at 140°C, DBTM synergised very effectively with BHBM at a 4:1 molar ratio (see Fig. 13) as measured by formation of colour and of unsaturation.[77]

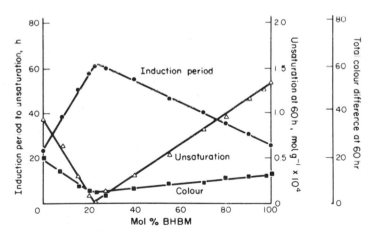

FIG. 13.   Synergistic optima for DBTM/BHBM mixtures (total concentration $5 \cdot 8 \times 10^{-3}$ mol/100 g) during oven ageing at 140°C, as measured by induction period to unsaturation formation, unsaturation at 60 h and colour formation at 60 h. (After ref. 79 with permission.)

7. SULPHIDES AS SYNERGISTS FOR ULTRAVIOLET STABILISERS

7.1. Ultraviolet Stabilisation of Polyolefins

Dialkyl and diaryl sulphides are not effective UV stabilisers when used alone but they do synergise effectively with 'UV absorbers', notably the 2-hydroxybenzophenones (**XXXIX**).[82]

$$\text{(structure)} \quad \text{—CO—} \quad \text{—OC}_8\text{H}_{17} \quad (\textbf{XXXIX}, \text{HOBP})$$

with HO group shown.

Sulphoxides and thiol sulphinates are very sensitive to the effects of UV light. Shelton has shown[27,80] that simple dialkyl sulphoxides photolyse to give radical products and Gilbert and co-workers[51] have demonstrated the photochemical formation of radicals by ESR from a wide variety of aromatic sulphoxides and thiol sulphinates related to **XXV**.

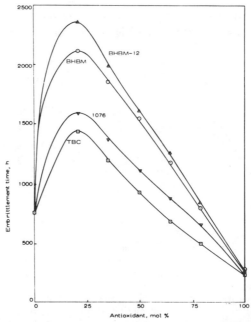

Fig. 14. Ultraviolet lifetimes of polypropylene films containing synergistic combinations of HOBP and phenolic antioxidants as a function of the molar proportion of antioxidants. (After ref. 83 with permission.)

4-Hydroxybenzylic sulphides (**XXVI**) as has been shown above are highly effective thermal antioxidants in polypropylene but, as might be expected, they are relatively weak light stabilisers.[81] However, they show an exceptionally high level of synergism with HOBP (**XXXIX**). This is shown typically for an antioxidant thiol (**XXVIa**) and its corresponding dodecyl sulphide (**XXVI**, $R = C_{12}H_{25}$) in Fig. 14. The monosulphide is particularly effective compared with one of the most effective CB–D antioxidants at the same molar concentration. A very sharp optimal molar ratio occurs at [HOBP]/[**XXVI**] = 3.[81]

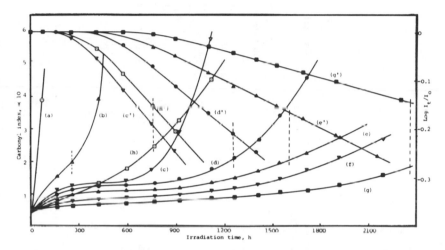

FIG. 15.   Relationship between the rate of photo-oxidation (a)–(h) of poly-propylene and the rate of decay of the UV absorbance of HOBP (a')–(h') in the presence of BHBM-12 (**XXVI**, $R = C_{12}H_{25}$). Total additive concentration, 10^{-3} g/100 g. (a) Control, no additive; (b) BHBM-12; (c), (c') 20% HOBP; (d), (d') 35% HOBP; (e), (e') 50% HOBP; (f) 65% HOBP; (g), (g') 80% HOBP; (h), (h') 100% HOBP. (After ref. 81 with permission.)

Figure 15 shows that the major function of the autosynergistic antioxidant is to protect the 2-hydroxybenzophenone against photo-oxidation. It has been demonstrated[82] that HOBP (**XXXIX**) is destroyed at least in part by hydroperoxide during photo-oxidation and it seems that a major function of the autosynergist is to inhibit the oxidative removal of HOBP during the induction period and to retard it during the first-order decay process. The protective action of sulphur antioxidants (**XXVI**) is due to their high peroxidolytic (PD–C) activity (see Section 6.1) which is superimposed on the chain-breaking (CB'–D)

TABLE 4

EFFECTIVENESS OF THIOGLYCOLLATE ULTRAVIOLET AND THERMAL STABILISERS (**XXVIII**) IN POLYPROPYLENE (Stabiliser concentration $= 1 \times 10^{-3}$ mol/100 g)[83]

Additive	UV embrittlement time (h)	Oven ageing at 120°C embrittlement time (h)	Solubility in hexane (g/100 g)	Relative UV absorbance
None	75	0·8	n.d.[a]	—
R=H	130	1·7	n.d.	0·11
R=C$_4$H$_9$	160	2·0	n.d.	0·15
R=C$_8$H$_{17}$	225	2·0	n.d.	0·18
R=C$_{12}$H$_{25}$	250	2·3	n.d.	0·18
R=C$_{18}$H$_{37}$	295	2·5	n.d.	0·22
XXXIX	750	0·9	41·0	1·0

[a] Abbreviation: n.d., not determined (too low to measure).

activity which they share with the commercial hindered phenols
(**XXXIIa** and **b**). It is significant that at a 4:1 molar excess of sulphur
antioxidant, the destruction of HOBP is actually sensitised (presumably by an excited state of the derived sulphoxide).

Autosynergistic UV stabilisers (e.g. **XXVIII**, $R = C_4H_9-C_{18}H_{37}$)
were not effective when these were used as conventional additives for
polypropylene.[83] Table 4 shows that the most effective autosynergist
(**XXVIII**, $R = C_{18}H_{37}$) was only about 40% as effective as HOBP
(**XXXIX**) on a molar basis although it was much more effective than
HOBP as a thermal oxidative stabiliser. Compound **XXVIII**, $R =
C_{18}H_{37}$ gave only 22% of the UV absorbance of HOBP. The absor-
bances of other members of the series were even lower, although the
extinction coefficients were all similar to HOBP in chlorobenzene,
indicating low solubility in the polymer. In fact, the solubilities of all
the benzophenone sulphides (**XXVIII**) were found to be extremely low
in aliphatic hydrocarbon solvents whereas HOBP was relatively soluble
(see Table 4).

7.2. Ultraviolet Stabilisation of ABS

Three-component synergism involving sulphide antioxidants has been
observed in the photostabilisation of acrylonitrile-butadiene–styrene
copolymers (ABS).[58,64] Table 5 shows the effectiveness of a simple

TABLE 5

EFFECTIVENESS OF COMMERCIAL ANTIOXIDANTS AND STABILISERS IN THE
ULTRAVIOLET STABILISATION OF ABS[58]

Additive	Concentration $\times 10^3$ (mol/100 g)	IP^a (h)	ET (h)
None	—	3	22
XXXIX	3	11	35
XXXIIa	4·9	10	34
III ($R = C_{12}H_{25}$)	1·5	5	25
XXV	4·9	10	42
XXXIX + **XXXIIa**	3 + 4·9	22	65
XXXIX + **XXV**	3·04 + 4·9	20	77
XXXIX + **XXXIIa** + **III** ($R = C_{12}H_{25}$)	3·04 + 4·9 + 1·5	25	85

^{a} Abbreviations: IP = induction period to carbonyl formation; ET = time to
embrittlement.

hindered phenol sulphide, **XXV**, and dilauryl thiodipropionate (DLTP; **III**, $R = C_{12}H_{25}$) both separately and in combination with HOBP, **XXXIX**.

The effect of a sulphide either in a separate molecule or as a phenolic autosynergist (**XXV**) is apparent from Table 5. However, quite substantial additive concentrations are required to achieve a four-fold increase in lifetime of ABS using multicomponent additive systems. Table 6 shows the effect of the autosynergistic UV stabiliser, EBHPT (**XXVIII**, $R = H$), as an adduct (**XXVIII**, $R = ABS$) on the stabilisation of ABS. The same Table shows the effect of introducing a further autosynergistic bound stabiliser derived from BHBM (**XXVIa**).

TABLE 6

EFFECT OF POLYMER-BOUND AUTOSYNERGISTS ON THE ULTRAVIOLET STABILITY OF ABS[58]

Bound antioxidant	Concentration $\times 10^3$ (mol/100 g)		Time to embrittlement (h)	
	U^a	E	U	E
XXVIII, R = ABS	3·0	1·9	52	48
XXVIII, R = ABS	3·75	2·4	53	51
XXVIII, R = ABS	4·5	2·7	57	53
XXVIII, R = ABS	6·0	3·6	62	60
XXVIII, R = ABS + **XXVI**, R = ABS	3·0 + 2·0	1·9 + 1·6	100	90
XXVIII, R = ABS + **XXVI**, R = ABS	3·0 + 6·0	1·9 + 4·8	160	140

^{a} Abbreviations: U = unextracted; E = exhaustively extracted.

The combination of autosynergistic bound stabilisers is found to be more effective than the single autosynergistic bound stabiliser, EBHPT, alone at the same molar concentration both before and after extraction to remove unbound side-reaction products. This cannot be due to compatibility or volatility effects since, by definition, the bound stabilisers are molecularly dispersed. This confirms the complementary stabilisation functions of the UV protective group (2-hydroxybenzophenone), the chain-breaking group (hindered phenol, CB–D) and the peroxidolytic sulphide function (PD–C) interacting in the same medium. Again it appears that the 'UV absorber' protects both the phenol and sulphide from oxidative photolysis, whereas the latter protects the 'UV absorber' from destruction by hydroperoxides both during thermal and photo-oxidation.

7.3. Ultraviolet Stabilisation of PVC

It has been shown[72,84,85] that peroxides formed during the processing of PVC are primarily responsible for subsequent photodegradation. In principle then, peroxidolytic alkyltin stabilisers should be effective UV stabilisers. In practice, when used alone, they are photo-pro-oxidants[71,74] in contrast to the tin maleates (**XXXV**, $R'R' =$ —CH=CH—) which are photostabilisers (Table 7). It has been shown that DOTG, unlike dibutyltin maleate (DBTM), is very photo-unstable in the presence of hydroperoxide due to the formation and subsequent rapid photolysis of the sulphoxide **XXXVII** (Scheme 16(c)).[74] As in the case of polypropylene and ABS (see previous section), HOBP (**XXXIX**) synergises effectively with the sulphide stabilisers[86] in PVC (Table 7) whereas the synergistic effect with the alkyltin maleate is much less marked.

TABLE 7

EFFECT OF THIOTIN STABILISERS AND SULPHUR ANTIOXIDANTS ON THE ULTRA-VIOLET STABILISATION OF PVC (The PVC was processed for 5 min at 170°C)[86]

Stabiliser	Concentration $\times 10^3$ (mol/100 g)	Embrittlement time (h)
None	—	670
DBTM	5·8	1000
DOTG	5·8	460
DBTM+HOBP	5·8+1·4	1400
DOTG+HOBP	5·8+1·4	1750
DBTM+BHBM	5·8[a]	1400
DOTG+BHBM	5·8[a]	400
DBTM+EBHPT	5·8[a]	1850
DOTG+EBHPT	5·8[a]	1000

[a] These formulations are the synergistic optima for thermal stabilisation shown in Fig. 12, Section 6.3.2.

Another way of achieving a similar synergistic effect is by the use of the polymer bound autosynergist, EBHPT[86] (**XXVIII**, R = PVC) in the presence of an alkyltin maleate. The sulphur autosynergists, BHBM (**XXVI**) and EBHPT (**XXVIII**), are ineffective photostabilisers alone

TABLE 8
EFFECT OF PARTIALLY REPLACING THE THERMAL SYNERGISTIC STABILISER (DBTM/BHBM, 4:1) BY THE UV STABILISER EBHPT (Total molar concentration was maintained constant at 5.8×10^{-3} mol/100 g. Processing time, 10 min)

Concentration $\times 10^3$ (mol/100 g)		UV embrittlement time (h)
DBTM/BHBM	EBHPT	
5·3	0·5	1360
4·8	1·0	1440
4·3	1·5	1500
3·8	2·0	1575
2·9	2·9	1660
2·0	3·8	1800
1·9	3·9	1975

due to their rapid destruction by HCl liberated in absence of an alkyltin thermal stabiliser but in the presence of DBTM they both become effective photostabilisers (Table 7). EBHPT which contains the 'UV absorbing' function is, however, more effective than the hindered phenolic autosynergist. It was shown earlier (Section 6.3.2) that BHBM is a more effective synergist than EBHPT in the stabilisa-

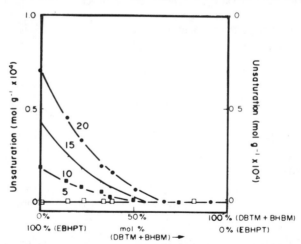

FIG. 16. Formation of unsaturation during the processing of PVC at 170°C as a function of DBTM/BHBM (4:1) to EBHPT. Numbers on curves are processing times in min. Total concentration, 5.8×10^{-3} mol/100 g. (After ref. 87 with permission.)

tion of PVC against the effects of processing and thermal oxidation and that the optimal molar [DBTM]/[BHBM] ratio in both respects is 4. Incremental replacement of this heat stabilising system by the UV stabiliser, EBHPT, maintaining the total molar concentration constant leads to a marked increase in UV stability (see Table 8) but with some sacrifice to the thermal stability of the polymer (see Fig. 16).[87] The commercial UV stabiliser HOBP (**XXXIX**) has a much more severe effect on thermal stability, showing that the deleterious effect is due to the 2-hydroxybenzophenone and not to the sulphide synergist. Figure 16 shows that although all the formulations are satisfactory at 5 min processing, with more severe processing (i.e. for longer times) less EBHPT can be tolerated in the system.[87] Table 8 and Fig. 14 clearly demonstrate that the choice of synergistic combination will depend on whether the main emphasis is on thermal or UV stability in the final product. Conflicting requirements for optimal antioxidant performance in different environments have been discussed in a recent review of antioxidant action.[42]

REFERENCES

1. SCOTT, G., *Atmospheric Oxidation and Antioxidants* (1965) Chapters 2 and 3. Elsevier, London and Amsterdam.
2. Ref. 1, Chapters 4 and 5.
3. DENISON, G. H., *Ind. Eng. Chem.*, **36** (1944) 477; DENISON, G. H. and CONDIT, P. C., *Ind. Eng. Chem.*, **37** (1945) 1102.
4. KENNERLEY, G. W. and PATTERSON, W. L., *Ind. Eng. Chem.*, **48** (1956) 1917.
5. Ref. 1, p. 188 *et seq.*
6. KHARASCH, M. S. and BART, J. G., *J. Org. Chem.*, **16** (1951) 128; KHARASCH, M. S., FONO, A. and NUDENBERG, W., *J. Org. Chem.*, **15** (1950), 748.
7. Ref. 1, p. 370.
8. ANON., *US Government Regulations Governing Meat Inspection* (1949). US Dept of Agriculture, Washington DC.
9. FARMER, E. H., *J. Farad. Soc.*, **38** (1942) 356.
10. FLETCHER, W. P. and FOGG, S. E., *Rubb. J. Ind. Plast.*, **134** (1958) 16; **135** (1958) 687.
11. DUNN, J. R. and SCANLAN, J., *Trans IRI*, **34** (1958) 228.
12. THOMAS, J. R., HARLE, O. L., RICHARDSON, W. L. and BOWMAN, L. O., *Additives in Lubricants Symposium* (1956) p. 138. ACS Div. Pet. Chem.
13. DUNN, J. R. and SCANLAN, J., *J. Appl. Polym. Sci.*, **35** (1959) 267.
14. BAUN, B. and PERUN, A. L., *ACS Meeting, Div. Org. Coatings Plast.*, **21** (2) (1961) 79.

15. Ref. 1, p. 295.
16. SHELTON, J. R. in *Polymer Stabilisation*, Ed. W. L. Hawkins (1972) Chapter 2. Wiley, New York.
17. HOLDSWORTH, J. D., SCOTT, G. and WILLIAMS, D., *J. Chem. Soc.* (1964) 4692.
18. HAWKINS, W. L. and SAUTTER, H., *J. Polym. Sci.*, A1, 7 (1963) 3499.
19. STAFFORD, R. L., *Proc. 3rd Rubb. Tech. Conf.*, (1954) 253.
20. Ref. 1, p. 433 *et seq.*
21. Hercules Powder Co., British Patent 851,670 (1960).
22. SCOTT, G. and SHEARN, P. A., *J. Appl. Polym. Sci.*, 13 (1969) 1329.
23. BATEMAN, L. and CUNNEEN, J. I., *J. Chem. Soc.* (1955) 1596.
24. BARNARD, D., BATEMAN, L., CAIN, M. E., COLCLOUGH, T. and CUNNEEN, J. I., *J. Chem. Soc.* (1961) 5339.
25. BATEMAN, L., CAIN, M. E., COLCLOUGH, T. and CUNNEEN, J. I., *J. Chem. Soc.* (1962) 3570.
26. BATEMAN, L. and HARGRAVE, K. R., *Proc. Roy. Soc.*, A224 (1954) 389, 399; BARNARD, D., HARGRAVE, K. R. and HIGGINS, G. M. C., *J. Chem. Soc.* (1956) 2845.
27. SHELTON, J. R., *Developments in Polymer Stabilisation—4*, Ed. G. Scott (1981) p. 23. Applied Science Publishers, London.
28. ARMSTRONG, C. and SCOTT, G., *J. Chem. Soc.* (1971) 1747.
29. ARMSTRONG, C., PLANT, M. A. and SCOTT, G., *Europ. Polym. J.*, 11 (1975) 161.
30. ARMSTRONG, C., HUSBANDS, M. J. and SCOTT, G., *Europ. Polym. J.*, 15 (1979) 241.
31. SCOTT, G., *Europ. Polym. J. Supplement* (1969) 189.
32. SCOTT, G., *Chem. Comm.* (1968) 1572.
33. NEUREITER, N. P. and BOWN, D. E., *Ind. Eng. Chem., Prod. Res. Dev.*, 1 (1962) 236.
34. SCOTT, G., *Pure Appl. Chem.*, 30 (1972) 267.
35. WIMS, A. M. and SWARN, S. J., *J. Appl. Polym. Sci.*, 19 (1975) 1243.
36. KOELEWIJN, P. and BERGER, H., *Rec. Trav. Chem.*, 91 (1972) 1275; 93 (1974) 63.
37. BLOCK, E., *J. Am. Chem. Soc.*, 94 (1972) 644.
38. BRIDGWATER, A. J. and SEXTON, M. D., *J. Chem. Soc., Perkin II* (1978) 530.
39. HUSBANDS, M. J. and SCOTT, G., *Europ. Polym. J.*, 15 (1979) 249.
40. FLOCKHART, B. D., IVIN, K. J., PINK, R. C. and SHARMAN, B. D., *Chem. Comm.* (1971) 339.
41. HENMAN, T. J. in *Developments in Polymer Stabilisation—1*, Ed G. Scott (1979) p. 79. Applied Science Publishers, London.
42. SCOTT, G., *S. Afr. J. Chem.*, 32 (1979) 137.
43. ARMSTRONG, C., INGHAM, F. A. A., PIMBLOTT, J. G., SCOTT, G. and STUCKEY, J. E., *Proc. Int. Rubb. Conf. Brighton* (May 1972) F2.1.
44. SCOTT, G. in *Developments in Polymer Stabilisation—4*, Ed G. Scott (1981) p. 10. Applied Science Publishers, London.
45. SCOTT, G. and SWEISS, A. A. A., unpublished work.
46. KATBAB, A. and SCOTT, G., *Chem. Ind.* (1980) 573.
47. Ref. 1, p. 396.

48. SCOTT, G., *Mechanisms of Reactions of Sulphur Compounds*, **4** (1980) 99.
49. HAWKINS, W. L. and SAUTTER, H., *J. Polym. Sci., A1* (1969) 3499.
50. KICE, J. C., PARHAM, F. M. and SIMONS, R. M., *J. Am. Chem. Soc.*, **82** (1960) 660.
51. CHATGILIALOGLU, C., GILBERT, B. C., GILL, B. and SEXTON, M. D., *J. Chem. Soc., Perkin II* (1980) 1141.
52. MILLER, E. G., RAYNER, D. R., THOMAS, H. T. and MISLOW, K., *J. Am. Chem. Soc.*, **90** (1968) 4861.
53. ABBOTT, D. J. and STIRLING, C. J. M., *Chem. Comm.* (1968) 165.
54. FARZALIEV, V. M., FERNANDO, W. S. E. and SCOTT, G., *Europ. Polym. J.*, **14** (1978) 785.
55. Ref. 1, p. 287 *et seq.*
56. JIRACKOVA, L. and POSPISIL, J., *Angew. Makromol. Chem.*, **66** (1968) 95.
57. JIRACKOVA, L. and POSPISIL, J., *Angew Makromol. Chem.*, **82** (1979) 197.
58. SCOTT, G. in *Developments in Polymer Stabilisation—1*, Ed G. Scott (1979) p. 309. Applied Science Publishers, London.
59. SCOTT, G., *J. Appl. Polym. Sci., Appl. Polym. Symp.*, **35** (1979) 123.
60. SCOTT, G. and YUSOFF, M. F., *Europ. Polym. J.*, **16** (1980) 497.
61. PLANT, M. A. and SCOTT, G., *Europ. Polym. J.*, **7** (1971) 1173.
62. SCOTT, G., *Plastics and Rubber: Processing*, **41** (June 1977) 41.
63. KULARATNE, K. W. S. and SCOTT, G., *Europ. Polym. J.*, **15** (1979) 827.
64. FERNANDO, W. S. E. and SCOTT, G., *Europ. Polym. J.*, **16** (1980) 971.
65. SCOTT, G. and YUSOFF, M. F., *Polym. Deg. Stab.*, **3** (1980–81) 53.
66. SCOTT, G., US Patent 4,213,892 (1980).
67. AJIBOYE, O. and SCOTT, G., unpublished work.
68. KATBAB, A. A. and SCOTT, G., *Polym. Deg. Stab.*, **3** (1981) 221.
69. Ref. 1, p. 313 *et seq.*
70. STAPFER, C. H. and GRANICK, J. D., *J. Polym. Sci.*, **A19** (1971) 2625.
71. COORAY, B. B. and SCOTT, G., *Developments in Polymer Stabilisation—2* Ed G. Scott, (1980) p. 53. Applied Science Publishers, London.
72. COORAY, B. B. and SCOTT, G., *Europ. Polym. J.*, **16** (1980) 169; *Chem Ind.* (1979) 741.
73. AYREY, G. and POLLER, R. C., *Developments in Polymer Stabilisation—2* Ed. G. Scott, (1980) p. 1. Applied Science Publishers, London.
74. COORAY, B. B. and SCOTT, G., *Polym. Deg. Stab.*, **2** (1980) 35.
75. COORAY, B. B. and SCOTT, G., *Europ. Polym. J.*, **17** (1981) 233.
76. COORAY, B. B. and SCOTT, G., (unpublished work).
77. COORAY, B. B. and SCOTT, G., *Europ. Polym. J.*, **16** (1980) 1145.
78. COORAY, B. B. and SCOTT, G., *Europ. Polym. J.*, **16** (1980) 379.
79. Ref. 1, p. 216.
80. SHELTON, J. R. and DAVIES, K. E., *Int. J. Sulphur Chem.*, **8** (1973) 217.
81. SCOTT, G. and YUSOFF, M. F., *Polym. Deg. Stab.*, **2** (1980) 309.
82. CHAKRABORTY, K. B. and SCOTT, G., *Europ. Polym. J.*, **15** (1979) 35.
83. SCOTT, G. and YUSOFF, M. F., *Europ. Polym. J.*, **16** (1980) 497.
84. SCOTT, G., TAHAN, M. and VYVODA, J., *Europ. Polym. J.*, **14** (1978) 1021.
85. COORAY, B. B. and SCOTT, G., *Polym. Deg. Stab.*, **3** (1980/1) 127.
86. COORAY, B. B. and SCOTT, G., *Europ. Polym. J.*, **17** (1981) 229.
87. COORAY, B. B. and SCOTT, G., *Europ. Polym. J.*, **17** (1981) 385.

Chapter 3

PEROXIDOLYTIC ANTIOXIDANTS: METAL COMPLEXES CONTAINING SULPHUR LIGANDS

Sahar Al-Malaika, Khirud B. Chakraborty and Gerald Scott

Department of Chemistry, University of Aston in Birmingham, UK

SUMMARY

The oxidation chemistry and technological behaviour in polyolefins of the major classes of metal thiolate antioxidants are reviewed. All the metal complexes are found to show a generally similar pattern of behaviour in that they are initially oxidised to the corresponding disulphides with the elimination of the metal and the formation of further oxidation products (sulphur acids) which behave as catalysts for the ionic decomposition of hydroperoxides.

It is concluded that a primary function of all classes in both thermal oxidation and photo-oxidation of polyolefins is to destroy hydroperoxides in an ionic process. However, in photo-oxidation, the effectiveness of metal thiolates as UV stabilisers is strongly dependent on their photo-stabilities since they are required to function as reservoirs for the formation of sulphur acids. The nickel complexes are in general most effective in this respect; however, the less stable complexes can be protected against photo-oxidative destruction by conventional UV absorbers.

1. INTRODUCTION

Transition metal complexes containing at least one sulphur ligand have been widely used as antioxidants in a variety of substrates. Technological media include lubricating oils, rubbers and plastics in applications

73

TABLE 1

STRUCTURE AND USES OF TRANSITION METAL COMPLEXES CONTAINING
SULPHUR LIGANDS

Metal complex	Code	Uses
	I, MDRC	Thermal and light stabilisers for polyolefins. Antioxidants and anticorrosives for lubricating oils. Time-controlled statilisation systems for polyolefins. *In situ* antioxidants and antiozonants for rubbers.
	II, MDRP	As for I.
	III, MRX	As for I.
	IV, MMBT	*In situ* antioxidants formed in the EV curing systems for rubber.
	V, MMBI	Non-staining antioxidants for rubbers and plastics.

which include UV and thermal stabilisation and controlled photodegradation. The most important structures and their uses are listed in
Table 1. For a review of the literature up to 1965, see ref. 1.

Mechanistically, these complexes have one feature in common; they
have all been shown to be oxidised during their antioxidant function to
sulphur acids. Consequently, they show a generally similar pattern of

behaviour as antioxidants to the sulphide antioxidants discussed in Chapter 2. They differ from the latter however in that the nickel, cobalt and copper complexes have the ability to function as UV stabilisers as well as thermal antioxidants. This appears to be associated with the strong UV absorbance of the metal ligand bond in the region 310–350 nm.

2. METAL MERCAPTOBENZTHIAZOLATES

2.1. Mercaptobenzthiazole Complexes as Thermal Antioxidants

The high oxidative stability of high accelerator, low sulphur vulcanisates (EV) has already been referred to in Chapter 2 and the need to develop practical curing systems at a cost acceptable to the rubber industry has led to an extensive evaluation of vulcanisates in which the antioxidant is produced as a result of the curing process. The technological performance of the EV rubbers has been reviewed in an earlier volume in this series.[2] The chemistry involved in the antioxidant action of the 'efficient vulcanisation' (EV) system involving the use of cyclohexylbenzthiazyl sulphenamide (**VI**) has been examined in some detail (Scheme 1). It is known that zinc mercaptobenzthiazolate (**IV**)

SCHEME 1

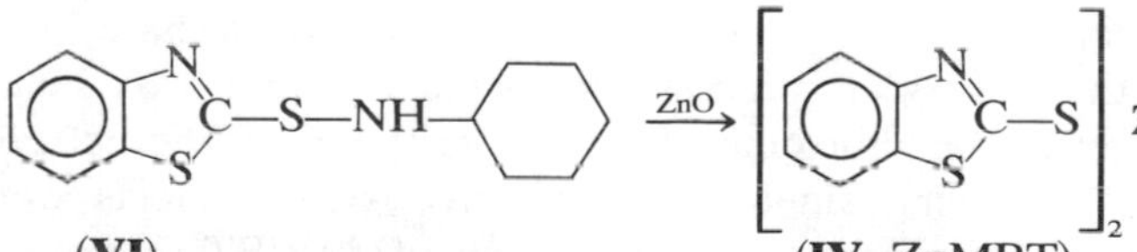

(**VI**)　　　　　　　　　　　　　　(**IV**, ZnMBT)

and its hydrate are the main products formed from this accelerator during vulcanisation (reaction 1)[2–4] and if these materials are removed from the vulcanisate by extraction, the very high level of oxidative stability disappears.[2,4]

Mercaptobenzthiazole (MBT; **VII**) and its zinc complex (ZnMBT; **IV**) are both very powerful catalysts for the decomposition of hydroperoxides.[5]

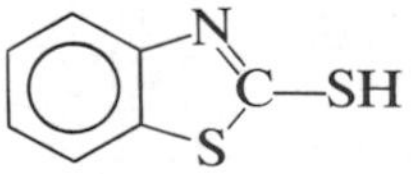

(**VII**, MBT)

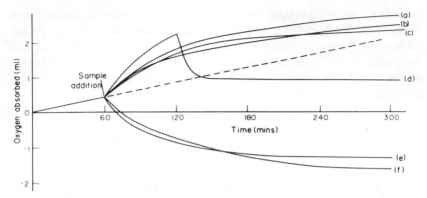

FIG. 1. Effect of ZnMBT on the oxidation of cumene initiated with cumene hydroperoxide (CHP) at varying molar ratios, [CHP]/[ZnMBT]. (a) $5:1$ (3×10^{-3} M); (b) $10:1$ ($1 \cdot 4 \times 10^{-3}$ M); (c) $5:1$ (9×10^{-3} M); (d) $10:1$ ($6 \cdot 7 \times 10^{-3}$ M); (e) $30:1$ ($1 \cdot 5 \times 10^{-3}$ M); (f) $30:1$ ($5 \cdot 1 \times 10^{-3}$ M). Numbers in parentheses are ZnMBT concentrations. (After ref. 5 with permission.)

In the presence of a molar excess of cumene hydroperoxide ([CHP]/[ZnMBT] = 30) the products formed were found to be predominantly those expected on the basis of an ionic decomposition in both cases. At this ratio, ZnMBT behaved as an effective antioxidant (with associated gas evolution) in an oxygen absorption test for cumene initiated by CHP (see Fig. 1) but at lower ratios pro-oxidant effects and autoretardation were observed. The addition of a base (pyridine) inhibited the antioxidant process and increased the pro-oxidant effect, confirming that the antioxidant is an acidic species.[5] The behaviour of MBT was somewhat similar but the pro-oxidant effects were less significant. The similarity of the behaviour of ZnMBT to that of other sulphur compounds discussed earlier (see, for example, Chapter 2, Fig. 4) is obvious. The products formed from both MBT and ZnMBT in the presence of tert-butyl hydroperoxide (TBH) at varying [ROOH]/[S] ratios proved to be much easier to determine than in the case of DLTP.[6] Both gave benzthiazolyl disulphide (MBTS, **VIII**) as the primary product, and at low molar ratios ([TBH]/[S] = 1) the main product (Fig. 2).

(VIII, MBTS)   (**IX**, BT)   (**X**, BTSO)

However, at [TBH]/[S] = 5 benzthiazole (BT, **IX**) and benzthiazole sulphonic acid (BTSO, **X**) became major products. At [TBH]/[S] = 20, the latter was found to be the only product.

A study of the antioxidant properties of the reaction products showed that all except BT were antioxidants in the presence of a hydroperoxide initiator and, more surprisingly, some, notably the

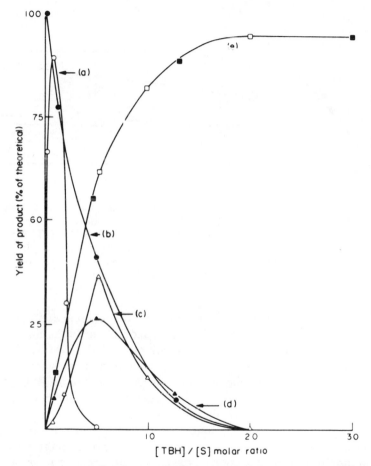

FIG. 2.   Effect of *tert*-butyl hydroperoxide/MBT molar ratio, [TBH]/[S], on the yield of products formed from MBT. (a) MBTS (MBT); (b) MBTS (MBTS); (c) BT (MBT); (d) BT (MBTS); (e) BTSO (MBT and MBTS). Compound in parentheses is the starting material. (After ref. 5 with permission.)

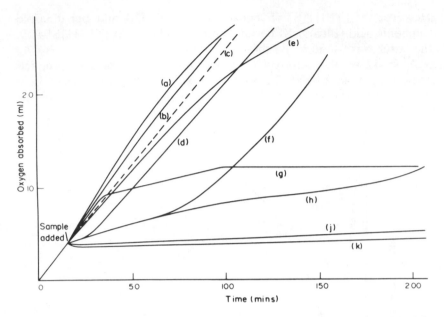

FIG. 3. Effect of MBT derivatives on the oxidation of cumene initiated by AIBN $(10^{-2}$ M) at 75°C. (a) MBTS $(20 \times 10^{-3}$ M)+pyridine $(20 \times 10^{-3}$ M); (b) MBTS $(1 \times 10^{-3}$ M); (c) control (no antioxidant); (d) MBT $(17 \times 10^{-3}$ M)+ pyridine $(17 \times 10^{-3}$ M); (e) MBT $(10^{-3}$ M); (f) ZnMBT $(13 \times 10^{-3}$ M); (g) MBTS $(20 \times 10^{-3}$ M); (h) MBT $(2 \cdot 5 \times 10^{-3}$ M); (j) BTSO $(1 \cdot 4 \times 10^{-3}$ M); (k) BTSO $(2 \cdot 4 \times 10^{-3}$ M).

sulphonic acid, were also antioxidants in the presence of an alkyl radical initiator (Fig. 3). This implies donor chain-breaking (CB–D) activity. MBTS is not a CB–D antioxidant unless oxidised further.[5] The sulphinic acid could not be isolated under the experimental conditions. Although the corresponding sulphonic acid was stable at 70°C, it was found to be unstable at 100°C and as reported previously[7] it eliminated SO$_2$ to give 2-hydroxybenzthiazole (**XI**). The competing pro-oxidant and antioxidant processes are summarised for MBT in Scheme 2. As in the case of the thiodipropionate esters, the pro-oxidant processes predominate during the early stages of the reaction or at low [ROOH]/[S] ratios, whereas the antioxidant processes predominate at extended times or at high [ROOH]/[S] ratios.

The antioxidant performance of ZnMBT is somewhat different from that of MBT. Zinc sulphinate (ZnBTS, **XII**) is formed in high yield in place of BTSO (**X**)[3] and this compound is considerably more stable than the free sulphinic acid.

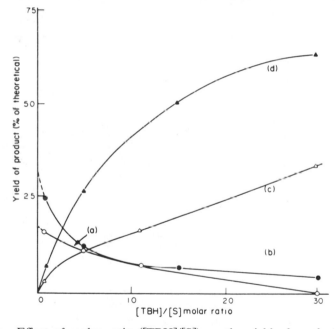

$$\left[\underset{S}{\overset{N}{\bigvee}} C - \overset{O}{\underset{\parallel}{S}} - O \right]_2 Zn \qquad (\textbf{XII, ZnBTS})$$

Benzthiazole is formed more slowly from it and the zinc sulphonate does not appear to be formed at all (see Fig. 4). ZnMBT and ZnBTS (**XII**) behave as reservoirs for the slow liberation of SO_3 by reaction with hydroperoxides and the non-volatility of the zinc complexes appears to be responsible for the relatively high effectiveness of ZnMBT (Fig. 6) compared with MBT (Fig. 5) at 140°C.

Figure 6 shows that an organic base (pyridine) has a favourable effect on the antioxidant activity of ZnMBT and this is consistent with

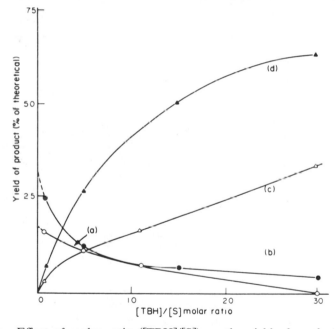

FIG. 4. Effect of molar ratio ([TBH]/[S]) on the yield of products on the reaction of TBH with ZnMBT and its pyridine complex (1:2). (a) MBTS (ZnMBT); (b) MBTS (ZnMBT/Pyr); (c) BT (ZnMBT); (d) BT (ZnMBT/Pyr). Compound in parentheses is the starting material. (After ref. 5 with permission.)

SCHEME 2. Transformation of mercaptobenzthiazole and its derivatives in the presence of a hydroperoxide. The boxes indicate the predominant process associated with each step.

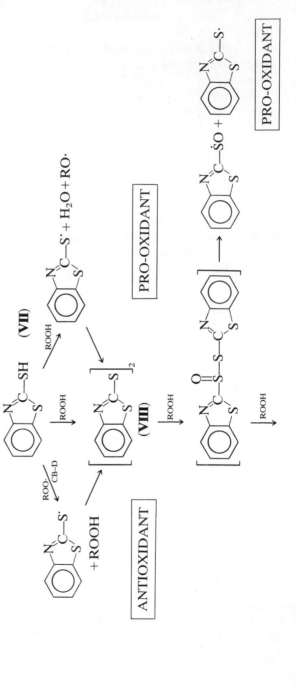

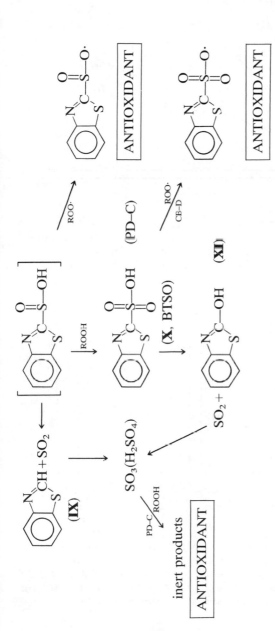

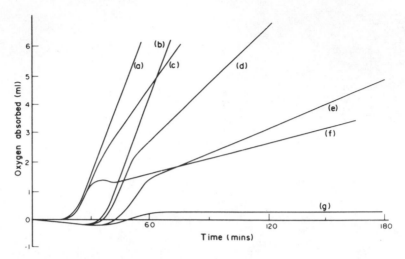

FIG. 5. Effect of MBT and MBTS on the oxidation of uninitiated paraffin oil at 140°C. (a) MBTS $(3 \times 10^{-3}$ M$)$ + pyridine $(3 \times 10^{-3}$ M$)$; (b) control (no antioxidant); (c) MBTS $(2 \times 10^{-3}$ M$)$; (d) MBT $(2 \times 10^{-3}$ M$)$; (e) MBT $(3 \times 10^{-3}$ M$)$; (f) MBTS $(3 \times 10^{-3}$ M$)$; (g) MBT $(4 \cdot 5 \times 10^{-3}$ M$)$. (After ref. 5 with permission.)

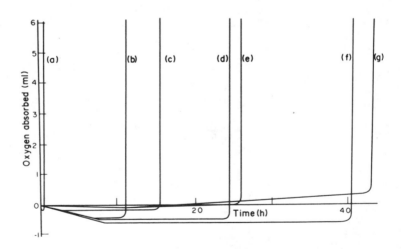

FIG. 6. Effect of MBT, ZnMBT and ZnMBT/Pyr $(1:2$ complex) on the oxidation of uninitiated paraffin oil at 140°C. (a) MBT $(1 \cdot 8 \times 10^{-3}$ M$)$; (b) ZnMBT $(0 \cdot 5 \times 10^{-3}$ M$)$; (c) ZnMBT/Pyr $(0 \cdot 5 \times 10^{-3}$ M$)$; (d) ZnMBT $(1 \cdot 2 \times 10^{-3}$ M$)$; (e) ZnMBT/Pyr $(0 \cdot 8 \times 10^{-3}$ M$)$; (f) ZnMBT $(2 \cdot 4 \times 10^{-3}$ M$)$; (g) ZnMBT/Pyr $(1 \cdot 6 \times 10^{-3}$ M$)$. (After ref. 5 with permission.)

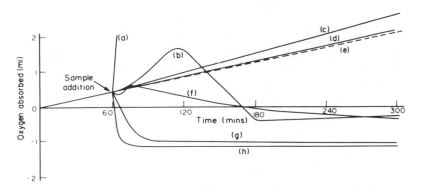

FIG. 7.   Effect of MBT and its oxidation products on the oxidation of cumene initiated by CHP $(10^{-1}$ M$)$ at 75°C. (a) MBTS $(4 \times 10^{-3}$ M$)$ + pyridine $(4 \times 10^{-3}$ M$)$; (b) MBTS $(10^{-3}$ M$)$ + pyridine $(4 \cdot 6 \times 10^{-3}$ M$)$; (c) MBT $(4 \cdot 6 \times 10^{-3}$ M$)$ + pyridine $(4 \cdot 6 \times 10^{-3}$ M$)$; (d) BTSO $(1 \times 10^{-3}$–6×10^{-3} M$)$ + pyridine (1:2); (e) control (no antioxidant); (f) MBT $(1 \times 10^{-3}$–10×10^{-3} M$)$; (g) BTSO $(0 \cdot 6 \times 10^{-3}$ M$)$; (h) BTSO $(1 \times 10^{-3}$ M$)$. (After ref. 5 with permission.)

the observation (see Fig. 4) that pyridine catalyses the formation of sulphur acids from the zinc sulphinate. A similar beneficial effect of bases on the peroxidolytic action of other zinc thiolates has been observed and will be discussed in Section 3.3. However, although pyridine was found to increase the activity of ZnMBT, it decreased the activity of benzthiazole sulphonic acid (BTSO, **X**). This compound was found to be a much more powerful antioxidant than MBT and MBTS since it did not show the characteristic pro-oxidant effects associated with these compounds in the presence of a hydroperoxide (Fig. 7). The sulphonic acid, BTSO (**X**), was found to be an effective antioxidant not only in the presence of a peroxide initiator but equally in the presence of an alkyl radical generator (see Fig. 3) indicating a high level of CB–D activity (see Scheme 2). Both MBT and MBTS develop this activity after oxidation, presumably to BTSO.

The importance of the lower molecular weight sulphur acids formed by the oxidation of both MBT and ZnMBT is shown in Fig. 8. The introduction of a molecular sieve above the surface of uninitiated paraffin oil at 140°C removed the small induction period associated with MBT and markedly reduced the much longer induction period associated with ZnMBT.

Although BTSO (**X**) is an effective antioxidant at temperatures below 100°C, it is much less effective than the zinc complexes at 140°C

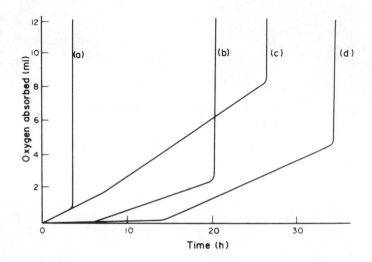

FIG. 8.   Effect of molecular sieve (5Å) on the antioxidant activity of MBT and ZnMBT in uninitiated paraffin oil at 140°C. (a) MBT (3×10^{-3} M); (b) ZnMBT (2×10^{-3} M); (c) MBT (5×10^{-3} M); (d) ZnMBT (3×10^{-3} M). (After ref. 5 with permission.)

TABLE 2

INDUCTION PERIODS (IP) FOR MBT, ZnMBT (**VII** AND **IV**) AND THEIR DERIVED OXIDATION PRODUCTS IN WHITE PARAFFIN OIL AT 140°C

Antioxidant	Concentration $\times 10^4$ (mol litre^{-1})	IP (h)
MBT (**VII**)	18	1
ZnMBT (**IV**, M = Zn)	5	12
	12	24
	24	41
ZnMBT + TBH (1:5)	12	40
ZnMBT/Pyr (1:2 complex)	5	16
	8	26
	16	43
BTSO (**X**)	50	4
	90	10
BTSO/Pyr (1:1)	90	1
ZnBTS (**XII**)	12	11

(see Table 2), due to its ready decomposition to give the volatile sulphur oxides. The chemistry of the antioxidant action of ZnMBT is summarised in Scheme 3.

SCHEME 3. Transformation of ZnMBT (**IV**) in the presence of a hydroperoxide.

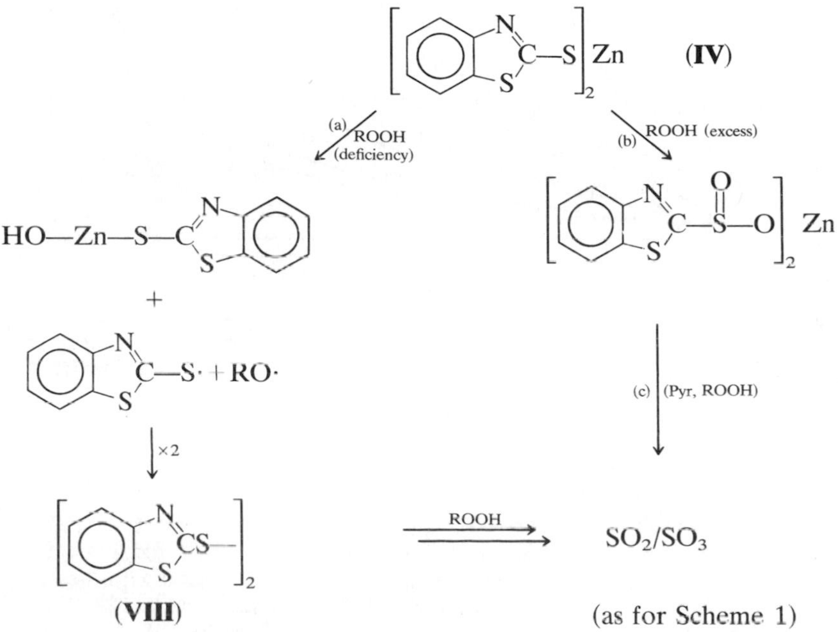

$$\left[\underset{S'}{\overset{N}{\bigcirc}}C-S\right]_2 Zn \qquad (IV)$$

(a) ROOH (deficiency)   (b) ROOH (excess)

$$HO-Zn-S-C\underset{S}{\overset{N}{<}}\bigcirc$$

$+$

$$\bigcirc\underset{S}{\overset{N}{<}}C-S\cdot + RO\cdot$$

$$\left[\underset{S'}{\overset{N}{\bigcirc}}C-\overset{O}{\underset{O}{\overset{\|}{S}}}-O\right]_2 Zn$$

(c) (Pyr, ROOH)

$\downarrow \times 2$

$$\left[\underset{S'}{\overset{N}{\bigcirc}}CS-\right]_2$$

(**VIII**)

$\xrightarrow{ROOH} SO_2/SO_3$

(as for Scheme 1)

2.2. Mercaptobenzthiazole and Mercaptobenzimidazole Complexes as Ultraviolet Stabilisers

Metal mercaptobenzimidazolates (**V**) are very closely related to the metal mercaptobenzthiazolates (**IV**) and have also been reported to be effective as non-discolouring and non-staining thermal and UV stabilisers for polyolefins and other polymers including rubbers.[8–10] The behaviour of these compounds as heat and light stabilisers in polyolefins has been found to be very similar to the metal mercaptobenzthiazoles (**IV**).[11,12] They again function primarily as peroxide decomposers;[11] the main difference between the two classes of compounds is that the MBI complexes (**V**) stabilise polyolefins against UV light more effectively than MBT complexes. The UV stabilisation

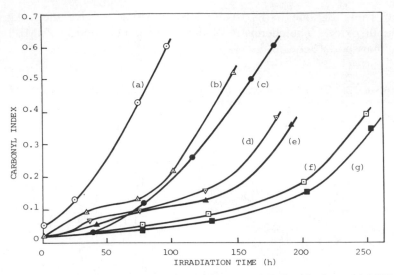

FIG. 9. Comparison of the metal complexes of MBT with those of MBI as UV stabilisers for polypropylene at the same molar concentration (6×10^{-4} mol/100 g). (a) Control (no antioxidant); (b) ZnMBT; (c) ZnMBI; (d) CuMBT; (e) NiMBT; (f) CuMBI; (g) NiMBI.

TABLE 3

EFFECT OF TRANSITION METAL COMPLEXES OF MBI (**V**) AND MBT (**IV**) ON THE ULTRAVIOLET LIFETIME OF POLYPROPYLENE AND RELATION TO ULTRAVIOLET ABSORBANCE (Polypropylene (PP) was processed with the additives in the RAPRA torque rheometer at 180°C/10 min prior to UV irradiation. Concentration of additives, 6×10^{-4} mol/100 g)

Additive	UV embrittlement time of PP (h)	Absorbance in PP (315–320 nm)	Extinction coefficient in benzene at 315–320 nm (litre/mol cm)
None (control)	90	—	—
ZnMBI	154	0·42	8×10^3
NiMBI	210	0·86	$1·5 \times 10^4$
CuMBI	255	1·02	$1·4 \times 10^4$
ZnMBT	140	0·16	$1·1 \times 10^3$
NiMBT	160	0·33	$1·3 \times 10^4$
CuMBT	163	0·30	8×10^3

PEROXYDOLYTIC ANTIOXIDANTS: METAL COMPLEXES 87

effectiveness of Zn, Ni and Cu complexes of MBI are compared with similar complexes of MBT in polypropylene at the same molar concentration in Fig. 9. The UV embrittlement times of polypropylene containing the above complexes are listed in Table 3. The relatively higher UV stabilisation effectiveness of the MBI complexes has been shown to be due to their higher UV stability compared with the MBT complexes.[11,12] Figure 9 and Table 3 also show that the Ni and Cu complexes of MBI are more effective UV stabilisers than the corresponding Zn complex. Moreover, the UV stabilising activity of the Ni and Cu complexes of MBI are concentration-dependent whereas the Zn complex is not (Fig. 10).

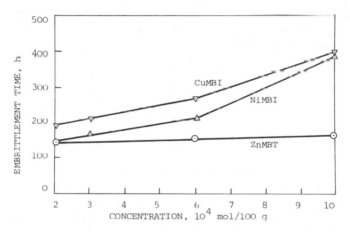

FIG. 10. Effect of MMBI concentration on photostabilising activity in polypropylene.

Although the Zn, Ni and Cu complexes of MBI have similar λ_{max} in the region 315–320 nm, the Zn complex was found to be destroyed much faster on UV irradiation than the Ni and Cu complexes, both in the polymer medium and in organic solution.[11] The strong UV absorbance of the Ni and Cu complexes at 315–320 nm and the slower decay of the UV absorbance under UV light suggests that these complexes might also function to some extent by UV screening in addition to peroxide decomposition (see last column, Table 3).

3. DITHIOCARBAMATE METAL COMPLEXES

As was mentioned in Chapter 2, the transition metal dithiocarbamates
(**I**) were among the earliest of the peroxidolytic antioxidants to be
investigated from a mechanistic point of view because of their importance in the 'sulphurless' vulcanisation of rubbers. It was recognised[1,7]
that sulphur dioxide is one of the products formed in their reaction
with hydroperoxides but the detailed chemistry by which they are
formed is much more complex than in the case of the corresponding
benzthiazolates.[1] Intermediate sulphinates (**XIII**) analogous to **XII**

$$R_2N\overset{\overset{S}{\|}}{C}-\overset{\overset{O}{\|}}{S}-O-Zn-O-\overset{\overset{O}{\|}}{S}-\overset{\overset{S}{\|}}{C}NR_2 \qquad (\textbf{XIII})$$

formed from ZnMBT have however been isolated[3] and it seems likely
that the overall mechanism is similar. Isothiocyanates have been identified among the products formed by reaction of hydroperoxides with
metal dithiocarbamates and there is evidence that under certain conditions the thiuram disulphide is a primary oxidation product. Scheme 4
summarises the probable chemical reactions occurring.

Recent interest in this class of PD–C antioxidant has centred on
their UV stabilising function in polyolefins.[6] NiDRC homologues (**I**,
M = Ni) have been singled out for particular attention.

3.1. Possible Ultraviolet Stabilisation Mechanisms of the Nickel Dithiocarbamates

It has been recognised for many years that the nickel dithiocarbamates
are effective UV stabilisers.[1] Since they have a high molar extinction
coefficient in the region 330 nm, effective screening of the incident
light was considered to be a primary function in addition to their PD–C
activity. The UV screening effect of these complexes is not in itself
sufficient to account for their very high activity and photophysical
investigations have led to the suggestion[13,14] that these compounds
may also deactivate photo-excited species formed in polymers under
the influence of light.

Although transition metal complexes do have this ability,[13–16] excited state 'quenching' appears to play a negligible role in the stabilisation of polyolefins by the dithiocarbamates[17–19] and it seems doubtful
whether photo-excitation of groups other than hydroperoxides, which
cannot be quenched by energy transfer, play a very significant role in
photo-initiation at least during the early stages of photo-oxidation.[20,21]

SCHEME 4

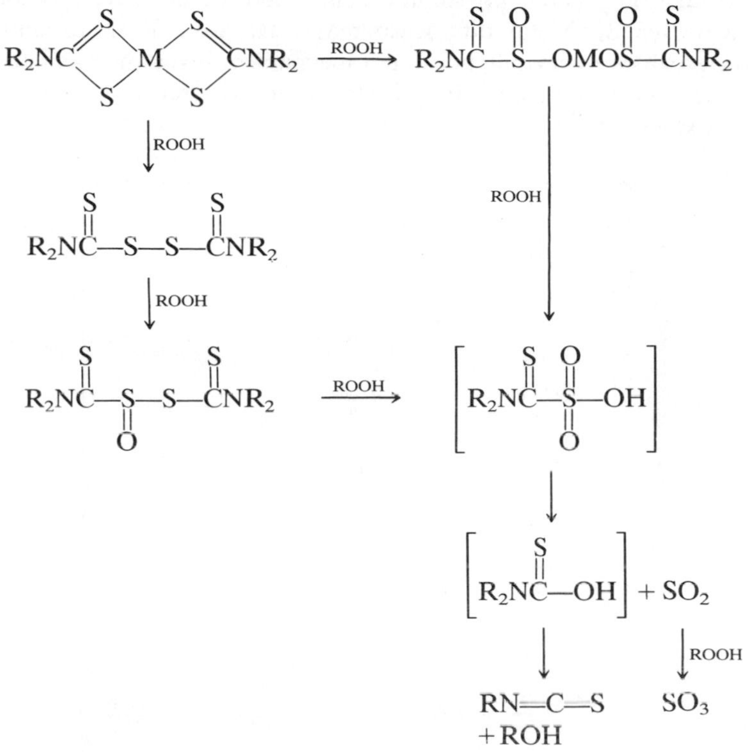

3.2. Ultraviolet Stabilisation Mechanism of the Metal Dithiocarbamates (MDRC)

The importance of hydroperoxides as initiators for the photo-oxidation of the polyolefins is now well established.[17-29] Not only is the rate of photo-oxidation a direct function of the initial (thermally produced) hydroperoxide concentration,[25,27,29] but the rate of photo-oxidation is proportional to [ROOH] during the initial stages of UV irradiation.[24,30] Moreover, removal of thermally formed peroxides in polyolefins by heating in the absence of oxygen leads to effective photostabilisation[25,26] comparable with that of unprocessed polymers. It follows then that additives which can remove peroxides during processing and UV exposure will be efficient antioxidants and UV stabilisers. It does not follow from this, however, that all effective thermal peroxide decomposers will be efficient light stabilisers,[31] since an important requirement of a UV stabiliser is its ability to withstand the effects of UV irradiation,[32] and many sulphur compounds do not have this ability (see Chapter 2). Metal dithiocarbamates (**I**) are known to destroy hydroperoxides by a catalytic (ionic) mechanism involving Lewis acids generated in a series of initial stoichiometric reactions with hydroperoxides.[7,33,34]

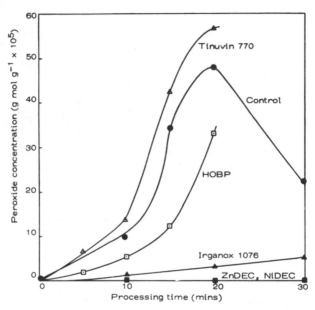

FIG. 11. Formation of hydroperoxides during the processing of polypropylene at 180°C in the presence and absence of antioxidants. (After ref. 6 with permission.)

Figure 11 shows that the formation of hydroperoxides in polypropylene is completely inhibited during processing in the presence of both ZnDEC and NiDEC. By contrast, a typical hindered phenolic antioxidant (Irganox 1076, **XIV**) slows down but does not inhibit peroxide formation and typical commercial UV stabilisers (**XV** and **XVI**) are virtually without effect under the same conditions.[6]

The effects of the same additives on the melt stability of polypropylene as measured by the change in melt flow index follow the same pattern (Fig. 12). Melt flow index (MFI), the amount of polymer extruded through a standard capiliary in a given time, is inversely

OH

tBu $\qquad$ tBu

CH$_2$CH$_2$COOC$_{18}$H$_{37}$

(**XIV**, Irganox 1076)

HO

$\qquad$—CO—$\qquad$—OC$_8$H$_{17}$

(**XV**, Cyasorb UV531, HOBP)

CH$_3$ CH$_3$

HN$\qquad$—OCO(CH$_2$)$_8$COO—$\qquad$NH   (**XVI**, Tinuvin 770)

CH$_3$ CH$_3$

CH$_3$ CH$_3$

related to molecular weight. In the case of the dithiocarbamates and Irganox 1076 (**XIV**), no change in melt viscosity occurs over 20 min,[6] whereas the conventional UV stabilisers have virtually no melt stabilising effect. Similar results have been found in low density polyethylene (LDPE) containing ZnDEC and NiDEC.[28] The stabilising effect of the

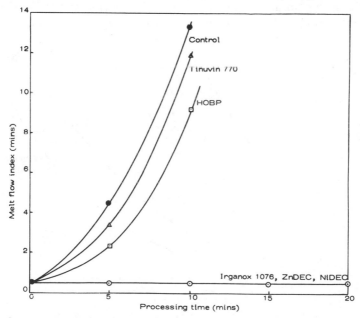

FIG. 12. Effects of antioxidants on the melt flow index (MFI) of polypropylene as a function of processing time at 180°C.

TABLE 4

THERMAL OXIDATIVE STABILITY OF POLYOLEFINS IN THE PRESENCE OF
ANTIOXIDANTS (Concentration of all additives, 3×10^{-4} mol/100 g)

Antioxidant	PP (140°C)[a] embrittlement time (h)	LDPE (110°C)[a] induction period (h)
None	0·5	10
ZnDEC (M = Zn, R = Et)	32	110
NiDEC (M = Ni, R = Et)	28	105
Irganox 1076 (**XIV**)	58	110
ZnDEC + Irganox 1076	165	300

[a] In the case of PP the embrittlement times follow the end of the
induction periods closely, whereas they are much longer than induction
periods for LDPE.

antioxidants was found to persist when polymer films were heated in
an air oven (Table 4).

Irganox 1076 (**XIV**), one of the most effective heat stabilisers,[35] has
a similar activity to ZnDEC and NiDEC as a thermal antioxidant in
LDPE, giving an almost identical induction period at the same molar
concentration at 110°C (Table 4). Combinations of 1076 and ZnDEC
or NiDEC are synergistic under thermal conditions[6,28] due to their

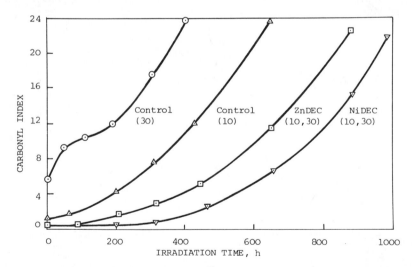

FIG. 13.   Effects of MDEC (3×10^{-4} mol/100 g) on the photostability of
polypropylene. (Numbers in parentheses are processing times (min) at 180°C.)

complementary mechanisms. The antioxidant action of the dithiocarbamates,[7,34] like other sulphur antioxidants,[5] involves an initial radical-generating step before the formation of the acidic species. An important function of chain-breaking phenolic antioxidants in this situation is to remove the initially formed radicals.[8,36]

The effect of processing on unstabilised polyolefins (LDPE and PP) is markedly to reduce UV stability (Fig. 13) due to the thermal oxidative formation of hydroperoxides (see Fig. 11) which are the primary photo-initiators. However, peroxides cannot be detected in thermally processed polypropylene or LDPE[28] containing metal dithiocarbamates. Figure 13 clearly shows that the build-up of carbonyl is inhibited not only during processing but during UV irradiation too, and the initial rate of carbonyl formation is much lower, giving an induction period to carbonyl formation. Moreover, both ZnDEC and NiDEC are as effective as UV stabilisers following a severe processing operation as they are in mildly processed samples (Table 5). This clearly indicates that the PD–C behaviour of ZnDEC or NiDEC begins in the polymer during heat treatment and is carried through to the UV

TABLE 5

EFFECT OF PROCESSING TIME ON THE ULTRAVIOLET LIFETIME OF LDPE FILMS CONTAINING ADDITIVES (Concentration of all additives, 3×10^{-4} mol/100 g)[28,32]

Antioxidant	Time to embrittlement (h)	
	10 min processed	30 min processed
None	1200	900
ZnDEC	1400	1400
NiDEC	1800	1800
NiDBC	2100	2100
HOBP	2200	1600
Irganox 1076	1800	1750
Tinuvin 770	2250	2400
Synergistic Systems		
NiDEC + UV531	4000	4000
ZnDEC + UV531	4000	4000
Antagonistic Systems		
NiDEC + Irganox 1076	1580	1580
ZnDEC + Irganox 1076	1250	1250
NiDEC + Tinuvin 770	1850	—

exposure stage. Although this antioxidant function continues during UV irradiation, both compounds are destroyed by light. The high efficiency of NiDEC compared with ZnDEC at the same molar concentration, particularly during the early stages of photo-oxidation, is not however a reflection of a difference in PD–C activity since under thermal oxidative conditions they have very similar antioxidant activity (Table 4). Their UV absorbance spectra and UV stabilities differ appreciably in both LDPE and PP.[30] NiDEC has a strong absorbance at 330 nm whereas ZnDEC absorbs much less strongly in this region but has a maximum at 285 nm whose intensity is only about 60% of that of NiDEC at 330 nm. This clearly affects their relative screening abilities in the region 300–350 nm. Both metal complexes decay in the first-order mode (Fig. 14)[6] but the half-life of ZnDEC is only about 25% of that of NiDEC in PP (see Fig. 14) and the length of the induction period to photo-oxidation corresponds almost exactly to the time for the disappearance of the metal complexes from the system. This clearly shows that the stability of the peroxide decomposer precursor is responsible under photo-oxidative conditions for its effectiveness as a UV stabiliser. As well as being a peroxide decomposer, NiDEC is an effective UV filter. However, this makes a relatively minor contribution to the overall stabilising function[18,28] (Fig. 15). ZnDEC does not

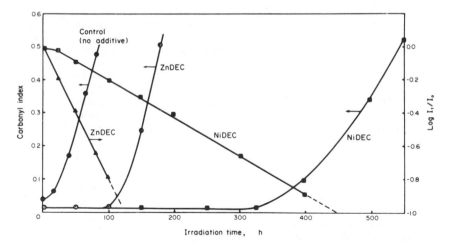

FIG. 14.   Relationship between the first-order decay of MDEC concentration and photo-oxidative induction period in polypropylene. (After ref. 6 with permission.)

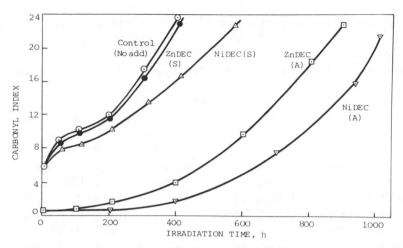

FIG. 15. Comparison of the effectiveness of MDEC as UV screens for unstabilised polymer (S) and as additives for LDPE screened with unstabilised LDPE (A) (processed 150°C/30 min)

seem to screen UV light at 290 nm at all, because the rate of photo-oxidation of the sample screened by a sample containing ZnDEC is almost the same as that of the control. The UV stabilising activity of NiDEC is strongly concentration-dependent whereas ZnDEC is not[38] and this is almost certainly associated with a difference in their abilities to absorb and deactivate light by a photophysical process at 290–350 nm, which is the critical region of the sun's spectrum responsible for sensitisation of photo-oxidation. Like the 2-hydroxybenzophenone UV stabiliser, NiDEC has an efficient method of internally deactivating the absorbed light.

On increasing the alkyl chain length in the nickel dithiocarbamates, the molar UV stabilising effect of NiDEC was found to increase both in LDPE (Table 5) and in PP.[37] The UV embrittlement time (1800 h) of LDPE observed with nickel diethyldithiocarbamate (NiDEC) increased to about 2100 h with nickel dibutyldithiocarbamate (NiDBC) (Table 5). The increased solubility of the additive in the polymer medium with the increase in chain length in the alkyl group of nickel dithiocarbamate has been found to be responsible for this effect.[38] Synergistic effects between peroxide decomposers and 'UV absorbers' have been known for many years and give rise to some of the most effective UV stabilising systems for polyolefins so far found.[33] Typical

'UV absorbers' such as 2-hydroxy-4-octoxybenzophenone (HOBP, **XV**) are unable to inhibit the formation of hydroperoxide in polyolefins[28,30] during processing and do not destroy it during photo-oxidation, although it is rapidly destroyed by hydroperoxide under these conditions.[39] HOBP is, therefore, an indifferent antioxidant during processing and appears to behave essentially as a screen during the early stages of photo-oxidation when hydroperoxides are the primary photo-initiators.[18] It behaves less effectively the more severe the processing operation[28,30] (Table 5). In the later stages of the photo-oxidation, however, when carbonyl photolysis is the main photodegradation process,[40] it has been shown that the UV stabiliser is sacrificially destroyed in the process of protecting the carbonyl group.[39]

Both ZnDEC and NiDEC show very powerful synergistic effects with HOBP in LDPE.[28] This is illustrated for ZnDEC in Fig. 16 and Table 5 and similar results have been obtained in PP[30] (Table 6). It is clear from Fig. 16 that the synergistic combination of metal dithiocarbamates with HOBP combines the desirable features of both

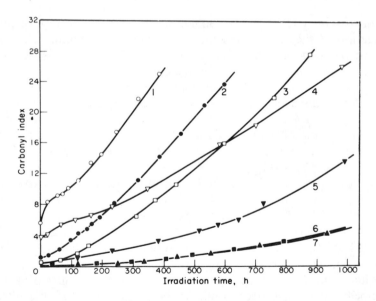

FIG. 16. Synergism between ZnDEC and HOBP under mild and severe processing conditions. 1, No additive (30 min); 2, no additive (10 min); 3, ZnDEC (10 and 30 min); 4, HOBP (30 min); 5, HOBP (10 min); 6, ZnDEC + HOBP (30 min); 7, ZnDEC + HOBP (10 min). (After ref. 6 with permission.)

TABLE 6
COMBINATION EFFECTS OF STABILISERS IN POLYPROPYLENE
(Concentration of all stabilisers, 3×10^{-4} mol/100 g)[30]

Stabiliser	UV embrittlement time (h)
Control (no additive)	90
ZnDEC	175
NiDEC	500
HOBP	245
ZnDEC+HOBP	700
NiDEC+HOBP	850

stabilisers, i.e. the well-defined induction period of the peroxide decomposer and the lower post-induction period rate of the UV absorber. A detailed study[28] of the mechanism of the synergistic action
has shown that the additives protect each other during thermal processing and on UV exposure and hence lengthen the period of their
independent but co-operative action. Figure 17 shows that metal
dithiocarbamates protect HOBP from destruction by peroxides during
processing and on UV exposure and in turn HOBP protects the metal
complexes from photolytic destruction both by screening of UV light
and by removing (sacrificially quenching) photo-excited reactive
species formed from the dithiocarbamates during irradiation. The

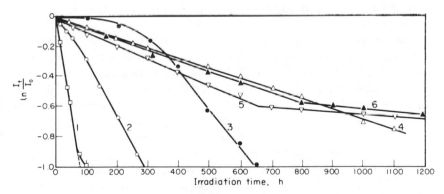

FIG. 17.   Change in UV and visible absorption spectra of additives during the
photo-oxidation of LDPE (processed 10 min at 150°C). Concentration of all
additives 3×10^{-4} mol/100 g. 1, ZnDEC (280 nm); 2, NiDEC (330 nm); 3,
NiDEC+HOBP (390 nm); 4, HOBP (330 nm); 5, ZnDEC+HOBP (330 nm);
6, NiDEC+HOBP (390 nm). Numbers in parentheses are λ_{max}. (After ref. 30
with permission.)

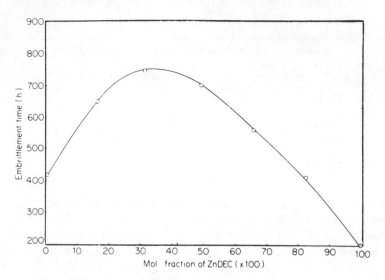

FIG. 18.   Synergism between ZnDEC and HOBP on the photostabilisation of polypropylene. Total additive concentration 6×10^{-4} mol/100 g.

synergistic effect varies with the molar ratio of the two types of stabilisers at constant total concentration and reaches a maximum at a molar [HOBP]/[ZnDEC] of $2:1$[30] (Fig. 18) indicating that the protection of the peroxide decomposer against photolysis is a major role of the synergist.

Combination of peroxide decomposers (ZnDEC and NiDEC) and a phenolic antioxidant, Irganox 1076, which are effective synergists during thermal oxidation (see Table 4), leads to antagonism during photo-oxidation (see Table 5). The antagonism between 1076 and metal dithiocarbamates has been shown[28] to be due to sensitisation of the photolytic destruction of the dithiocarbamates by oxidation products of phenol (particularly stilbenequinones) leading to a decrease in their effectiveness. PD–C antioxidants (ZnDEC and NiDEC) are also antagonistic towards the hindered piperidine UV stabiliser, Tinuvin 770 (**XVI**) in LDPE during photo-oxidation[36] (see Table 5). It is known[30,42] that the hindered piperidine is not itself the effective UV stabiliser and that it has to be first oxidised to the nitroxyl radical. This conversion is brought about, at least in part, by oxidation by hydroperoxides.[36] The more severe the processing conditions, leading to increased hydroperoxide concentration, the higher is the conversion of amine to nitroxyl[41] and the more effective is the hindered piperidine as a UV stabiliser. PD–C antioxidants (metal dithiocarbamates) thus inhibit the conversion of amine to nitroxyl radical by removing peroxide from the system, thereby reducing the apparent effectiveness of hindered amine as a UV stabiliser.

3.3. Amine Complexes of Zinc Dialkyldithiocarbamates

It has been shown in the previous section that zinc diethyldithiocarbamate is a powerful antioxidant in polyolefins and an effective UV stabiliser in the initial photo-oxidation stage. The major limitation to its effectiveness as a UV stabiliser is that it is light-unstable and photolyses rapidly under conditions of UV exposure. Two major requirements of an effective photo-antioxidant are compatibility and UV stability. The UV stabilisation effectiveness of zinc diethyldithiocarbamate is considerably improved on complexing with both aliphatic and aromatic amines.[37] The piperidine and diazabicyclo-octane (DABCO) complexes of zinc diethyldithiocarbamate are seen to be more effective photostabilisers for polypropylene than the parent zinc diethyldithiocarbamate in Fig. 19. It is also apparent from Fig. 19 that amine complexes of zinc diethyldithiocarbamate gave sharp induction periods to the onset of carbonyl formation during the photo-oxidation of

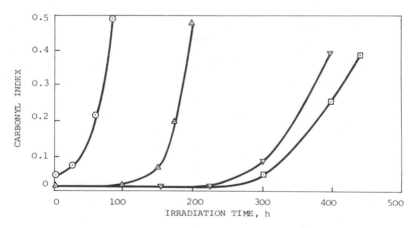

FIG. 19. Effect of amine complexing of zinc diethyldithiocarbamate on UV stabilising effectiveness. Processed 180°C/10 min; concentration 6× 10^{-4} mol/100 g in each case. ⊙ Control (no additive); △ ZnDEC; ▽ ZnDEC–piperidine; ▢ ZnDEC–DABCO.

polypropylene. This is a characteristic of peroxide-decomposing antioxidants. The amines alone do not provide any significant photo-stabilising effect compared with the control. As already seen, ZnDRC complexes exert their antioxidant and UV stabilisation action in polymers by catalytically decomposing peroxides into non-radical products (see Scheme 3, reaction (b)). The catalytic decomposition of peroxides is not however due to the metal dithiocarbamates themselves but to sulphur oxides produced by an initial reaction between metal dithiocarbamates and hydroperoxides involving free radicals. It is clearly seen from Fig. 20 that both the initial rate and the catalytic decomposition rate of *tert*-butyl hydroperoxide in chlorobenzene under UV light are increased by complexing zinc diethyldithiocarbamates with amines. The UV stability of a zinc diethyldithiocarbamate–DABCO complex in PP films is compared with the parent zinc diethyldithiocarbamate in Fig. 21 by following the decay of the UV absorption maxima at 285 nm. ZnDEC–DABCO has an initial induction period before starting to decay and the rate of decay is much slower compared with ZnDEC alone. However, the UV stabilities of the amine complexes of ZnDEC in solution are almost identical to that of the parent ZnDEC.[38] This seems to suggest that the higher UV stability of ZnDEC on complex formation with the amines is associated with their enhanced compatibility in the polymeric medium.

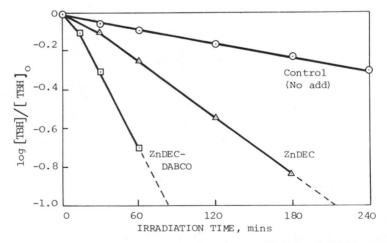

FIG. 20.   First-order plot for the decomposition of TBH by ZnDEC and its DABCO complex in chlorobenzene under UV irradiation. Metal complex concentrations 10^{-4} M; TBH 10^{-2} M.

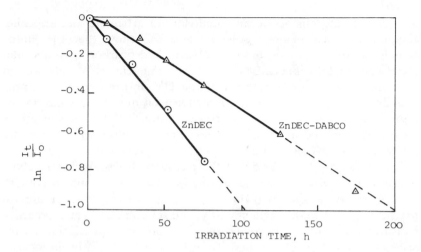

FIG. 21. Decay of UV absorbance of ZnDEC (285 nm) and ZnDEC-DABCO (285 nm) in polypropylene during photo-oxidation. Concentration 6×10^{-4} mol/100 g.

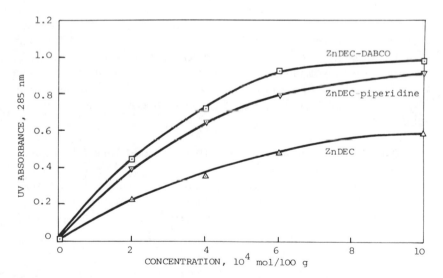

FIG. 22. Ultraviolet absorbance at 285 nm of ZnDEC and its amine complexes as a function of additive concentration.

Solubility measurements of zinc diethyldithiocarbamate and its amine complexes in hydrocarbon solvents at 30°C show that the amine complexes are in general more soluble in hydrocarbon solvents than the parent metal dithiocarbamates.[38] The importance of solubility is also supported by measuring the initial UV absorbance of the compounds in polypropylene after subjection to a mild processing operation. ZnDEC shows lower absorbance values at all concentrations compared with the amine complexes (Fig. 22) although their molar extinction coefficients in organic solution are identical.

The above results suggest that the increased UV stabilisation effectiveness of ZnDEC on complex formation with amines may be due to one or more of the following factors: (a) increased solubility in polymers; (b) increased light stability; and (c) enhanced catalytic activity for hydroperoxides. However, the contribution of physical or chemical quenching of photo-excited ZnDEC by the amines in the overall stabilisation process cannot be ruled out.[6]

4. DITHIOPHOSPHATE AND XANTHATE METAL COMPLEXES

Two other closely related classes of sulphur-containing metal complexes which exhibit powerful thermal antioxidant activity are the metal dialkyldithiophosphates (II)[43-45] and metal xanthates (III).[45] The nickel complexes like the nickel dithiocarbamates are effective light stabilisers for polyolefins,[45-48] the barium diaryl dithiophosphates have been reported[44] to be stabilisers for halogen-containing polymers, e.g. PVC, and the nickel complexes are also effective antiozonants for rubbers.[49]

4.1. Nickel Dialkyldithiophosphates (NiDRP) and Nickel Xanthates (NiRX) as Stabilisers for Polyolefins

The behaviour of the transition metal dithiophosphates (II) and metal xanthates (III) as thermal and UV stabilisers in polyolefins generally resembles that of the dithiocarbamates (I).[48] Figure 23 compares the

(XVII, R = isoPr)   (XVIII, R = isoPr)

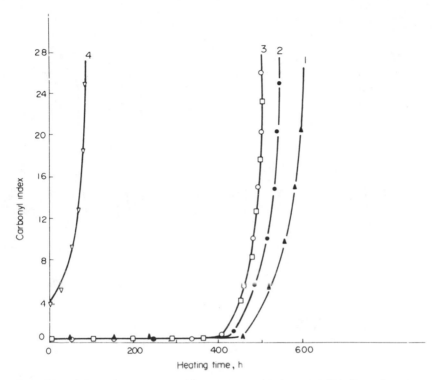

FIG. 23.   Effect of thermal oxidation (at 110°C) on LDPE films containing additives $(2 \cdot 5 \times 10^{-4} \text{ mol}/100 \text{ g})$. 1, NiDBP; 2, NiDiPP; 3, NiDiPP extracted before oven ageing (□) and DiPDiS (○); 4, control (no additive). (After ref. 48, with permission.)

effectiveness as thermal antioxidants of typical nickel dialkyl-dithiophosphates with a derived disulphide (**XVII**). The similar performance of the disulphide at the same molar concentration suggests that the latter is involved in the antioxidant function of the metal complex. A similar behaviour is observed with the xanthates (Fig. 24) and in this case it was established that the thiocarbonyl function became chemically attached to the polymer backbone through the dixanthogen intermediate (**XVIII**).

In spite of the similar behaviour of the metal complexes and the related disulphides during thermal oxidation, they behave quite differently during photo-oxidation.[48] Whereas the nickel dithiophosphates and nickel xanthates are effective UV stabilisers, the corresponding

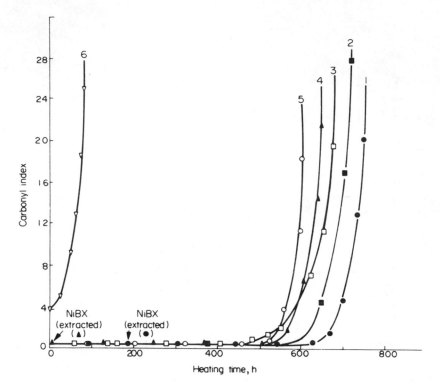

FIG. 24. Effect of thermal oxidation (at 110°C) in the presence of air on LDPE films containing additives (2.5×10^{-4} mol/100 g). 1, NiBX extracted after the destruction of the complex (200 h at 110°C); 2, NiBX untreated; 3, NiEX untreated; 4, NiBX extracted prior to oven ageing; 5, NiBX; 6, control (no additive). (After ref. 48, with permission.)

disulphides are not. Figure 25 shows that extraction of an LDPE film containing nickel dibutyldithiophosphate effectively removes the UV stabilising component, namely the nickel complex itself. The thiophosphoryl disulphides, although formed, were found to be relatively weak UV stabilisers. Likewise the dixanthogens (**XVIII**) are weak UV stabilisers compared with the corresponding nickel complexes and the UV stability of the nickel xanthate is reflected in its long photo-oxidative induction period (Fig. 26).[45]

Figure 27 shows that, as in the case of the dithiocarbamates (see Section 3), there is a close correlation between loss of the metal

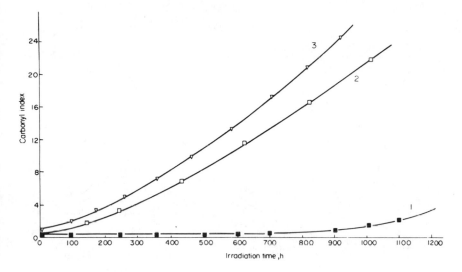

FIG. 25. Effect of NiDBP (extracted and unextracted) on the photo-oxidation of LDPE (processed at 150°C/30 min in a closed mixer, concentration $2 \cdot 5 \times 10^{-4}$ mol/100 g). 1, NiDBP; 2, NiDBP (extracted); 3, control (no additive). (After ref. 48, with permission.)

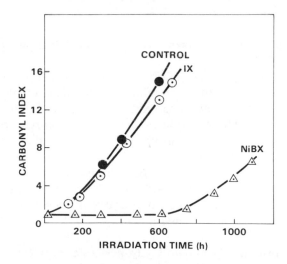

FIG. 26. Effect of NiBX and **IX** on the photo-oxidation of LDPE (processed at 150°C/30 min in a closed mixer; concentration $2 \cdot 5 \times 10^{-4}$ mol/100 g).

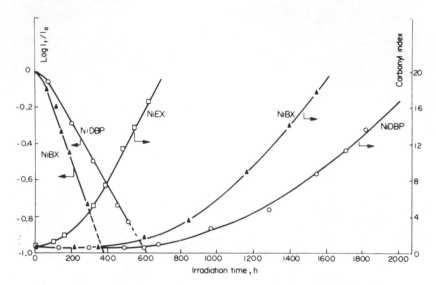

FIG. 27.   Kinetics of the disappearance of NiEX, NiBX and NiDBP (316 nm) in LDPE during photo-oxidation compared with the changes of carbonyl index under the same conditions (concentration of additives $2 \cdot 5 \times 10^{-4}$ mol/100 g). (After ref. 48, with permission.)

complex by photolysis or photo-oxidation and the end of the photo-oxidative induction period.

4.2. Mechanisms of Antioxidant Action of the Dithiophosphates and Xanthates

The generally accepted mechanism of action of dithiophosphates and xanthates in common with other sulphur-containing complexes is that they function both as hydroperoxide decomposers and as radical scavengers.[6,45,50–56] The question of the relative contribution of these processes to the overall stabilising effect[52,53] and the actual intermediates involved[54] however has not been completely resolved. The dominating importance of the peroxidolytic function of metal dithiophosphates and xanthates and their derived products during the stabilisation of polymers has become apparent from recent studies in model systems at elevated temperatures.[50,57]

4.3. Decomposition of Cumene Hydroperoxide by NiDRP and NiRX

Reactions of nickel dithiolates (NiDRP, NiRX, NiDRC) with cumene hydroperoxide (CHP) at different temperatures (30–150°C) in the

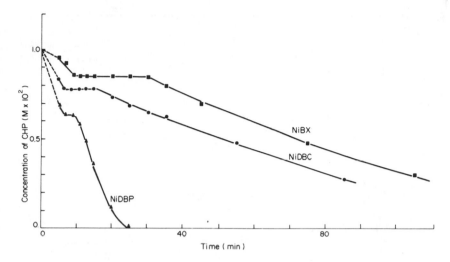

FIG. 28. Decomposition of CHP (1×10^{-2} M) in chlorobenzene at 110°C in the presence of nickel dithiolates (2×10^{-4} M). (After ref. 50, with permission.)

presence and absence of an oxidisable substrate showed[50,57] that the nickel complexes decompose CHP in two distinct stages whose relative importance is a function of the molar ratio of the two components. Figure 28 shows that all the three nickel complexes show the same general behaviour. In all cases, the rapid initial catalytic stage is favoured by a low molar ratio of hydroperoxide to complex and gives rise to homolytic products. The second catalytic stage, which follows an induction period, assumes greater significance at higher molar ratios of hydroperoxide to complex, and gives rise to an ionic decomposition of hydroperoxides. Figure 29 shows that α-cumyl alcohol and acetophenone (the homolytic products of hydroperoxide decomposition[6,37]) are exclusively formed at, and below, a molar ratio [CHP]/[NiDBP] = 10.[50] This clearly demonstrates the importance of the first catalytic free radical pathway at low molar ratios, during the reaction of CHP with NiDBP at elevated temperatures.

The relative contribution of the initial catalytic process to the overall stabilising effect of NiDRP was found[57] to decrease with increase in the complex concentration. This is demonstrated in Fig. 30, which shows a linear relationship between the amount of hydroperoxide decomposed during the first catalytic stage and the concentration of the nickel complex.

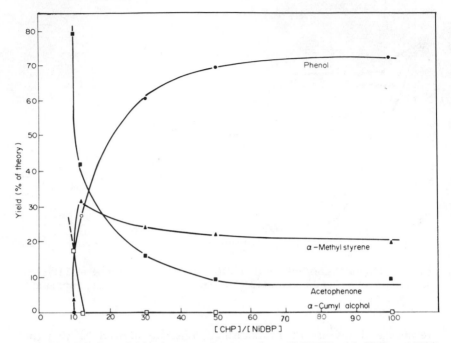

FIG. 29.   Product yields after complete reaction of NiDBP with CHP at 110°C in chlorobenzene at various molar ratios [CHP]/[NiDBP]. (After ref. 50, with permission.)

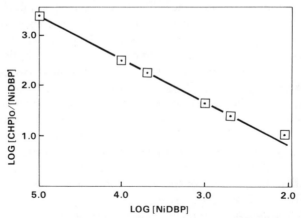

FIG. 30.   Relationship between number of moles of CHP decomposed during the first stage, per mole of NiDBP at different additive concentrations. (CHP concentration in chlorobenzene, 1×10^{-2} M).

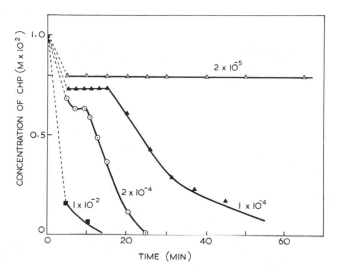

Fig. 31. Decomposition of CHP (1×10^{-2} M) in chlorobenzene at 110°C in the presence of NiDBP at the molar concentrations shown.

In contrast to the above, an inverse relationship was found[50] between the length of the secondary induction period and the concentration of the nickel complex at constant hydroperoxide concentration (Fig. 31). This implies that the induction period corresponds to the time required for the formation of a critical concentration of the active catalyst from the nickel complex. The opposite relationship has been reported[54] for ZnDIP, where the length of the induction period was found to increase with increase in the initial zinc complex concentration. Consequently, the formation of an active catalyst from the zinc complex was discounted in this case.[54]

A product study[57] at intervals during the decomposition of CHP in the presence of NiDBP at 110°C and at molar [CHP]/[NiDBP] = 100 (see Fig. 32) demonstrated the changeover from an essentially homolytic to an essentially heterolytic reaction. Clearly heterolytic decomposition predominates at the end of the reaction.

The fact that nickel dithiolates (e.g. NiDRC) are not themselves responsible for the catalytic decomposition of hydroperoxides has been known for many years[7] and has recently been confirmed for the case of nickel dithiophosphate.[50] During the reaction of NiDRP with hydroperoxides, the nickel complex is always destroyed before the onset of

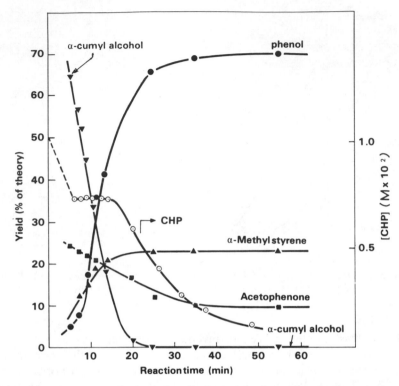

FIG. 32.   Kinetics of ionic and radical product formation in the reaction of CHP $(1 \times 10^{-2} \text{M})$ and NiDBP $(1 \times 10^{-4} \text{M})$ in chlorobenzene at 110°C. CHP decomposition curve for the same reaction is also shown.

the secondary catalytic stage (Fig. 33). This evidence, coupled with the fact that the length of the secondary induction period observed during reaction of NiDBP with hydroperoxides increases with decreasing complex concentration, provides further confirmation for the formation and accumulation of an active catalyst or catalysts from the nickel complex during the induction period.

4.4. Decomposition of CHP by NiDRP-derived Intermediates
There appears to be some conflict in the published literature on the nature of the catalyst(s) that lead to the ionic peroxidolytic action of the metal dithiolates. The formation of disulphides and their intermediacy during the peroxide-decomposing action of ZnDIP has been

questioned.[54,56,59] There is unequivocal evidence[60] that the corresponding disulphide is one of the transformation products formed during the early stages of the reaction of CHP with NiDBP at high temperatures. This process is quite analogous to that involved in the formation of an ionic catalyst for the decomposition of hydroperoxides from zinc mercaptobenzthiazole and its metal complexes through the intermediacy of the disulphide (see Section 2.1) and is also consistent with the mechanism proposed for the antioxidant function of MDRC in hydroperoxide-initiated systems (see Section 3.1).

A kinetic investigation of the reaction of CHP with di-isopropyl thiophosphoryl disulphide (DiPDiS) showed[50] that under the same conditions as those used in the investigation of NiDBP, its behaviour was similar (Fig. 34). However, two major differences were observed.

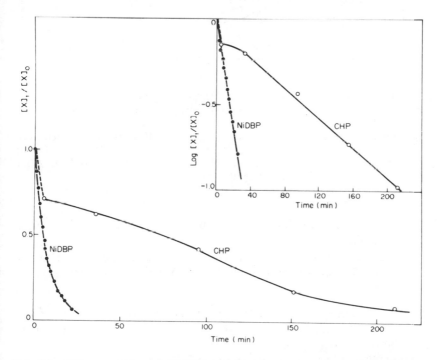

FIG. 33.   Decomposition of CHP $(1 \times 10^{-2}$ M) in chlorobenzene at 70°C in the presence of NiDBP $(8 \times 10^{-4}$ M) and the associated decay of the NiDBP UV absorbance. (The inset shows first-order kinetic plots for the same reaction.) (After ref. 50, with permission.)

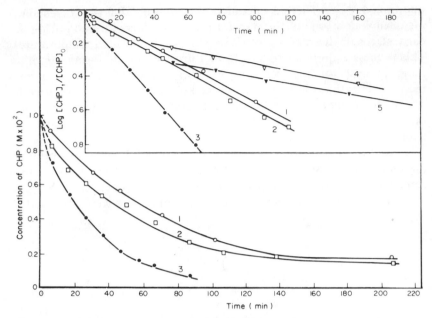

FIG. 34.   Decomposition of CHP $(1 \times 10^{-2} M)$ in dodecane at 150°C in the presence of DiPDiS at the molar concentrations indicated below. (Inset compares first-order kinetics for the reactions of DiPDiS and NiDBP with CHP under identical conditions.) The plots for NiDBP are those obtained from the third stage of the reaction. 1, DiPDiS $(1 \times 10^{-4} M)$; 2, DiPDiS $(2 \times 10^{-4} M)$; 3, DiPDiS $(10 \times 10^{-4} M)$; 4, NiDBP $(1 \times 10^{-4} M)$; 5. NiDBP $(2 \times 10^{-4} M)$. (After ref. 50, with permission.)

(1)   There was no induction period before the second catalytic stage (cf. Fig. 28).

(2)   Homolytic breakdown of hydroperoxides made only a minor contribution to the overall decomposition process (cf. Figs 29 and 35).

Figure 35 shows that at molar [CHP]/[DiPDiS] = 10, mainly ionic products are found. Even at a 1:1 molar ratio, 36% of the products are heterolytic (e.g. phenol).

Alkyl disulphides and monosulphides[61,62] have been shown to be ineffective antioxidants, and inhibition was attributed to products formed from them as hydroperoxides accumulate in the substrate.[4,63] It has been suggested[64,65] that hydroperoxides formed in the substrates at high temperatures oxidise disulphides to thiolsulphinates and that effective inhibition then results from peroxide decompositions by the thiolsulphinate or subsequent reaction products (Scheme 5). Disulphides of thiophosphoric acid (**XVII**) and xanthic acid (**XVIII**) have

SCHEME 5

$$\text{RSSR} \xrightarrow{\text{ROOH}} \overset{\overset{\displaystyle O}{\parallel}}{\text{RS}}\text{—SR} \xrightarrow{\text{ROOH}} \text{non-radical products}$$

recently been found[57] to be unable to bring about thermal decomposition of hydroperoxides below 70°C. Furthermore, they are not destroyed by hydroperoxides unless initiated, for example by UV light.

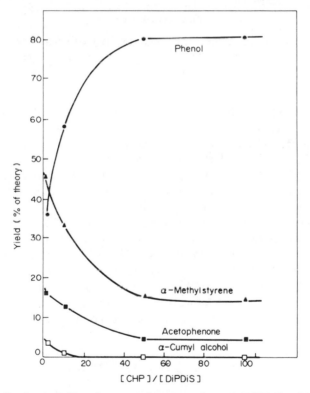

FIG. 35. Product yields after complete reaction of DiPDiS with CHP at 110°C in chlorobenzene at various molar ratios ([CHP]/[DiPDiS]). (After ref. 50, with permission.)

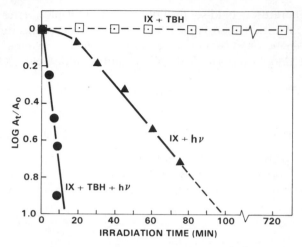

FIG. 36. Disappearance of xanthogen (**IX**, 0.5×10^{-4} M) (285 nm) in the presence of *tert*-butyl hydroperoxide (0.5×10^{-2} M) in the presence and absence of light.

Figure 36 compares the rate of disappearance of the 285 nm band of dixanthogen in the presence of *tert*-butyl hydroperoxides (TBH) at molar [TBH]/[X] = 100 at room temperature, in the presence and absence of UV light. The temperature required for the reaction of disulphides (e.g. thiophosphoryl disulphide with hydroperoxide) is significantly higher than that required for similar reactions of the corresponding metal complexes (e.g. ZnDRP).

The nature of the ionic catalysts(s) for peroxide decomposition formed from NiDBP and its corresponding disulphide has recently been investigated.[60] It was found (see Fig. 37) that in the reaction of CHP with NiDBP in chlorobenzene at 110°C at molar [CHP]/[NiDBP] = 100, sulphonic acid (**XIX**, Scheme 6) and thionophosphoric acid (**XX**) were formed during the induction period. Whereas no phenol (a diagnostic ionic decomposition product) was

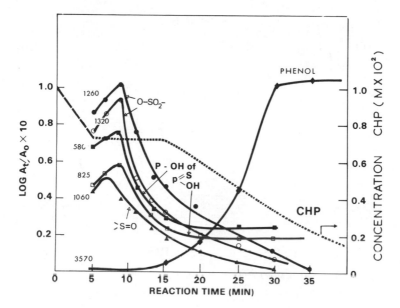

FIG. 37. Formation and decay of products formed in the reaction between cumene hydroperoxide (CHP) and nickel dibutyldithiophosphate (NiDBP) at molar [CHP]/[NiDBP] = 100 in chlorobenzene at 110°C. Numbers on curves are IR absorbances, cm^{-1}. The decay of CHP is superimposed.

found during the build-up of the disulphide-derived oxidation products, phenol began to be formed after about 10 min from the start of the reaction, coincident with the onset of decomposition of oxidation intermediates **XIX** and **XX**. The production of SO_3 and H_2SO_4 (Scheme 6) would therefore account for the formation of phenol and acetone during the decomposition of CHP in the presence of nickel dithiophosphates. The mechanism outlined in Scheme 6 is broadly analogous to that proposed for the formation of acidic species from mercaptobenzthiazolate complexes (see Section 2.1) and from the dithiocarbamates (see Section 3). Furthermore, the product distribution from CHP in the presence of NiDBP is strikingly similar to that observed with sulphur dioxide,[67] confirming the role of sulphur acids in the peroxidolytic function of nickel dithiophosphates.

Nickel xanthates and dithiocarbamates show quite similar behaviour to that of nickel dithiophosphates in decomposing hydroperoxides (see Fig. 28). A comparison of their contribution to the catalytic ionic

SCHEME 6

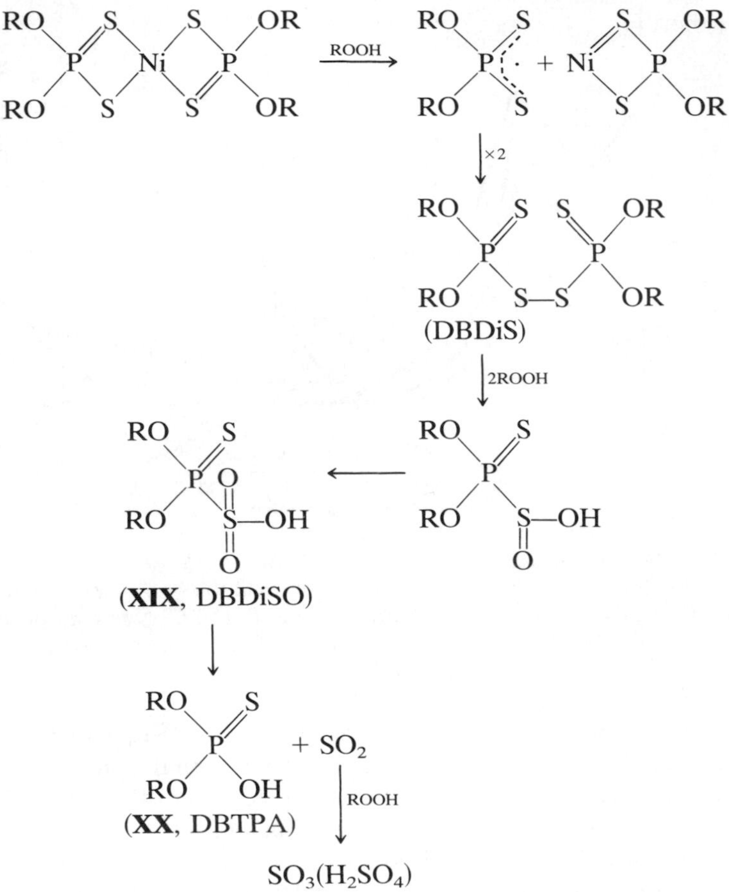

TABLE 7

RELATIVE CONTRIBUTIONS TO THE CONCENTRATION AND
HOMOLYTIC DECOMPOSITION OF CUMENE HYDROPEROXIDE BY
NICKEL DITHIOLATES AT $110°C$ (Molar [CHP]/[complex] = 125)

Complex	Percentage of initial CHP		
	Phenol	α-Methylstyrene	Acetophenone
NiDBP	90·8	2·4	6·8
NiDBC	89·0	5·3	5·7
NiBX	76·6	18·5	5·4

decomposition of CHP at molar [CHP]/[complex] = 125 at 110°C is shown in Table 7. These results suggest a slightly larger contribution from the ionic pathway in the case of NiDRP.

5. OVERVIEW OF THE METAL THIOLATE ANTIOXIDANT MECHANISMS

It is clear from a consideration of Schemes 2, 3, 4 and 6 that catalytic peroxidolysis (PD–C) is the most important mechanism involved in the antioxidant action of the metal thiolate antioxidants. It is also evident that, although other mechanisms may be involved in specific cases (see below), this mechanism is common to all the sulphur complexes and is able to account for the salient technological characteristics of the metal thiolates.

5.1. Catalytic Peroxidolysis (PD–C)

The main distinguishing features of this mechanism are as follows:

(a) The effective antioxidants are not the initial metal complexes but acidic products formed from them by oxidation. The metal thiolates function as reservoirs for the active catalytic species which is liberated slowly and in a controlled way by hydroperoxides formed in the polymer. In this respect, the metal thiolates behave very similarly to the alkyl sulphide antioxidants discussed in Chapter 2 and they are in general as effective as thermal antioxidants as the alkyl sulphides, if not more so. As melt stabilisers, they are considerably more effective than the alkyl sulphides and do not require auxiliary synergists.

(b) The metal thiolates vary considerably in their ability to function as UV stabilisers. The determining feature in this case appears to be the nature of the metal ion which in turn determines the UV stability of the metal complex. In general the transition metal complexes are more UV stable and hence better UV stabilisers than the Group II metal complexes. However, iron is an exception to this and although its complexes function as UV stabilisers at relatively high concentrations, they are very concentration-sensitive and are used as the basis of time-controlled stabilisers in polymers (see *Devolopments in Polymer Stabilisation—5*, Chapter 4).

(c)   In spite of the fact that some metal thiolates are indifferent UV stabilisers, they all show powerful synergistic effects with commercial 'UV absorbers' (cf. the dialkyl sulphides, Chapter 2). Two effects appear to be responsible for this. The metal thiolates are able to protect the UV absorbers from destruction by peroxides both during processing and UV exposure and the UV stabilisers protect the sulphur components from photolysis.

5.2. Chain-breaking Donor (CB–D) Activity

There is evidence that both the parent metal complexes and many of their derived oxidation products (e.g. sulphinic acids, sulphonic acids, sulphur dioxide, etc.) have some activity as alkylperoxyl radical scavengers (CB–D). In this respect they resemble the alkyl sulphide antioxidants and this mechanism probably makes a subsidiary although ill-defined contribution to their overall activity.

5.3. Ultraviolet Absorption

The transition metal complexes generally have a high molar extinction coefficient, in the region 300–330 nm, and this generally parallels their stability to UV light in this region. Ultraviolet filtering appears to make some contribution to their UV stabilising ability but their effectiveness as UV screens does not by itself account for their high activity.

5.4. Excited State Quenching

In spite of the fact that many transition metal thiolates have been shown to be effective quenchers for photo-excited states of reagents (e.g. oxygen) or oxidation products (carbonyl) present during photooxidation, no clear evidence has emerged to suggest that these processes make a significant contribution to their photostabilising effect in polymers. There are good theoretical arguments against their functioning in this way and much of the published data advocating this mechanism can be equally well explained on the basis of the peroxidolysis mechanism.

REFERENCES

1. SCOTT, G., *Atmospheric Oxidation and Antioxidants* (1965) pp. 186, 290. Elsevier, London and Amsterdam.

2. STUCKEY, J. E. in *Developments in Polymer Stabilisation—1*, Ed. G. Scott, (1979), p. 101. Applied Science Publishers, London.
3. BROOKS, L. A., *Rubb. Chem. Tech.*, **36** (1963) 887.
4. INGHAM, F. A. A., SCOTT, G. and STUCKEY, J. E., *Europ. Polym. J.*, **11** (1975) 783.
5. HUSBANDS, M. J. and SCOTT, G., *Europ. Polym. J.*, **15** (1979) 879.
6. SCOTT, G., *Pure Appl. Chem.*, **52** (1980) 365.
7. HOLDSWORTH, J. D., SCOTT, G. and WILLIAMS, D., *J. Chem. Soc.* (1964) 4692.
8. SCOTT, G. and SHEARN, P. A., *J. Appl. Polym. Sci.*, **13** (1969) 1329.
9. LE BRAS, J., *Rev. Gen. Caout.*, **21** (1944) 3; LE BRAS, J. and VIGER, L., *Rev. Gen. Caout.*, **21** (1944) 89.
10. SHELTON, J. R. and COX, W. L., *Ind. Eng. Chem.*, **43** (1957) 456.
11. DWEIK, H. S., CHAKRABORTY, K. B. and SCOTT, G., unpublished work.
12. ABED-ALI, S., CHAKRABORTY, K. B. and SCOTT, G., unpublished work.
13. CARLSSON, D. J., SUPRUNCHUK, T. and WILES, D. M., *J. Polym. Sci.*, **B11** (1973) 61.
14. GUILLORY, J. P. and BECKER, R. S., *J. Polym. Sci.*, *A1*, **12** (1974) 993.
15. HAMMOND, G. S. and FOSS, R. P., *J. Phys. Chem.*, **68** (1969) 3739.
16. CHIEN, J. C. W. and CONNOR, W. J., *J. Am. Chem. Soc.*, **90** (1968) 1001.
17. RANAWEERA, R. P. R. and SCOTT, G., *J. Polym. Sci., Polym. Lett. Edn*, **13** (1975) 71.
18. RANAWEERA, R. P. R. and SCOTT, G., *Europ. Polym. J.*, **12** (1976) 591.
19. RANAWEERA, R. P. R. and SCOTT, G., *Europ. Polym. J.*, **12** (1976) 825.
20. RANBY, B. and RABEK, J., Eds., *Reactions of Singlet Oxygen with Polymers* (1978), p. 230, Wiley London.
21. WILES, D. M. and CARLSSON, D. J., *Polym. Deg. Stab.*, **3** (1980–81) 61.
22. SCOTT, G., *Macromol. Chem.*, **8** (1972) 319.
23. MELLOR, D. C., MOIR, A. B. and SCOTT, G., *Europ. Polym. J.*, **9** (1973) 291.
24. CARLSSON, D. J., GARTON, A. and WILES, D. M., *Macromolecules*, **9** (1976) 695.
25. CHAKRABORTY, K. B. and SCOTT, G., *Polymer*, **18** (1977) 98.
26. AMIN, M. U., SCOTT, G. and TILLEKERATNE, L. M. K., *Europ. Polym. J.*, **11** (1975) 85.
27. CHAKRABORTY, K. B. and SCOTT, G., *Europ. Polym. J.*, **13** (1977) 731.
28. CHAKRABORTY, K. B. and SCOTT, G., *Europ. Polym. J.*, **13** (1977) 1007.
29. SCOTT, G., *Developments in Polymer Stabilisation—1*, Ed. G. Scott (1979) p. 309. Applied Science Publishers, London.
30. CHAKRABORTY, K. B. and SCOTT, G., *Polym. Deg. Stab.*, **1** (1979) 37.
31. HUTSON, G. V. and SCOTT, G., *Europ. Polym. J.*, **16** (1974) 45.
32. SCOTT, G., *ACS Symp. Series*, **25** (1976) 340.
33. Ref. 1, Chapter 5.
34. CHENIER, J. H. B. and HOWARD, J. A., *Can. J. Chem.*, **54** (1976) 390.
35. PLANT, M. A. and SCOTT, G., *Europ. Polym. J.*, **7** (1971) 1173.
36. CHAKRABORTY, K. B. and SCOTT, G., *Chem. Ind.*, (1978) 237.
37. CHAKRABORTY, K. B., PhD Thesis, University of Aston in Birmingham (1977).

38. CHAKRABORTY, K. B. and SCOTT, G., unpublished work.
39. CHAKRABORTY, K. B. and SCOTT, G., Europ. Polym. J., 15 (1979) 35.
40. CHEW, C. H., GAN, L. M. and SCOTT, G., Europ. Polym. J., 13 (1977) 361.
41. BAGHERI, R., CHAKRABORTY, K. B. and SCOTT, G., Polym. Deg. Stab, in press.
42. CARLSSON, D. J., GARTON, A. and WILES, D. M., Developments in Polymer Stabilisation—1, Ed. G. Scott (1979) p. 219. Applied Science Publishers, London.
43. Ref. 1, p. 271.
44. CHADWICK, D. H. and WATT, R. S., Phosphorus and its Compounds, Ed. J. R. V. WAZER, (1961) Vol. 2, p. 1264. Interscience Publishers, New York.
45. AL-MALAIKA, S., PhD Thesis, University of Aston in Birmingham (1981).
46. CARLSSON, D. J. and WILES, D. M., J. Macromol. Sci., Rev. Macromol. Chem., C14(1) (1976) 65.
47. CARLSSON, D. J. and WILES, D. M., J. Macromol. Sci., Rev. Macromol. Chem., C14(2) (1976) 155.
48. AL-MALAIKA, S. and SCOTT, G., Europ. Polym. J., 16 (1980) 709.
49. Ref. 1, p. 506.
50. AL-MALAIKA, S. and SCOTT, G., Europ. Polym. J., 16 (1980) 503.
51. HOWARD, J. A., OHKATSU, Y., CHENIER, J. H. B. and INGOLD, K. U., Can. J. Chem., 51 (1973) 1543.
52. HOWARD, J. A. and CHENIER, J. H. B., Can. J. Chem., 54 (1976) 382.
53. HOWARD, J. A. and CHENIER, J. H. B., Can. J. Chem., 54 (1976) 390.
54. BURN, A. J., CECIL, R. and YOUNG, V. U., J. Inst. Pet., 57 (1971) 319.
55. BURN, A. J., Tetrahedron, 22 (1966) 2153.
56. IVANOV, S. K., Developments in Polymer Stabilisation—3, Ed. G. SCOTT (1980) p. 55 et seq. Applied Science Publishers, London.
57. AL-MALAIKA, S. and SCOTT, G., unpublished work.
58. LAVER, H. S., Developments in Polymer Stabilisation—1, Ed. G. SCOTT (1979) p. 167. Applied Science Publishers, London.
59. OHKATSU, Y., KIKKAWA, K. and OSA, T., Bull. Chem. Soc. (Jap.), 51(12) (1978) 3606.
60. AL-MALAIKA, S. and SCOTT, G., Polymer, in press.
61. BARNARD, D., BATEMAN, L., CAIN, M. E., COLCLOUGH, T. and CUN-NEEN, J. I., J. Chem. Soc. (1961) 5339.
62. HAWKINS, W. L. and SAUTTER, H., J. Polym. Sci., Part A, 1 (1963) 3499.
63. BARNARD, D., BATEMAN, L., CUNNEEN, J. I. and SMITH, J. F., Chemistry and Physics of Rubber-like Substances, Ed. L. BATEMAN (1963) p. 593. MacLaren, London.
64. SHELTON, J. R., Polymer Stabilisation, Ed. W. L. HAWKINS (1972) Chapter 2. Wiley, New York.
65. BARNARD, D., BATEMAN, L., COLE, E. R. and CUNNEEN, J. I., Chem. Ind., (1958) 918.
66. BRIDGWATER, A. J., DEVER, J. R. and SEXTON, M. D., J. Chem. Soc., Perkin II (1980) 1006.
67. HUSBANDS, M. J. and SCOTT, G., Europ. Polym. J., 15 (1979) 249.

Chapter 4

EFFECT OF VARIOUS PHOTOSTABILISERS ON THE PHOTO-OXIDATION OF POLYPROPYLENE

F. Tudos, G. Bálint and T. Kelen

Central Research Institute for Chemistry of the Hungarian Academy of Sciences, Budapest, H-1025

SUMMARY

Hydroperoxide groups and double bonds present as impurities in poly-propylene are the main photo-initiators. The rupture of C–C bonds present in the polymer in high concentration may also play a part. The initiating roles of carbonyl and titanium impurities initially present are not significant.

The effectivity of the photostabilisers increases with concentration but in the case of a 2-hydroxybenzophenone, a 2-hydroxybenzotriazole and metal complexes, specific effectivity reaches a maximum in the concentration range 7–12 mmol/litre. Specific effectivity of the hindered piperidines increases gradually at least up to 20 mmol/litre.

A study of the kinetics of stabiliser consumption points to different stabilisation mechanisms. The zero-order kinetics of the hydroxybenzophenone can be explained by its role as a UV-absorber and quencher; part of the energy that it absorbs does not dissipate harmlessly, and leads to the decomposition of the molecule. The hydroxybenzotriazole derivative acts mainly as a UV-absorber, with very high initial efficiency. Later, probably due to its reaction with oxidation products, a drastic decrease in concentration occurs, and its overall efficiency is not as high as that of the hydroxybenzophenone. The metal dibutyldithiocarbamates do not have a significant role as UV-absorbers. These compounds cause ionic decomposition of hydroperoxides. Nickel dibutyldithiocarbamates

121

may play a role in quenching excited oxygen and they also take part in radical reactions.

Nitroxyl radicals are formed from sterically hindered piperidine deriva-tives during processing and during irradiation. Their concentration changes during photo-oxidation can be described by a curve reaching a maximum with time. As these additives cannot be considered as UV-absorbers, quenchers or hydroperoxide decomposers, their excellent stabilising efficiency must be interpreted in terms of a radical scavenging mechanism. The nitroxyl radical, formed in a reaction with oxygen or peroxy radicals, reacts with chain-propagating alkyl radicals. The nit-roxyl radicals regenerated from the adduct in the course of oxidation then continue the stabilising process. Termination of the chain-propagating R· and ROO· radicals seems to be a very important process in the course of photo-oxidation. The high efficiency of TMPS can be explained by the fact that the nitroxyl radical is continuously regenerated, thus leading to a quasi-stationary radical concentration.

1.   INTRODUCTION

An important factor determining the durability of plastic products prepared from polyolefins for outdoor applications is their response to exposure to climatic conditions. One of the most harmful effects is caused by sunlight which causes oxidative chain scission and finally leads to total deterioration of the polymer.

Products prepared from polyolefins can be partially protected from the harmful effect of UV light by reduction or elimination of the defects initiating photo-oxidation in the polymer, and partially by the application of appropriate stabilisers.

Photo-oxidation can be retarded by various additives, which inhibit the individual steps of the overall process. Burchill et al.[1] and Carlsson et al.[2] have proposed detailed reaction schemes for photodegradation inhibited or terminated by different stabilisers.

However, the published literature contains conflicting views on the initiation step and mode of action of the various photostabilisers, probably due to differences in experimental conditions. In this chapter a systematic attempt is made to clarify the most probable step of photo-initiation and to elucidate the potential mode of action of some highly effective stabilisers.

2. POSSIBLE SITES OF INITIATION AND POSSIBLE METHODS OF STABILISATION DURING PHOTO-OXIDATION

Photo-oxidation of saturated polyolefins is generally assumed to involve a free-radical chain reaction following the mechanism formulated by Semenov[3] for the autoxidation of hydrocarbons and adapted by Bolland and Gee[4] to the thermal autoxidation of oil and rubber. The main steps of this mechanism are:

$$RH \xrightarrow{\Delta H,\, h\nu} radicals \tag{1}$$

$$R^{\cdot} + O_2 \longrightarrow ROO^{\cdot} \tag{2}$$

$$ROO^{\cdot} + RH \longrightarrow ROOH + R^{\cdot} \tag{3}$$

$$ROOH \xrightarrow{\Delta H,\, h\nu} RO^{\cdot} + {}^{\cdot}OH \tag{4}$$

$$2ROOH \longrightarrow ROO^{\cdot} + RO^{\cdot} + H_2O \tag{5}$$

$$RO^{\cdot} + RH \longrightarrow ROH + R^{\cdot} \tag{6}$$

$${}^{\cdot}OH + RH \longrightarrow HOH + R^{\cdot} \tag{7}$$

$$2ROO^{\cdot} \longrightarrow products\ (ketones, \tag{8}$$
$$alcohols,\ etc.)$$

Radicals are formed during initiation which react with oxygen leading to degenerated chain reactions. Reaction 2 is faster than reaction 3. The decomposition of hydroperoxides by heat or UV light (reaction 4) causes the formation of alkyloxyl and hydroxyl radicals leading to degenerate chain branching.

2.1. Hydroperoxide Initiation

Hydroperoxide decomposition may also take place via reaction 5, resulting in alkyloxyl, alkylperoxyl radicals and water. The reaction of alkyloxyl and hydroxyl radicals with the polymer (reactions 6, 7) gives alcohol and/or water and alkyl radicals, which further react according to reaction 2. In the course of chain termination peroxyl radicals are transformed into molecular products, with the formation of mainly alcoholic and ketonic groups (reaction 8).

In the case of polypropylene (PP), reaction 3 can take place via intermolecular or intramolecular steps. In the intermolecular case isolated hydroperoxides are formed whereas in intramolecular hydrogen abstraction adjacent hydroperoxides are formed. The latter process results in hydroperoxides associated by hydrogen bonding. According to literature data,[5] 90% of hydroperoxides formed in oxidised PP are intramolecular with hydrogen bonds.

On the basis of the above mechanism, the initiation step would be the breaking of a C–H bond under the influence of UV light (reaction 1). Polyolefins are, however, alkanes with long hydrocarbon chains and do not absorb in the actinic range of sunlight reaching earth (290–400 nm). It is generally assumed therefore that the initiation of polyolefin photo-oxidation is due to defect sites or impurities present in the original polymer or formed during processing. These may be oxygen-containing groups, or traces of initiators which can be excited by the absorption of UV light followed by decomposition into radicals, or by energy transfer to the polymer.

The energy distribution of sunlight on the surface of the earth and the approximate dissociation energies of some bonds in polyolefins are presented in Table 1.[6] According to the data, the most probable photo-initiation step seems to be the homolytic decomposition of the O–O bond in hydroperoxides present in the polymer as impurities. The relatively low-energy light required for decomposition comprises about 60% of the sunlight on the surface of the earth. Although the decomposition of C–O and C–C bonds is a theoretical possibility, these are much less probable processes since they absorb only weakly in the UV range of sunlight reaching the earth's surface. However, the high concentration of the C–C bonds may make their scission more probable.

TABLE 1

ENERGY DISTRIBUTION OF SUNLIGHT ON THE SURFACE OF EARTH[6]

Percentage with energy greater than E	Wavelength (nm)	$-E$		Type of bonds in polyolefins
		$(kcal/mol)$	(kJ/mol)	
80	~1226	~21	~88	—
60	~820	~35	~147	O–O
40	~610	~47	~197	—
20	~470	~62	~260	—
10	~390	~73	~306	—
5	~350	~80	~335	C–O

Hydroperoxides can also be generated, although with lower probability and only at advanced stages of photo-oxidation, by other UV excitation processes. For example, vinyl unsaturation formed by Norrish-II decomposition of carbonyl can react with singlet oxygen resulting in the formation of allylic hydroperoxide by reaction 9.[6a] In

$$
\text{H—}\overset{\underset{\displaystyle O—O}{|}}{\underset{}{C}}\text{—}\overset{\underset{\displaystyle}{|H|}}{C}\text{—}\overset{\underset{\displaystyle}{CH_3}}{CH}\text{—CH}_2\text{—} \longrightarrow \left[\ \text{H—}\overset{\underset{\displaystyle O—O}{|}}{\underset{}{C}}\text{—}\overset{\underset{\displaystyle}{|H|}}{C}\text{—}\overset{\underset{\displaystyle}{CH_3}}{CH}\text{—CH}_2\text{—}\ \right]^{*} \longrightarrow
$$

$$
\text{H—}\overset{\underset{\displaystyle O—O}{|}}{\underset{}{C}}\text{—}\overset{\underset{\displaystyle}{|H|}}{C}\text{—}\overset{\underset{\displaystyle}{CH_3}}{CH}\text{—CH}_2\text{—} \longrightarrow \text{H—}\overset{\underset{\displaystyle OOH}{|}}{\underset{}{C}}\text{—}\overset{\underset{\displaystyle}{CH_3}}{C}\text{=}\overset{\underset{\displaystyle}{CH_3}}{C}\text{—CH}_2\text{—} \quad (9)
$$

the course of addition the delocalised electrons of the excited oxygen form a bond with the antiparallel electrons of the double bond, thus producing an intermediate complex with a four-membered ring. Splitting of the C–O bond in the ring leads to the formation of a biradical, which can be stabilised by hydrogen abstraction from the tertiary carbon atom in the β position, resulting in allyl hydroperoxide.

2.2. Carbonyl Initiation

Photochemical decomposition of carbonyl groups formed during storage and processing takes place via the well-known Norrish-I and Norrish-II processes.[6b] Ketones show relatively low absorptions in the range 270–290 nm, due to $n-\pi^*$ transition. This is the electronic transition of lowest energy and is characteristic of groups containing non-bonding electron-pairs which, as in the case of C=O groups, falls in the region of 350 kJ/mol. Light with this energy is, however, present in the sunlight on the surface of the earth only in a very low percentage. The probability of such processes is therefore expected to be very low, and may take place only at an advanced stage of oxidation when the carbonyl concentration is high.

The structure of unsaturated carbonyl groups was studied by Allen et al. before and after irradiation by a luminescence method.[7–9] These authors found that phosphorescence characteristic of α,β-unsaturated ketones, present in the original polymer, disappeared after irradiation, and another spectrum characteristic of saturated ketones was obtained. They concluded that the α,β-unsaturated ketones are converted by light into β,γ-unsaturated ketones and by repeated light absorption they ultimately decompose via Norrish-I or -II type reactions. A parallel process is the conversion of α,β-unsaturated carbonyls into saturated ketones by cross-linking. A correlation has been found between the phosphorescence intensity of the original polymer and photo-oxidation rate,[9] which, however, does not prove unambiguously the role of ketones in initiation. A possible reaction between the double bonds in unsaturated ketones and ground-state or excited oxygen as discussed above must also be taken into account. In the course of this reaction unsaturated hydroperoxide may be formed by a Norrish-II reaction. Recently, Allen[10] has found that irradiation in an inert atmosphere (which destroys α,β-unsaturated carbonyls and hydroperoxide impurities) does not significantly affect the photo-oxidation rate.

Chakraborty and Scott[11] have shown experimentally that the photo-oxidation rate of polypropylene containing only carbonyl groups, without hydroperoxide, is similar to that of the impurity-free sample, while the initial hydroperoxide content increases the rate of the oxidation. We obtained similar results in studying the effect of pretreatment on the photo-oxidation of polypropylene without the addition of a stabiliser. A detailed description of the experimental conditions has been provided elsewhere;[12–14] therefore only a short review is given here.

The polypropylene (PP) studied was a laboratory product (free of additives), Daplen AT 10, in the form of 100 μm-thick films at 170°C. Irradiation of the samples was performed in a xenon-lamp apparatus in which the light was focused by elliptical mirrors on to the samples mounted on a rotating drum in a thermostat. The temperature of the films was kept at 70°C, corresponding to the temperature of samples exposed to sunshine. At the sample surface the average intensity of illumination was 4×10^{-7} mol photon s^{-1} cm^{-2}. Allowing for the alternating light and dark periods—due to the rotation of the drum with the samples—the dosage of irradiation on the samples in 24 h was about 2·5–3·5 Mlxh (megalux-hours). The concentration of carbonyl (C=O) and associated hydroperoxide groups (HP) was determined by use of infrared spectroscopy of the film samples, by measuring the integrated absorption of the bands at 1715 and 3400 cm^{-1}.[15]

The samples were processed in a Rheocord (Haake) apparatus at 175°C. The concentration of the hydroperoxides was similar to that of the original sample after processing for 5 h, while the carbonyl concentration was three times as high as the original ($\sim 0 \cdot 1$ mol/litre). This concentration is assigned to the end of the induction period of carbonyl formation during the irradiation of PP without additives and processing, so a reduction in the induction period could be expected in the case of the processed sample. In spite of this, the induction period of samples with different initial amounts of carbonyls did not show any significant change (Table 2). The number-average molecular mass, determined by gel permeation chromatography, however, decreased from the original value of 92 500 to 17 000 after processing for 5 h. (This means that the main degradation process in a closed chamber is chain scission.)

TABLE 2

INDUCTION PERIODS TO CARBONYL FORMATION WITH DIFFERENT
PROCESSING TIMES

Processing time (min)	Induction period (Mlxh)
0	4·8
30	4·0
60	4·0
150	3·2
300	3·2

2.3. Initiation by Catalyst Residues

Some authors consider catalyst residues to be initiators for the photo-oxidation of polymers.[16-18] In polypropylene polymerised by the Ti/Al-complex catalyst, catalyst residues are probably present as TiO_2, $OTiCl_2$, $OTiCO(OR)$, $TiCl_2(OR)_2$, $TiCl_3$ or $Ti(OR)_4$, where R denotes alkyl groups or polymer chain. There are, however, conflicting views in the literature regarding the initiating effect of catalyst residues. According to Carlsson and Wiles[16] the role of titanium residues from the catalyst is similar to the effect of hydroperoxides. Cicchetti and Gratani[17] found tetra- and tri-butyl titanate to accelerate (in 0·01 mol/litre concentration) the photodegradation of 2,4,6,8-tetrabutylnonane (liquid model compound of PP). On the other hand no correlation was found, e.g. by Gugumus,[18] between the titanium content (7–49 ppm) and the photostability of PP as determined by 50% tensile strength retention.

Our experimental results related to this problem are shown in Table 3, where induction periods to carbonyl and hydroperoxide formation ($t_{i,CO}$ and $t_{i,HP}$, respectively), as well as the maximum formation rate (W_{CO} and W_{HP}, respectively) are given for an irradiated PP film free of impurities, and PP containing titanium residues at a concentration of 200 ppm. All samples also contained 0·2 mass % Topanol OC antioxidant. Polypropylene samples and analytical assessment of titanium content were kindly supplied by Montedison, Italy.

TABLE 3

INDUCTION PERIODS (t_i) AND MAXIMUM RATES (W) FOR CARBONYL AND HYDROPEROXIDE FORMATION FOR PP SAMPLES WITH AND WITHOUT TITANIUM

Sample	Carbonyl formation		Hydroperoxide formation	
	$t_{i,CO}$ (Mlxh)	W_{CO} (mol/litre Mlxh)	$t_{i,HP}$ (Mlxh)	W_{HP} (mol/litre Mlxh)
PP	13·2	0·090	11·1	0·011
PP + 200 ppm Ti	14·0	0·094	11·2	0·013

As can be seen from the data, a Ti concentration up to 200 ppm does not significantly affect the induction period and the rate of formation of carbonyl and hydroperoxide groups.

2.4. Polycyclic Aromatic Compounds

In addition to these initiation processes, some papers refer to the photo-initiating role of condensed aromatic compounds in the air;[19] the practical importance of this is, however, negligible.

In our opinion, the initiating sites of photo-oxidation are mainly hydroperoxides and double bonds occurring in the processed polymer as defects, as well as those arising from the scission of C–C bonds present in the polymer in high concentration. The photo-initiating role of carbonyl, titanium and other impurities is of minor importance.

3.   POSSIBLE PHOTOSTABILISING MECHANISMS

Considering the initiating steps detailed above, the following possibilities can be taken into account for intervention in the photo-oxidation process, or for stabilisation of the polymer system.

3.1. Ultraviolet Screening
The light falling on the surface of the polymer can be screened by non-transparent filters (screens), e.g. carbon black, various pigments, etc. These, however, can be used only in non-transparent pigmented polymer.

3.2. Ultraviolet Absorption
Ultraviolet-absorbers blended into the polymer can absorb actinic light which is harmful to the polymer. The absorbed excitation energy is dissipated harmlessly, the stabilisers being re-converted to their original ground states.

3.3. Quenching
The excess excitation energy can be transferred to quenchers from the excited carbonyl or singlet oxygen, followed by harmless dissipation.

3.4. Radical Scavenging
Radicals formed during oxidation can react with radical scavengers to given non-radical products.

3.5. Peroxide Decomposers
Hydroperoxides causing degenerate chain branching can be decomposed by peroxide decomposers in a non-radical process.

3.6. Metal Ion Deactivators
Finally, complexing agents may deactivate catalyst residues. (These stabilisers are of no practical significance.)

4. EFFECT OF DIFFERENT TYPES OF STABILISERS ON PP PHOTO-OXIDATION

4.1. Types of Photostabilisers Investigated
Based on the above considerations, the following types of additives have been studied as photostabilisers in PP; hydroxybenzophenone derivatives, substituted benzotriazole derivatives, metal complexes of different ligands and different metals, and sterically hindered piperidine derivatives (detailed data are given in Table 4).

The additives were selected on the basis of the following criteria. In

TABLE 4

PHOTOSTABILISERS STUDIED IN POLYPROPYLENE

Compound (designation)	Structure	Commercial product	Produced by
2-Hydroxy-4-octoxy benzophenone (Cy)		Cyasorb UV 531	Cynamid
2[2′-Hydroxy-3′-tert-butyl-5′-methylphenyl] 5-chlorobenzotriazole (Tin 326)		Tinuvin 326	Ciba–Geigy
Metal acetyl acetonates (Me-AcAc)	Me=Zn(II), Ni(II), Al(III)	—	Synthesis
Metal dialkyldithio-carbamates (Me-DETC) (Me-DBTC)	R = Et, Bu; Me = Cd(II), Bi(III), Co(II), Ni(II), Pb(II), Zn(II)	—	Synthesis
Tetramethylpiperidine-N-oxyl (TMP–NO)		—	Synthesis
Bis(tetramethyl-piperidine sebacate) (TMPS)		Tinuvin 770	Ciba–Geigy

addition to commercially available hydroxybenzophenone and benzotriazole derivatives used as reference materials, some organic complexes containing metals with varying valency were examined, taking
into consideration literature data on their antioxidant characteristics.

The antioxidant behaviour of Ni, Zn, Co and Cd chelates in the
oxidation of hydrocarbons has been observed by several authors.[20-24]
This effect is generally attributed to the reaction of chelates with
alkylperoxy radicals formed during oxidation, and to non-radical decomposition of chain-propagating hydroperoxides, by chelates. For
earlier reviews, see *Developments in Polymer Stabilisation—1*, Chapter
5, and Chapter 3 in the present volume.

In the oxidation of squalene Burn[25] found that the antioxidant
activity of complexes containing different kinds of metals decreases in
the following order: $Zn(II) > Sb(III) > Cd(II) > Pb(II) > Bi(III)$. According to Gervits et al.[26] the inhibiting efficiency of Zn, Pb and
Bi dibutyldithiocarbamates in the oxidation of cumene initiated by
azobisisobutyronitrile varies with the metal. On the other hand the
experimental results of Baum and Perun[27] showed that the nature of
the metal does not affect antioxidant activity, as only the ligand takes
part in the hydroperoxide-decomposing reaction of the complexes.

Some reference has been made in the literature to the triplet
carbonyl and singlet oxygen quenching effect of the Ni, Co and Zn
complexes.[16,23,28-33] According to Carlsson et al.,[34] the singlet oxygen
quenching effect of the metal complexes increases in the following
order: $Zn < Co < Ni < Fe$. Briggs and McKellar[35,36] observed a correlation between the triplet quenching efficiency of the nickel complexes
and UV stabiliser effectivity in PP. On the basis of these observations
it might be expected that complexes of other metals with varying
valency should also show a stabilising effect in the photo-oxidation of
polyolefins.

Several papers have been published recently dealing with the excellent photostabilising efficiency of sterically hindered piperidine derivatives in polyolefins;[37-46] in spite of this, however, no generally accepted
ideas have been formulated regarding the mode of action of such
compounds. According to most authors[18,37,41-45] a nitroxyl radical is
formed from the piperidine, and this is responsible for the stabilising
effect. Assumptions regarding the generation of the nitroxyl radical
are, however, different.

According to Chakraborty and Scott,[37] the original amine reacts

with hydroperoxide during processing and the N˙ radical formed produces a —NOOH group, in a reaction with oxygen, and by hydrogen
abstraction from the polymer. The latter decomposes into nitroxyl and
hydroxyl radicals. During this process, however, alkoxyl, macro alkyl
and hydroxyl radicals are formed, which may promote the propagation
of oxidation. Grattan et al.,[41] Sedlar et al.[43] and Allen et al.[44] have
suggested the formation of a complex between the nitroxyl radical and
hydroperoxide, while Gugumus[18] assumes the formation of a complex
between the nitroxyl and alkylperoxy radicals.

Some authors presume a reaction between the nitroxyl and alkyl
radical in the course of stabilisation,[39,42,44,46] but according to
Carlsson[45] and Sedlar et al.[43] this reaction is not probable, due to the
higher reactivity of the alkyl radical with oxygen.

These conflicting views justify further investigation of the photostabilising effect of piperidine derivatives.

4.2. The Effect of Additives on the Initial Carbonyl and Hydroperoxide Content of Polymer Films

In studying the different types of photostabilisers, it seems feasible to
examine first the effect of additives on oxidation in the course of
processing. The initial carbonyl contents of PP films compression
moulded in a nitrogen atmosphere at 170°C for stabilisers added at
0·2, 0·4, 0·5 and 1·0 mass % are presented in Fig. 1. In each case
photostabilisers were used together with 0·2 mass % Topanol OC
antioxidant. The initial carbonyl content of films containing Cy and

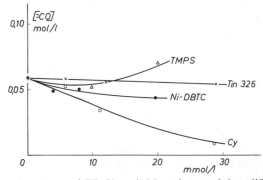

FIG. 1.	Carbonyl content of PP films (100 μm) containing different types of
stabilisers after compression moulding, as a function of stabiliser concentration.

TMPS was calculated by subtracting the number of carbonyl and/or ester groups introduced by the stabilisers. (The data in Fig. 1 show mean values, within ±10% experimental error.)

As is evident from Fig. 1, an appreciable antioxidant effect is obtained only in the case of the hydroxybenzophenone Cy. A similar antioxidant effect was observed by Hutson and Scott,[47] while TMPS shows some slight pro-oxidant features especially at higher concentrations. Chakraborty and Scott also found pro-oxidant behaviour in the course of processing experiments with LDPE containing Tinuvin 770.[37] In spite of this, Tinuvin 770 was found to give the highest protection against photo-oxidation.

Initial hydroperoxide concentration does not show any significant change as a function of stabiliser concentration.

4.3. Photo-oxidation of PP Containing Photostabilisers

4.3.1 Effect of Photostabiliser Concentration on the Formation of Carbonyl and Hydroperoxide Groups

The rate of formation of carbonyl and hydroperoxide groups during UV irradiation does not change significantly with the concentration of the stabiliser (in the range 0·2–1·0 mass %, i.e. 4–30 mmol/litre); higher concentrations of the stabiliser only increase the length of the induction period. The kinetics of carbonyl and hydroperoxide formation are shown in Figs. 2 and 3 in the presence of Cy. It can be seen

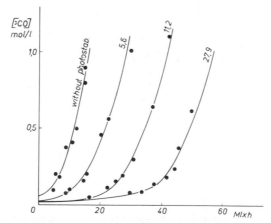

FIG. 2.   Kinetics of carbonyl group formation in PP films containing Cyasorb UV 531. The numbers on the curves denote concentration in mmol/litre.

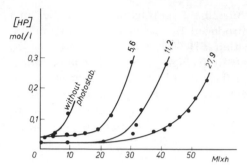

FIG. 3. Kinetics of hydroperoxide group formation in PP films containing Cyasorb UV 531 during UV irradiation. The numbers on the curves denote concentrations in mmol/litre.

that the carbonyl induction period is only slightly shorter than that for hydroperoxide. In other words, the formation of both oxygen-containing groups starts at almost the same UV dosage.

Similar curves were obtained for PP containing 0·2–1·0 mass % of the hydroxybenzotriazole, Tin 326, as a function of the dosage (Fig. 4), but the induction periods were much shorter than in the case of Cy.

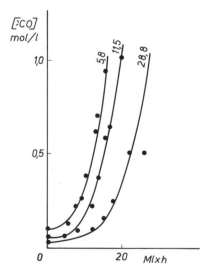

FIG. 4. Kinetics of carbonyl group formation in PP films containing Tinuvin 326 during UV irradiation. The numbers on the curves denote concentrations in mmol/litre.

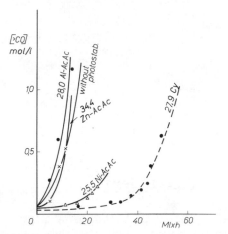

FIG. 5. Kinetics of carbonyl group formation in PP films containing 1·0 mass % metal acetylacetonates during UV irradiation. The numbers on the curves denote concentrations in mmol/litre.

Photo-oxidation kinetics for the metal complexes are shown in Figs. 5–8 (carbonyl formation curves for Cy of 1·0 mass % (28·8 mmol/litre) are also plotted in the Figures, for comparison). Significant stabilising effects were obtained only with samples containing 0·4 mass % Ni-DBTC (7·8 mmol/litre), and 1·0 mass % Co-DBTC

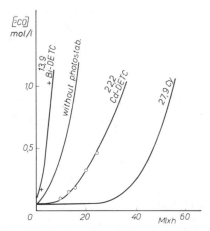

FIG. 6. Kinetics of carbonyl group formation in PP films containing 1·0 mass % metal diethyldithiocarbamates during UV irradiation. The numbers on the curves denote concentrations in mmol/litre.

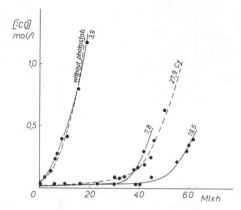

Fig. 7. Kinetics of carbonyl group formation in PP films containing Co-dibutyldithiocarbamates during UV irradiation. The numbers on the curves denote concentrations in mmol/litre.

and Ni-DBTC (13·55 and 19·48 mmol/litre, respectively); see Figs. 7 and 8. Zn-DBTC and Pb-DBTC were similar to the control without any photostabiliser. One reason for this may be that these stabilisers undergo thermal or photochemical decomposition during compression moulding or during the early stages of irradiation. (The UV-absorption characteristic of these additives shows a dramatic decrease after compression moulding or in the first stage of irradiation.) On the other hand, it appears that Al- and Bi-acetylacetonates (or their transforma-

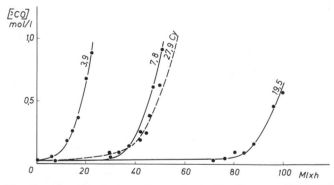

Fig. 8. Kinetics of carbonyl group formation in PP films containing Ni-DBTC during UV irradiation. The numbers on the curves denote concentrations in mmol/litre.

tion products) exert some photosensitising effect. Similar initiating effects have also been described in the literature with other dithiocarbamates.[29,48,49]

Kamiya and Ingold[21] studied the autoxidation of tetralin containing Mn(II) and Co(II) decanoates and found that oxygen absorption increases with the increase of the amount of additive up to a given concentration (e.g. in the case of the Mn complex up to 0.1 mmol/litre), and after achieving a maximum rate a sudden decrease was observed. That is, the catalytic activity of the additive turns into antioxidant behaviour.

Black[50] simulated the catalyst–inhibitor conversion by computer for the above system. It has been shown that if the concentration of Co^{2+} is lower than that of hydroperoxide, then the hydroperoxide-decomposing reaction is suppressed and the catalysed process predominates. At higher metal-ion concentrations ionic decomposition of the hydroperoxide or peroxyl radical takes place. According to the computer-calculated results, the rapid fall of oxygen absorption after the maximum (catalyst–inhibitor conversion) occurs at approximately 10^{-2} mol/litre metal concentration.

Similar results were obtained in the case of metal complexes by plotting the induction periods of carbonyl formation (t_i) as a function of the initial concentration of the stabiliser (c_0), as in Fig. 9. It can be seen that the use of low concentrations of the stabilisers (below 2–3 mmol/litre) gives practically no stabilising effect. On the other

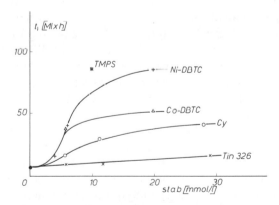

Fig. 9. Dependence of the induction periods to carbonyl formation on stabiliser concentrations.

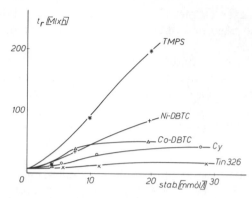

FIG. 10.   Relation between the UV dosage and the occurrence of spontaneous rupture of PP films containing various stabiliser concentrations.

hand, high concentrations (above 20 mmol/litre) offer no economic advantage. (No carbonyl formation can be detected in samples containing TMPS at 1·0 mass % up to the time of spontaneous rupture of the films; thus the induction period is given in Fig. 9 only in the case of 0·5 mass % TMPS.) Similar curves (Fig. 10) can be constructed for the concentrations associated with the spontaneous rupture of the samples (t_r). The high effectivity of TMPS appears also in this plot.

4.3.2. Change in Stabiliser Concentration During Irradiation

In order to elucidate the mode of action of the stabilisers, the kinetics of stabiliser consumption were measured. Ultraviolet stabilisers generally have an absorption peak in the UV wavelength range and their concentrations can be determined by measuring the UV spectra of PP samples containing them. In the case of PP without photostabilisers, UV absorption increases at wavelengths less than 275 nm due to the absorption of oxygen-containing groups formed during irradiation. Above this range, however, no significant change can be observed. Thus, the concentration of additives can be followed if they absorb above 275 nm. The concentration of TMPS was assessed at 268 nm. Since carbonyl and hydroperoxide groups are formed in its presence, the TMPS concentration may be estimated although with a relatively high margin of error ($\approx 25\%$). The wavelengths of the absorption peaks used for determination of the stabilisers as well as their absorption coefficients determined in benzene solution are listed in Table 5.

TABLE 5

WAVELENGTH ABSORPTION PEAKS USED FOR DETERMINATION OF STABILISERS
AND THEIR CALCULATED ABSORPTION COEFFICIENTS

Stabiliser	Absorption band (nm)	Absorption coefficient, $\varepsilon \times 10^{-3}$ ($litre\ mol^{-1}\ cm^{-1}$)
Cy	327	10·29
	360	1·85
Tin 326	354	16·62
	390	3·98
Zn-AcAc	275	17·30
	332	0·47
Ni-AcAc	295	16·69
Cd-DETC	266	33·80
	286	13·00
Bi-DETC	258	47·20
	323	6·83
Co-DBTC	323	27·86
Ni-DBTC	328	33·30
	396	5·43
TMP-NO	245	0·66
TMPS	268	1·71

Stabilisers show different types of consumption kinetics during irradiation. The hydroxybenzophenone derivative, Cy, is consumed during irradiation according to linear kinetics, and carbonyl formation begins when the concentration of the stabiliser drops to 7·0 mmol/litre (Fig. 11). Similar results were found by Vink,[51] who measured oxygen consumption and stabiliser concentration during irradiation.

The consumption of the benzotriazole derivative, Tin 326, is initially slow, similar to that of Cy, but after the appearance of carbonyl groups (when there is still 12·5 mmol/litre stabiliser present in the system) the rate of consumption increases (Fig. 12). Vink reports a kinetic curve of similar character for the consumption of 2(2'-hydroxy-5'-methylphenyl)benzotriazole in the course of irradiation.[51]

The consumption of metal complexes is shown during irradiation for Ni-AcAc (Fig. 13), Co-DBTC and Ni-DBTC (Figs. 14 and 15). (Diethyldithiocarbamates, as well as Zn and Pb dibutyldithiocarbamates, were found to decompose during processing or in the initial stage of irradiation: characteristic absorption peaks in the UV spectra were found to decrease quickly.)

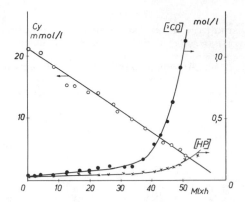

FIG. 11.   The change in Cyasorb UV 531 and carbonyl concentrations during UV irradiation of PP films. Initial stabiliser concentration, 27·88 mmol/litre.

Decomposition of metal complexes can be described by an exponential kinetic curve (Figs. 14 and 15), and carbonyl formation starts when concentration of the stabiliser falls below 1·0 mmol/litre. This correlates with the induction periods in Fig. 9 (i.e. below this low concentration, Co-DBTC and Ni-DBTC do not exert any inhibiting effect on photo-oxidation). This is in agreement with the results of Chakraborty and Scott.[65] A similar exponential concentration decrease can be observed in the case of TMP-NO and TMPS (Fig. 16).

It should be noted that the concentration of piperidine derivatives

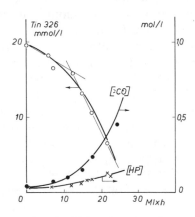

FIG. 12.   The change in Tinuvin 326 and carbonyl concentrations during UV irradiation of PP films. Initial stabiliser concentration, 28·81 mmol/litre.

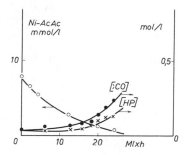

Fig. 13.   The change of Ni-AcAc and carbonyl concentrations during UV irradiation of PP films. Initial concentration of stabiliser, 25·50 mmol/litre.

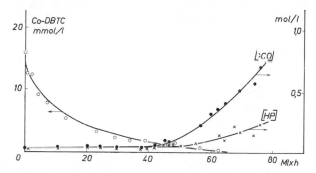

Fig. 14.   The change of Co-DBTC and carbonyl concentrations during UV irradiation of PP films. Initial stabiliser concentration, 19·47 mmol/litre.

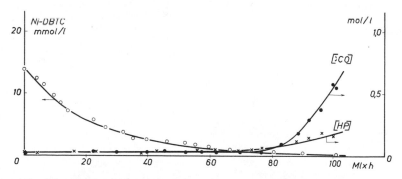

Fig. 15.   The change of Ni-DBTC and carbonyl concentrations during irradiation of PP films. Initial stabiliser concentration, 19·48 mmol/litre.

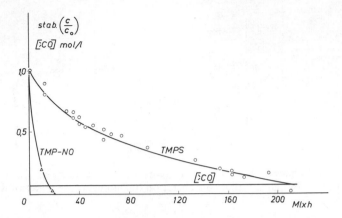

FIG. 16.   The change of TMP-NO, TMPS and carbonyl group concentrations during irradiation of PP films.

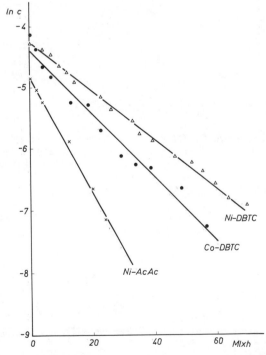

FIG. 17.   Logarithmic plot of metal complex concentration change during UV irradiation of PP films.

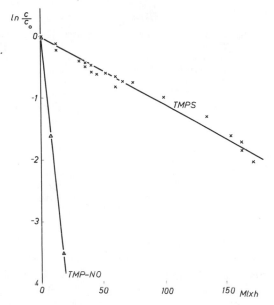

FIG. 18. Logarithmic plot of the change in concentration of piperidine derivatives during UV irradiation.

cannot be measured accurately from their UV spectra. (However, absorption coefficients calculated from calibration curves of solutions differ from the value obtained in the polymer.) The relative concentration change of piperidine derivatives is therefore given by the ratio of absorption at 268 nm to the absorption of the sample before irradiation at the same wavelength ($D/D_0 = c/c_0$).

The logarithm of concentrations of metal complexes and piperidine derivatives are plotted as a function of UV dosage in Figs. 17 and 18. Rate constants (k) of consumption were zero order in the case of Cy and Tin 326, and first order in the case of metal complexes and piperidines. These values together with the half lives, $\tau_{1/2}$, are presented in Table 6.

From these data there is clearly an inverse correlation between the efficiency and decomposition rate of the stabilisers.

4.3.3. Change in Molecular Mass and Chain Scission During Irradiation

The effect of stabilisers on chain scission was studied by measuring the molecular mass and molecular mass distribution during irradiation, by gel permeation chromatography and viscometry.

TABLE 6

RATE CONSTANTS (k) AND HALF LIVES $(\tau_{1/2})$ FOR THE CONSUMPTION OF STABILISERS

Stabiliser	$k(mmol/litre\ Mlxh)^a$	$k(Mlxh^{-1})^b$	$\tau_{1/2}(Mlxh)$
Cyasorb UV 531	0·35	—	—
Tin 326			
(first period)	0·31	—	—
(second period)	1·02	—	—
Ni-AcAc	—	0·0924	7·50
Co-DBTC	—	0·0512	13·54
Ni-DBTC	—	0·0395	17·55
TMP-NO	—	0·206	3·36
TMPS	—	0·0106	65·39

^{a} First order: $-dc/dt = k$.
^{b} Second order: $-dc/dt = kc$.

Molecular mass change measured by viscometry $(\bar{M}_V)$ and number-average molecular mass obtained by GPC $(\bar{M}_n)$ show a similar decrease as a function of UV dosage (Fig. 19). Curves of carbonyl formation are also plotted in the Figure. Values of $\bar{M}_V$ and $\bar{M}_n$ show a rapid decrease in the case of PP without photostabiliser (a similarly fast decrease was found by Adams[52] upon irradiation of PP films of 500 μm thickness).

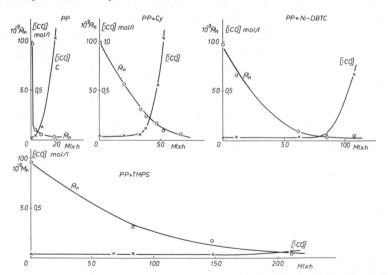

FIG. 19.   Relation between number-average molecular mass and carbonyl formation during irradiation of PP containing different types of stabilisers.

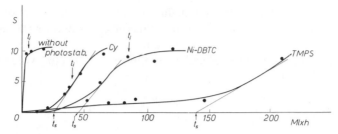

FIG. 20. Dependence of the average number of chain scissions on UV exposure in the presence of different types of stabilisers (1·0 mass % concentration).

The rate of decrease is, however, slower in the case of Cy, Ni-DBTC, and TMPS in the order indicated. At the end of the induction period $\bar{M}_n$ decreases to one-tenth of its original value—except for the sample containing Cy, where the decrease is about one-fifth of the original value at the end of the induction period.

Similar curves of inverse character are obtained by plotting the number of scissions $\left(S - \dfrac{\bar{M}_{n,0}}{\bar{M}_n} - 1\right)$ as a function of dosage, as in Fig. 20. In this plot, well-defined induction periods appear in the case of polymers of different composition, which can be detected earlier than the induction period of carbonyl formation (t_i). These induction periods obtained by determining the number of chain scissions (t_s) represent new independent data for the estimation of the efficiency of different stabilisers. This gives information on the degradation of polymers at very low irradiation dosages.

Except for the sample containing Cy, the induction period to carbonyl formation (t_i) occurs at about $S = 9$, while in the presence of Cy, carbonyl formation starts at $S = 5$. This indicates that the hydroxybenzophenone derivative retards chain scission to a greater extent than it retards the incorporation of oxygen into the polymer chain. In the case of the unstabilised control and samples containing Ni-DBTC, the rate of carbonyl formation is initially slower than chain scission, but subsequently chain scission slows down compared with carbonyl formation. In the presence of Cy, however, carbonyl formation parallels chain scission. The quantum yield of the chain scission process can be calculated from the slope of the chain scission curve:

$$\phi = \frac{\text{number of chain scissions in the film}}{\text{number of photons absorbed in the film}} = \frac{w_s}{I_{abs}}$$

The value of ϕ has been calculated from the number of chain scissions in a $0 \cdot 01$ cm^3 film of 100 μm thickness divided by the number of photons absorbed in the same volume.

The number of macromolecules in films of $0 \cdot 01$ cm^3 volume was calculated as follows:

$$\frac{WN}{M_n} = \frac{0 \cdot 0091 \times 6 \cdot 10^{23}}{92\,500} = 5 \cdot 9 \times 10^{16}$$

where W is the mass of a $0 \cdot 01$ cm^3 film, N is Avogadro's number and $\bar{M}_n$ is the number-average molecular mass of the polymer.

The number of chain scissions taking place in the same volume upon a dosage of 1 Mlxh was found to be:

$$w_s = \frac{WN}{\bar{M}_n} \cdot \frac{\Delta s}{D} \,(\text{Mlxh}^{-1}\,\text{cm}^{-2})$$

where D is the UV dosage in Mlxh.

For determination of the number of photons absorbed in the film, average illumination (I_0) was expressed in terms of time required for 1 Mlxh dosage (average illumination, 4×10^{-7} mol photon s^{-1} cm^{-2}; actual exposure of films, 1/15th of average illumination; with 1 Mlxh corresponding to 10 h):

$$I_0 = \frac{4}{15} \times 10^{-7} \times 3600 \times 10 \times 6 \times 10^{23} = 5 \cdot 76 \times 10^{20} \,(\text{photon Mlxh}^{-1}\,\text{cm}^{-2})$$

The number of photons absorbed by the films of different composition (I_{abs}) can be obtained by multiplying the value of I_0 by the average absorption ratio of the films in the range of 250–350 nm (A):

$$I_{abs} = I_0 \cdot A$$

TABLE 7

CHAIN SCISSION DATA FOR PP SAMPLES CONTAINING $1 \cdot 0$ MASS % STABILISERS

Sample	Chain scission rate $w_s \times 10^{-16}$ $(mlxh^{-1}\,cm^{-2})$	Absorption ratio A	Number of photons absorbed by the films $I_{abs} \times 10^{-20}$ $(photon\,Mlxh^{-1}\,cm^{-2})$	Quantum efficiency of chain scission ϕ
PP	18·47	0·32	1·84	$1 \cdot 00 \times 10^{-3}$
PP+Cy	1·48	0·99	5·70	$2 \cdot 59 \times 10^{-5}$
PP+Ni-DBTC	1·24	0·80	4·61	$2 \cdot 69 \times 10^{-5}$
PP+TMPS	0·57	0·34	1·84	$3 \cdot 10 \times 10^{-5}$

The data for samples containing different stabilisers are presented in Table 7.

As can be seen, the quantum efficiency of chain scission is about one and a half orders of magnitude lower in the presence of photostabilisers. The lowest value is obtained with the hydroxybenzophenone, Cyasorb UV 531 (1·0 mass%).

5. THE MECHANISM OF ACTION OF ULTRAVIOLET STABILISERS

5.1. Ultraviolet-absorber Character (Diffusivity)

In order to investigate the screening effect of UV stabilisers, Guillory et al.[53] irradiated PP samples behind stabiliser solutions with a mercury lamp and compared the irradiation time required for the rupture of samples on bending through 180° with the irradiation time required for rupture of samples containing stabilisers. It was found that only a benzotriazole derivative acted as a pure absorber; addition of the 2-hydroxybenzophenone to the polymer resulted in a higher activity than could be accounted for by UV-absorption, as evidenced by the fact that when a hydroxybenzophenone was blended into the polymer, sample rupture required longer irradiation time than when it was used as a screen in front of the polymer.

In another process, the authors plotted rupture times as a function of light fraction absorbed by the film (LAP %), which was calculated as follows:

$$\% \text{ LAP} = 100 \sum_{\lambda_2}^{\lambda_1} (1 - T(\lambda)) \frac{A_p(\lambda)}{A_p(\lambda) + A_s(\lambda)}$$

where $T(\lambda)$ denotes the light transmitted by the polymer containing the stabiliser, at wavelength λ, and $A_p(\lambda)$ and $A_s(\lambda)$ are the absorbances of the polymer and stabiliser respectively at wavelength λ. (This method, however, does not consider the change in the absorption spectrum of the stabilised polymer during irradiation.) It was concluded that the benzotriazole derivative behaved simply as a UV screen, whilst the hydroxybenzophenone and the nickel chelate showed higher stabilising effects than expected on the basis of their UV-absorptions.

Zannuci and Lappin[54] plotted the ratio of chain scission of PP films containing benzophenone, benzotriazole and metal-complex UV stabilisers to the chain scission of polymer without stabiliser (S/S_0), as a function of the ratio of the fraction of light absorbed by the polymer with and without stabiliser (L/L_0). If the additive exerts a protective effect merely as a UV-absorber, the S/S_0 values should give a straight line as a function of L/L_0. This straight line was obtained by screening the sample with the hydroxybenzophenone in a separate film. When incorporated in the polymer, however, all three additives give points below the straight line. The correct interpretation of the phenomena may be that the stabilisers investigated have some other stabilising effect, in addition to UV absorption. Unambiguous evidence for exclusively UV-absorber effect has been shown, however, only in the case of 4-(dodecoxy)-2-hydroxybenzophenone stabiliser.[54] The higher efficiency of the other benzophenone stabilisers may be due to additional mechanisms.

The UV absorbing characteristics of benzotriazole derivative were studied by Vink et al.[55]—see Developments in Polymer Stabilisation—3, Chapter 4. They found that the rate of oxygen absorption by the polymer with additive in the course of irradiation is proportional to the

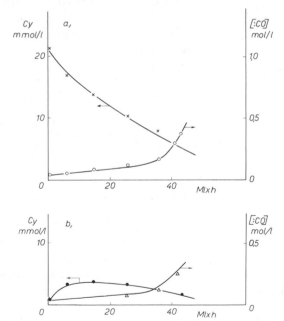

FIG. 21.   Photo-oxidation of PP film (a) containing 1·0 mass % Cy and (b) without any photostabiliser, placed behind (a).

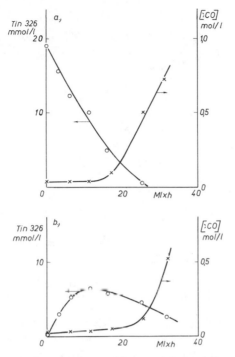

FIG. 22.   Photo-oxidation of PP film (a) containing 1·0 mass % Tin 326, and
(b) behind (a) without any photostabiliser.

square root of the light intensity in the presence of a benzotriazole,
while in the case of a benzophenone it is proportional to the first
power of the light intensity. The former stabiliser has therefore only a
UV-absorber effect, while in the latter case some other additional
effect—probably radical scavenging—has to be taken into account.

Similar results have been reported by Heller and Blattman,[56] who
plotted the ratio of the light absorbed by PP with and without stabiliser
characteristic of the pure screening effect (the so-called protective
factor) as a function of the logarithm of film thickness. With increase in
film thickness the protective factor increased only in the presence of
the benzophenone and benzotriazole, implying that such stabilisers
exert a UV-absorber effect.

The results of our investigations into the UV-absorbing characteristics of various stabilisers are shown in Figs. 21–24. The kinetics of
carbonyl formation in films containing 1·0 mass % additives (a) and in

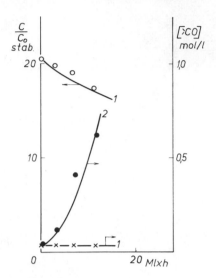

FIG. 23.   Photo-oxidation of PP film (1) containing Ni-DBTC (1·0 mass %), and (2) behind (1) without any photostabiliser.

films without any additive placed right behind the former (b) are presented. Stabiliser concentration is also plotted in the Figures for both samples. On the basis of the curves it can be stated that, although Cy, Tin 326 and Ni-DBTC have significant absorption in the actinic range (see Table 5), the UV-absorber character is predominant only in

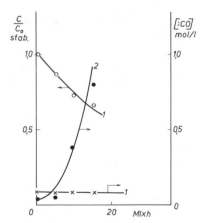

FIG. 24.   Photo-oxidation of PP film (1) containing 1·0 mass % TMPS and (2) behind (1) without photostabiliser.

the first two stabilisers. In the case of a front film with Cy (Fig. 21), carbonyl formation begins at the same time in both samples; in the case of a front sample with Tin 326, oxidation of the back film is retarded relative to that of the front film. The reason for this is that a considerable amount of the stabiliser has migrated into the back film without additive, and consumption is slower than in the front film due to the additional UV-absorber effect on the front film, and the lower availability of oxygen. Cyasorb also shows diffusion, although much less than Tin 326.

Ni-DBTC, on the other hand, exerts no significant screening effect when placed in front of a photo-oxidising film with no additive (Fig. 23). The same holds true for TMPS (Fig. 24). The two latter stabilisers do not show any migration ability.

On the basis of the above results and in accordance with literature data the following picture has emerged for the relative importance of UV absorption by the stabilisers:

> Hydroxybenzophenone derivatives   UV-absorber
> Hydroxybenzotriazole derivatives   UV-absorber
> Ni-DBTC   weak UV-absorber
> TMPS   non UV-absorber

5.2. Changes in Molecular Mass Distribution of Films Containing Different Photostabilisers during Irradiation

The change in molecular mass of polyolefins during irradiation is the topic of several papers;[52,57-59] no studies have been reported, however, regarding the effect of different photostabilisers on the change of molecular mass.

In this section, we present the molecular mass distributions of PP films irradiated in the presence of highly efficient photostabilisers, which show differing behaviour.

The molecular mass distribution of a stabiliser-free film, Fig. 25, and that of PP with 1·0 mass % TMPS, Fig. 26, gradually shift toward lower molecular mass, whilst in the presence of Cy and Ni-DBTC the distribution becomes bimodal at the beginning of irradiation (Figs. 27 and 28). In the latter case the peak in the range of higher molecular mass continuously decreased, and a new peak appeared in the region of the lower molecular mass, then with gradual decrease of the former, the distribution shifted continuously into the range of low molecular mass.

152   F. TUDOS, G. BÁLINT AND T. KELEN

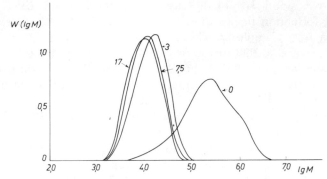

FIG. 25.   Molecular mass distribution on PP films without photostabiliser.
Numbers denote exposure time in Mlxh.

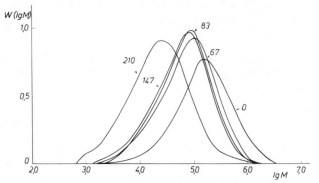

FIG. 26.   Molecular mass distribution in PP films with 1·0 mass % TMPS.
Numbers on curves denote exposure time in Mlxh.

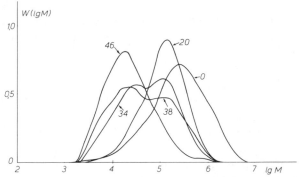

FIG. 27.   Molecular mass distribution in PP films with 1·0 mass % Cy.
Numbers on curves denote exposure time in Mlxh.

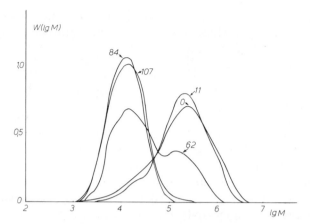

FIG. 28. Molecular mass distribution in PP films containing 1·0 mass % Ni-DBTC. Numbers on curves denote exposure time in Mlxh.

In interpreting the above observations, we assume that Cy and Ni-DBTC act in a different way in the surface layers from that in the back of the polymer. In order to verify this assumption we superimposed three layers of film (of 30 μm thickness) on each other, and studied chain scission after 34 Mlxh dosage. It can be seen from the data of Table 8 that chain scission is higher in films nearer the light source, in the presence of both Cy and Ni-DBTC.

TABLE 8

NUMBER-AVERAGE MOLECULAR MASS AND CHAIN SCISSION OF PP FILMS CONTAINING STABILISERS AFTER IRRADIATION (Radiation dosage, 34 Mlxh; thickness of each film 30 μm)

Film layer	1·0 mass % Cy additive		1·0 mass % Ni-DBTC additive	
	$\bar{M}_n$	S	$\bar{M}_n$	S
Initial value	92 500	—	92 500	—
First	12 300	6·52	17 200	4·83
Second	27 500	2·361	28 900	2·20
Third	31 900	1·901	23 900	2·87

5.3. Reaction with Hydroperoxides

The reaction between organic metal complexes containing sulphur and hydroperoxides has been dealt with in several publications.[24,29,45,60,66]

It is generally accepted that reaction between metals with variable valency and hydroperoxides is first-order for both components, and takes place via radical formation:

$$M^n + ROOH \rightarrow M^{n+1} + RO^{\cdot} + {}^{-}OH$$

$$M^{n+1} + ROOH \rightarrow M^n + ROO^{\cdot} + H^+$$

This mechanism does not explain, however, the stabilising effect of metal complexes. The radicals formed initiate the decomposition of hydroperoxide and also decompose the metal complex. According to Howard et al.[61] the reaction of metal complexes and the peroxyl radical can take place in three different ways: via peroxyl anion and cation radical formation (reaction 10(a)); via peroxyl anion, cation with metal content and dithiocarbamate radical formation (reaction 10(b)); or the peroxyl radical is added to the metal complex, and the unpaired electron is shifted to the carbon atom (reaction 10(c)).

(10)

In the case of dialkyldithiophosphates, Liston and co-workers[62] obtained ESR spectra characteristic of the radicals of pathways 10(a) and 10(c), while Howard et al.[61] found pathway 10(b) to be more probable in the case of the dithiocarbamates. Since there was no experimental evidence to indicate that peroxyl radicals reacted with organosulphur compounds at the sulphur atom,[61] the authors assumed that the reaction occurs at a metal centre rather than at sulphur. This is supported by the low exponential factor.[61]

$$ROO^{\cdot} + X{-}Zn{-}\dot{X} \rightarrow ROO{-}Zn{-}X + X^{\cdot} \qquad (11)$$

Chemical interaction between the metal complexes and hydroperoxides takes place initially in a stoichiometric reaction. According to Chien and Boss,[24] during this stage six molecules of hydroperoxide are converted to alcohol while compounds of higher oxidation state are formed from the metal complex. Finally, sulphur trioxide, formed from sulphur dioxide in a stoichiometric reaction with hydroperoxides,[61a] decomposes hydroperoxides ionically.[29]

In our opinion the probability of the latter mechanism is higher, because hydroperoxide decomposition via radical formation would accelerate photo-oxidation chain propagation. Since, in the case of PP, hydroperoxide sequences appear along the molecule backbone, Chien postulates that these might be transferred into five-membered ring peroxides, with evolution of hydrogen peroxide,[24] by reaction 12.

$$\text{(structure)} \xrightarrow{\text{stabiliser}} \text{(structure)} + H_2O_2 \qquad (12)$$

The hydroperoxide-decomposing ability of metal dithiocarbamates has been discussed in several publications.[29,45,63–66] Scott et al.[29,63,64] presume that intermediate products are formed from dithiocarbamates in the reaction with hydroperoxides (see Chapter 3). Dithiocarbamates completely inhibit hydroperoxide formation even during compression moulding, and the hydroperoxide-decomposing efficiency is higher at elevated temperatures.[65] On irradiating processed PP containing hydroperoxide after immersion in a solution of Ni-DBTC and tetramethylhydroxypiperidine, Carlsson and Wiles[45,66] observed a rapid decrease in polypropylene hydroperoxide concentration.

The reactivity of different kinds of photostabilisers toward hydroperoxides was therefore studied in a reaction with tert-butyl hydroperoxide and polypropylene hydroperoxides.

In Fig. 29 the concentration of tert-butyl hydroperoxide is followed in the reaction with Ni-DBTC in chloroform solution, in air, at room temperature. Solutions were mixed in a 1:1 ratio. Hydroperoxide concentration decreased very rapidly after reaction with Ni-DBTC.

Figure 30 shows the decrease in hydroperoxide concentration measured after reaction of the stabilisers with tert-butyl hydroperoxide in

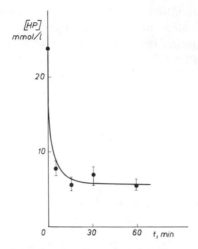

FIG. 29. Concentration change of *tert*-butyl hydroperoxide during reaction with Ni-DBTC in chloroform solution, at room temperature. Initial Ni-DBTC concentration, 36·9 mmol/litre.

chloroform solution, in air, at room temperature for 3 min. Appreciable hydroperoxide decomposing efficiency was observed only in the case of Ni-DBTC, although TMPS also showed some slight degree of reactivity.

Similar results were obtained by reacting the stabilisers with PP hydroperoxide in the dark at 70°C (Fig. 31). Stabilisers were incorp-

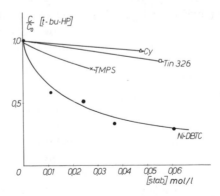

FIG. 30. Relative concentration change of *tert*-butyl hydroperoxide in chloroform solution during reaction with different concentrations of photostabilisers at room temperature (concentration measured after 30 min). Initial *t*-bu-HP concentration, 21·9 mmol/litre.

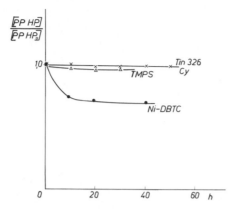

FIG. 31. Hydroperoxide concentration change in processed PP film (PPHP) swollen in solutions of different photostabilisers. (Data in text.)

orated by swelling benzene solution into films of processed PP of 100 μm thickness. After storage for 10 min the films were dried. The amounts of stabilisers migrating into the films were as follows: Cy, 85 mmol/litre; Tin 326, 35 mmol/litre; Ni-DBTC, 72 mmol/litre; TMPS, 92 mmol/litre. It can be seen from the Figure that, in the case of Ni-DBTC, at the beginning of the reaction the concentration of the PP hydroperoxide decreased rapidly and then fell more gradually.

On the basis of these experiments and in agreement with the literature, the following statements can be established regarding the hydroperoxide-decomposing activity of the photostabilisers investigated:

(a) 2-Hydroxybenzophenone and 2-hydroxybenzotriazole do not react with hydroperoxide.
(b) Ni-DBTC reacts with hydroperoxide and TMPS reacts with hydroperoxide to some extent.

5.4. Radical Formation

Only sterically hindered piperidine derivatives give rise to appreciable concentrations of radicals in a PP film at room temperature or at the temperature of irradiation. The generally accepted view[38,39,45] is that the nitroxyl radical is formed from the piperidine derivatives during irradiation[67,68] or in reaction with the oxidation products.[37,42] Other authors, however, doubt the presence and stabilising role of the N-oxyl

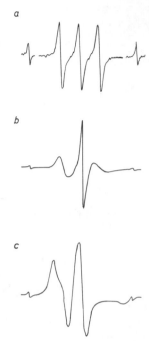

Fig. 32.   (a) ESR spectrum of TMP-NO radical in toluene. Amplification (A), 450; modulation (M), 0·05 G. (b) ESR spectrum of TMP-NO radical in PP (A, 50; M, 1 G). (c) ESR spectrum of radicals formed from TMPS in PP (A, 710; M, 1 G).

radical[69] since the sample containing the piperidine derivative does not show characteristic discolouration, and the shape of the ESR signal differs from that of the nitroxyl radical.

In order to throw light on this problem, the ESR spectra of the nitroxyl radical in toluene solution (Fig. 32(a)) and in PP (Fig. 32(b)) is presented for comparison with the ESR spectrum of TMPS in PP (Fig. 32(c)). It can be seen that the ESR spectrum of the latter does differ from the spectrum of the nitroxyl radical (Fig. 32(a) and (b)). This difference can be explained by the assumption that radicals of different mobility are present in the polymer and thus, instead of three signals of identical width, signals with different widths appear in the ESR spectrum of TMP-NO in PP (Fig. 32(b)). In addition, a sharp singlet is superimposed on the middle line of the spectrum, indicating that some of the radicals are present in high local concentration, and are thus in close paramagnetic interaction. In Fig. 32(c) a non-equidistant aniso-tropic signal can be observed, due to the fact that mobility of the radicals formed from TMPS in PP is much lower than that of TMP-NO.[14]

Chakraborty and Scott[37] and Sedlar et al.[43] attribute the formation of the nitroxyl radical to the reaction of the piperidine with hy-droperoxides during processing according to reaction 13:

$$RO-\underset{}{\bigcirc}N-H+HOOR \longrightarrow RO-\underset{}{\bigcirc}N^{\cdot}+H_2O+RO^{\cdot}$$

$$\downarrow \qquad (13)$$

$$RO-\underset{}{\bigcirc}N-\dot{O} + \dot{O}H \longleftarrow RO-\underset{}{\bigcirc}NOOH+R^{\cdot}$$

In the course of this process, however, $R^{\cdot}$, $RO^{\cdot}$, and $^{\cdot}OH$ radicals are formed, which can initiate the oxidation of the polymer. The occurrence of reaction 13, if it takes place under photo-oxidative conditions, would not explain the excellent stabilising effect of this compound since the piperidine is not a stabiliser in the melt and indeed the hydroperoxide concentration in PP containing TMPS was found to be actually higher than in PP without this stabiliser.[37]

The ESR spectra of TMPS in powder form were studied as well as in solution (Fig. 33). A spectrum similar to that obtained for radicals formed from TMPS in PP (Fig. 32(c)) and characteristic of nitroxyl radicals with low mobility is obtained in the case of TMPS powder (Fig. 33(a)). The ESR spectrum of TMPS in toluene solution (Fig. 33(b)) is similar to that of TMP-NO in the same solution (Fig. 32(a)). This means that nitroxyl radicals are formed from TMPS, and this is confirmed by the agreement between the g-values of both spectra. The radical concentration is about one order of magnitude higher in the polymer containing TMP-NO than in the polymer with TMPS, and is about three orders of magnitude lower in the case of TMPS powder than in the polymer containing 1·0% TMPS. (At the wings of the spectra the third and fourth lines of Mn:MgO can also be seen.)

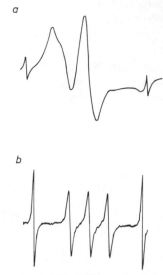

FIG. 33. (a) ESR spectrum of TMPS powder (A, 180; M, 2 G). (b) ESR spectrum of TMPS in toluene (A, 1400; M, 0·5 G).

The formation of the nitroxyl radical thus cannot be simply the result of reaction with hydroperoxides. The presence of radicals in TMPS powder, as well as the formation of the radicals in PP containing TMPS after compression moulding lead to the conclusion that radicals are formed due to the effect of oxygen and heat or irradiation. This assumption is supported by the results of Ivanov et al.[67] and Anisimova

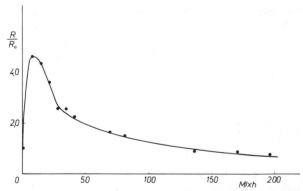

FIG. 34. Change in radical concentration in PP containing 1·0 mass % TMPS, during UV irradiation.

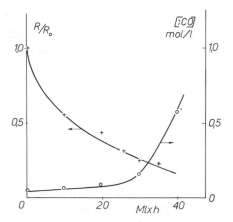

FIG. 35. Change in TMP-NO radical concentration in PP containing 1·0 mass % during UV irradiation.

et al.[68] According to these authors secondary and tertiary amines are readily oxidised to N-oxyl radicals in a reaction with excited singlet oxygen (see *Developments in Polymer Stabilisation—5*, Chapter 3).

The change of concentration of radicals formed from TMPS in PP during irradiation is shown in Fig. 34 (arbitrary units). The amount of radicals first increases and then, after a maximum, slowly decreases. In the case of TMP-NO, a monotonic decrease can be observed in the intensity of the ESR signal after irradiation (Fig. 35). This compound is not an effective photostabiliser due to its very rapid loss during irradiation. On the other hand, N-oxyl radicals are continuously formed from the TMPS, which remains in the polymer due to its high molecular mass and serves as a radical reservoir. The maximum in the curve (Fig. 34) may be attributed to the fact that the radicals formed continuously from the TMPS are subsequently consumed in reaction with R˙ or ROO˙ formed in the polymer.[42,70]

6. CO-OPERATIVE EFFECTS OF STABILISERS

6.1. Combination of a 2-Hydroxybenzophenone and a Nickel Complex

The advantage of co-addition of different types of stabilisers was first mentioned by Melchore.[71] Savides[72] found the stabilising effect of

Cyasorb UV 531 (0·25 mass %) with the 2,2-thiobis(4-*tert*-octylphenolato)-*n*-butylamine–Ni complex to be higher in PP film (375 μm thick) than the effect of the two components used alone, at 0·5 mass %.

Later a thorough study was made by Chakraborty and Scott[73] on the effect of co-addition of 2-hydroxy-4-octoxybenzophenone (Cy) with both Zn-DETC and Ni-DETC in PE films of 200 μm thickness on carbonyl formation during irradiation. (Total amount of additives was 3·3 mmol/litre.) These authors found that the mixture when used as a screen was not so effective as when used as an additive. They suggested that the formation of radicals from the Ni-DETC is inhibited by Cyasorb which acts both as a UV-absorber and as a quencher. The authors also investigated the effect of co-addition of Zn-DETC with Cyasorb UV 531 (similarly at 3·3 mmol/litre total concentration) in PP and found a synergistic effect on carbonyl formation[65] (see Chapter 3).

In order to study the synergistic effect, we investigated the co-addition of Cy and Ni-DBTC in 0·5% concentration (total concentration, 23·86 mmol/litre). Induction periods to carbonyl formation ($t_{i,CO}$) in the case of PP containing this mixture are compared in Table 9, with those for samples containing Cy and Ni-DBTC alone at 0·5 and 1·0 mass %.

TABLE 9

INDUCTION PERIODS TO CARBONYL FORMATION IN PHOTO-OXIDATION OF PP IN PRESENCE OF ONE OR TWO PHOTOSTABILISERS

Additive	$t_{i,CO}$ (Mlxh)
None	6·0
0·5% Cy	32·5
0·5% Ni-DBTC	51·5
0·5% Cy + 0·5% Ni-DBTC	55·0
1·0% Cy	41·0
1·0% Ni-DBTC	86·0
1·0% Cy + 1·0% Ni-DBTC	99·0

If the addition of about 0·4–0·5 mass % gives the highest specific efficiency (Fig. 9), it might be expected that the combination of both stabilisers at this concentration should result in an additive effect. The expected maximal efficiency was, however, not achieved.

The difference between Scott's results and our own may be due to the difference in concentration of the additives since there is evidence that above 0·4 mass % Ni-DBTC is lost by blooming (see Chapter 3).

Similarly, no additive effect can be attained by joint addition of Cy and Ni-DBTC at 1·0 mass % (Table 9). The induction period in this case was found to be 99 Mlxh, in contrast to the additive value of 127 Mlxh.

6.2. Combinations of Benzotriazole and Piperidine Derivatives

The synergistic effect of a 1:1 combination of a 2-hydroxybenzotriazole and a hindered piperidine was reported by Gugumus,[18] who measured the exposure time up to 50% retention of impact strength, in the case of 2 mm PE plaques. The higher stabilising effect was explained by the assumption that the benzotriazole screens the bulk of the thick plaques due to its UV-absorber character, thus hindering the decomposition of piperidine derivatives by light. The effect of co-addition of piperidine derivatives with other types of stabilisers was recently investigated by Allen.[40] No additive effect could be achieved by co-addition of 0·1% of hindered piperidine with hydroxybenzotriazoles, hydroxybenzophenones or nickel complexes on measuring the

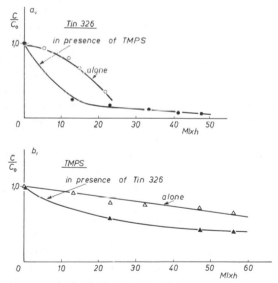

FIG. 36. (a) Change of Tin 326 concentration both alone (1·0 mass %) and with 1·0 mass % TMPS. (b) Change of TMPS concentration both alone (1·0 mass %) and with 1·0 mass % Tin 326.

embrittlement time and carbonyl index after irradiation of 200 μm PP films in a Xenotest.

The co-addition of TMPS together with Tin 326, Ni-DBTC and Cy (1·0+1·0%) was studied with the following results. Carbonyl and hydroperoxide concentrations remained unchanged in the presence of TMPS+Tin 326 during irradiation—similar to the case when TMPS was used alone—up to spontaneous rupture of the film at 171 Mlxh. The change in concentration of the stabiliser, however, showed a different behaviour from that obtained by use of the stabilisers separately. As shown in Fig. 36(a), in the presence of TMPS the concentration of Tin 326 drops steeply at the beginning of irradiation. Similarly, the consumption of TMPS is somewhat faster in the presence of Tin 326 (Fig. 36(b)). This phenomenon can be explained by assuming reaction between Tinuvin 326 and the [=CO] groups of TMPS.[74]

6.3. Combinations of a Metal Complex and Sterically Hindered Piperidine Derivative

Using combinations of these two kinds of stabilisers, Chakraborty and Scott[37] found an antagonistic effect. They explained this as being due to the hydroperoxide-decomposing activity of the metal complexes which reduced the formation of the nitroxyl radical by reaction of the piperidine with hydroperoxide during processing. Allen[40] also explains the decreased effect of a combination of a nickel complex of a sterically hindered phenol type (Irgastab 2002) with a piperidine derivative by the above mechanism. This author suggests that the antagonistic effect of the piperidine derivative in the presence of DABCO—a singlet oxygen quencher—is due to the inhibiting behaviour of DABCO on hydroperoxide formation which thus suppresses the formation of the nitroxyl radical.

In our opinion, however, nitroxyl radicals are formed in other reactions besides that of TMPS with hydroperoxides (see Section 5.4).

The change in concentration of carbonyl and hydroperoxide groups together with stabiliser consumption are shown in Fig. 37, during the photo-oxidation of a PP film containing 1·0+1·0 mass % of Ni-DBTC and TMPS. The consumption of Ni-DBTC, at least at the beginning of irradiation, approaches the consumption curve obtained in the presence of the stabiliser alone (dotted line in Fig. 37). No ESR signal characteristic of the nitroxyl radical could be detected, however, in the presence of Ni-DBTC. This leads to the conclusion that nickel complexes, being singlet oxygen quenchers,[32,34,75,76] will hinder the forma-

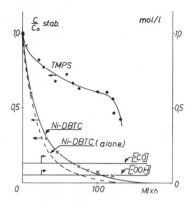

FIG. 37. Photo-oxidation of PP films with 1·0+1·0 mass % TMPS+Ni-DBTC.

tion of nitroxyl radicals by this mechanism. If this assumption is valid, it would indicate that the nitroxyl radical is partly formed by interaction between TMPS and excited singlet oxygen.

6.4. Combinations of Hydroxybenzophenone and Piperidine Derivative

The effect of a combination of 1·0+1·0 mass % Cy+TMPS is shown in Fig. 38. As can be seen, the consumption of Cy is slower than when it is used alone and the concentration change of TMPS is similarly

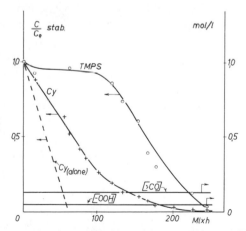

FIG. 38. Photo-oxidation of PP film with 1·0+1·0 mass % TMPS+Cy.

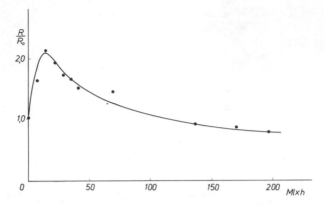

FIG. 39.   Change of nitroxyl radical concentration in PP film with $1\cdot0+1\cdot0$ mass % TMPS+Cy, during UV irradiation.

slower (compare Fig. 36(b)) at least as long as Cy is present in the system. Exponential consumption of TMPS begins only after exposure to 120 Mlxh, when about 80% of the Cy had been consumed.

At the beginning of irradiation the radical concentration is about twice as high as its initial value when each stabiliser is used alone (Fig. 39).

Ultraviolet dosage values corresponding to the spontaneous rupture of the films (t_r) upon the addition of different stabilisers ($1\cdot0$ or $1\cdot0+1\cdot0$ mass%) are summarised in Table 10.

The highest efficiency is achieved by using Cy and TMPS ($1\cdot0+1\cdot0$ mass %). The dosage required for spontaneous rupture of the film of the above composition is practically additive in this case. Considering

TABLE 10

ULTRAVIOLET DOSAGE VALUES (Mlxh) CORRESPONDING TO THE SPONTANEOUS RUPTURE (t_r) OF PP FILMS OF 100 μm THICKNESS FOR COMBINATIONS OF ULTRAVIOLET STABILISERS (Concentration of mixture of stabilisers, $1\cdot0+1\cdot0$ mass %. Concentration of single stabiliser, 1 mass %)

	Tin 326	Cyasorb UV 531	Ni-DBTC	TMPS
Tin 326	22	—	—	171
Cyasorb UV 531	—	44	99	246
Ni-DBTC	—	99	101	148
TMPS	171	246	148	210

the shape of t_r–c curves shown in Fig. 10, this indicates synergism, since in the case of Cy, the combination gives a higher efficiency than might be expected from the t_r values at 2% Cy alone.

7. CONCLUSIONS CONCERNING THE MODE OF ACTION OF PHOTOSTABILISERS

7.1. Hydroxybenzotriazole Derivatives

The results of our investigations on the screening ability and other supplementary experiments have led to the conclusion, in agreement with other authors, that this type of stabiliser acts primarily as a UV-absorber. Reactivity to hydroperoxides is negligible, and consumption begins when oxygen-containing products (ketones and in particular acetone) are formed from the polymer in the course of oxidation. This suggests that the benzotriazole decomposes by reaction with these products.

The reaction of benzotriazole derivatives with ketonic products has been established by Pac et al.[74] in experiments involving the reaction between Tinuvin 326 and acetone. Since in the course of PP photo-oxidation acetone is formed by the Norrish-II decomposition of ketone at the end-group of the polymer molecule, the above assumption seems well based. The fact that the concentrations of both Tin 326 and TMPS decrease faster in combination also points to a reaction with the C=O group. In addition the possibility of hydroxybenzotriazole reaction with peroxyl radicals[77] cannot be excluded.

7.2. Hydroxybenzophenone Derivatives

This type of additive is known to exert a complex effect. On the basis of screening investigations it has been stated that these derivatives may act as UV-absorbers. The bimodal character of the molecular mass distribution can also be explained by this effect. Vink[51] suggests that the zero-order consumption of hydroxybenzophenone during irradiation is due to its quenching effect: after quenching, part of the excited hydroxybenzophenone returns to its ground state, dissipating harmlessly the excess energy, whereas the other part of the additive decomposes.

Similarly the higher effectivity of hydroxybenzophenones in slowing polymer chain scission can be interpreted by excited carbonyl quenching, by retarding the photolysis of carbonyl compounds.

The combination of hydroxybenzophenones and sterically hindered piperidine derivatives retards the development of the maximum nitroxyl radical concentration and increases its stationary concentration compared with the values obtained by the use of TMPS alone. This may be attributed to the slight quenching effect exerted by hydroxybenzophenones on the excited oxygen necessary for the formation of nitroxyl radicals.

7.3. Metal Complexes

In spite of the fact that many metal complexes have significant absorption in the range 300–350 nm, screening experiments show that these complexes transmit light in other parts of the actinic range of the UV spectrum, and therefore exert no significant UV screening effect. Their stabilising effect is partly due to ionic decomposition of the hydroperoxide formed, and partly to reaction with excited carbonyls and with oxygen.[66,78] This may also explain the higher rate of decomposition of Ni-DBTC upon irradiation in powder form than in the polymer, as well as its faster decomposition in oxygen than in nitrogen.[45,79] Similarly, the experimental observation that Ni-DBTC retards the formation of the nitroxyl radical from TMPS can also be explained in terms of singlet oxygen quenching. (After consumption of Ni-DBTC, radicals formed from TMPS can be detected.)

The bimodal character of molecular mass distribution of the polymer in the presence of Ni-DBTC is probably related to the fact that in the surface layers the additive is readily accessible to oxygen and therefore decomposes faster in these layers; it can be supplemented with difficulty from the inside due to its relatively low diffusivity.

Possible participation of Ni complexes in radical reactions cannot be excluded either. Howard et al.[61] reported a reaction between DETC complex and ROO˙ radical in solution, and Ranaweera and Scott[23] showed that these complexes can interrupt kinetic chains.

7.4. Sterically Hindered Piperidine Derivatives

Some authors assume that the amine associates with the hydroperoxides in the polymer[41,80] and thus practically neutralises the main initiation sites. This assumption is, however, in contradiction with the experimental observation that hydroperoxide content can be measured also in the presence of TMPS. Hydrogen abstraction by the peroxyl radical propagating oxidation probably occurs more readily from the amine than from the polymer.[81] The iminyl radical or the original amine is converted into a nitroxyl radical in a reaction with oxygen.[67] After an initial increase, a decrease in concentration of the nitroxyl radical can be observed during irradiation. This decrease is a result of the reaction with the radicals formed during photo-oxidation of the polymer, in the course of which reaction chain propagation is retarded.

The reaction of aromatic amines with R˙ and ROO˙ radicals is discussed by Dubinskaya et al.[82] The fast reaction between nitroxyl radicals and alkyl radicals has been well known for many years.[83,84] Shilove et al.[42] have proved experimentally that nitroxyl radicals react with alkyl radicals to give oxygen-substituted adducts of the hydroxylamine type, from which again a nitroxyl radical is formed by oxidation. (For fuller discussion of the cyclical regeneration mechanism, see *Developments in Polymer Stabilisation—3*, Chapter 1, and *ibid.—5*, Chapter 3.)

In our opinion, the nitroxyl radical is formed from the piperidine during UV irradiation by direct reaction with oxygen, reaction 14(a), or by hydrogen abstraction by the ROO˙ radical followed by oxygen absorption (reaction 14(b), (c)). The nitroxyl radical can be built into the polymer upon reaction with the macro-alkyl radicals formed from it during irradiation (reaction 14(d)).

$$ (14) $$

In reaction 14(b) alkylperoxyl radicals are terminated, although at high oxygen concentration or upon the decrease of the initial amine concentration, the regenerated radicals result in new ROO˙ radicals (14(e)). A stationary radical concentration is set up due partly to the nitroxyl radical formed from TMPS, and partly to those regenerated from the NO–R bond in the polymer.

170   F. TUDOS, G. BÁLINT AND T. KELEN

REFERENCES

1. BURCHILL, B. J. and PINKERTON, D. M., *J. Polym. Sci. Polym. Symp.*, **55** (1976) 185.
2. CARLSSON, D. J., GARTON, A. and WILES, D. M. in *Developments in Polymer Stabilisation—1*, Ed. G. Scott (1979), p. 219. Applied Science, London.
3. SEMENOV, N. N., *Chain Reactions* (1934) (in Russian). Goskhimtechizdat, Moscow.
4. BOLLAND, J. L. and GEE, G., *Trans. Faraday Soc.*, **42** (1946) 236, 244.
5. CHIEN, J. C. W., VANDENBERG, E. J. and JABLONER, H., *J. Polym. Sci.* A-1, **6** (1968) 381.
6. GUILLET, J. E., DHANRAJ, J., GOLEMBA, F. J. and HARTLEY, G. H. in *ACS Adv. in Chem. Ser.* **85** (1968) 272.
6a. TROZZOLO, A. M. and WINSLOW, F. H. *Macromol.*, **1** (1968) 98.
6b. CALVERT, S. G. and PITTS, N. S. (1966) *Photochemistry*, Wiley & Sons, New York.
7. ALLEN, N. S. and MCKELLAR, J. F., *Chem. Ind.* (1977) 537.
8. ALLEN, N. S. and MCKELLAR, J. F. in *Developments in Polymer Degradation—2*, Ed. N. Grassie (1979), p. 129. Applied Science, London.
9. ALLEN, N. S., HOMER, J. and MCKELLAR, J. F., *Chem. Ind.*, (1976), 692.
10. ALLEN, N. S., *Polym. Deg. Stab.*, **2** (1980) 155.
11. CHAKRABORTY, K. B. and SCOTT, G., *Polymer*, **18** (1977) 98.
12. BÁLINT, G., KELEN, T., REHÁK, Á. and TÜDŐS, F., *React. Kin. Cat. Lett.*, **4** (1976) 467.
13. BÁLINT, G., KELEN, T., TÜDŐS, F. and REHÁK, Á., *Polym. Bull.*, **1** (1979) 647.
14. BÁLINT, G., ROCKENBAUER, A., KELEN, T., TÜDŐS, F. and JÓKAY, L., *Polym. Photochem.*, **1** (1981) 139.
15. LÁSZLÓ-HEDVIG, S., IRING, M., BÁLINT, G., KELEN, T. and TÜDŐS, F., *J. Appl. Polym. Sci., Appl. Polym. Symp.*, **35** (1979) 161.
16. CARLSSON, D. J. and WILES, D. M., *J. Macromol. Sci., Rev. Macromol. Chem.*, **C14** (1976) 65.
17. CICCHETTI, O. and GRATANI, F., *Europ. Polym. J.*, **8** (1972) 561.
18. GUGUMUS, F. in *Developments in Polymer Stabilisation—1*, Ed. G. Scott (1979) p. 261. Applied Science, London.
19. ASPLER, J., CARLSSON, D. J. and WILES, D. M., *Macromol.* **9** (1976) 691.
20. EMANUEL, N. M., *Izv. Akad. Nauk, SSSR Ser. Khim.*, 1974, 1056.
21. KAMIYA, Y. and INGOLD, K. U., *Can. J. Chem.*, **42** (1964) 2424.
22. URI, N., *Chem. Ind.* (1967) 2060.
23. RANAWEERA, R. P. R. and SCOTT, G., *Europ. Polym. J.*, **12** (1976) 825.
24. CHIEN, J. C. W. and BOSS, C. R., *J. Polym. Sci.* A-1, **10** (1972) 1579.
25. BURN, A. J., *Adv. Chem. Ser.*, **75** (1968) 323.
26. GERVITS, L. L., ZOLOTOVA, N. V. and DENISOV, E. T., *Vysokomol. Soed.*, **A17** (1975) 2112.
27. BAUM, B. and PERUN, A. L., *Am. Chem. Soc., Div. Org. Coatings Plast. Chem.*, **21** (1961) 79.

28. CARLSSON, D. J., SPROULE, D. E. and WILES, D. M., *Macromol.*, **5** (1972) 659.
29. HOLDSWORTH, J. D., SCOTT, G. and WILLIAMS, D., *J. Chem. Soc.*, (1964) 4692.
30. MELLOR, D. C., MOIR, A. B. and SCOTT, G., *Europ. Polym. J.*, **9** (1973) 219.
31. ADAMCHYK, A. and WILKINSON, F., *J. Chem. Soc. Farad. Trans. II.* **68** (1972) 2031.
32. ZWEIG, A. and HENDERSON, W. A., *J. Polym. Sci., Polym. Chem. Edn.*, **13** (1975) 717.
33. GUILLORY, J. P. and COOK, C. F., *J. Am. Chem. Soc.*, **95** (1973) 4885.
34. CARLSSON, D. J., SUPRUNCHUK, T. and WILES, D. M., *J. Polym. Sci., Polym. Lett.*, **11** (1973) 61.
35. BRIGGS, P. J. and MCKELLAR, J. F., *Chem. Ind.*, (1967) 622.
36. BRIGGS, P. J. and MCKELLAR, J. F., *J. Appl. Polym. Sci.*, **12** (1968) 1825.
37. CHAKRABORTY, K. B. and SCOTT, G., *Chem. Ind.*, (1978) 237.
38. SHLYAPINTOHK, V. Ia., IVANOV, V. B., HVOSTACH, O. M., SHAPIRO, A. B. and ROZANTSEV, E. G. *Dokl Akad. Nauk. SSSR*, **225** (1975) 1132.
39. CHAKRABORTY, K. B. and SCOTT, G., *Polymer*, **21** (1980) 252.
40. ALLEN, N. S., *Polym. Deg. Stab.*, **2** (1980) 129.
41. GRATTAN, D. W., REDDOCH, A. H., CARLSSON, D. J. and WILES, D. M., *J. Polym. Sci., Polym. Lett.*, **16** (1978) 143.
42. SHILOV, JU. B. and DENISOV, E. T., *Vysokomol. Soed.*, **A16** (1974) 2313.
43. SEDLAR, J., PETRUJ, J. and PAC, J., communication at *Int. Symp. Mechanism of Degradation and Stabilisation of Hydrocarbon Polymers, Prague*, (1979), to be published.
44. ALLEN, N. S., MCKELLAR, J. F. and WILSON, D., *Polym. Deg. Stab.*, **1** (1979) 205.
45. CARLSSON, D. J., *J. Appl. Polym. Sci.*, **22** (1978) 2217.
46. SOHMA, J., in *Developments in Polymer Degradation—2*, Ed. N. Grassie (1979), p. 99. Applied Science Publishers Ltd., London.
47. HUTSON, G. V. and SCOTT, G., *Europ. Polym. J.*, **10** (1974) 45.
48. AMIN, M. U. and SCOTT, G., *Europ. Polym. J.*, **10** (1974) 1019.
49. OSAWA, Z. and SAITO, T., *Polym. Prepr. (New Orleans Meeting, 1977)*, **18** (1977) 420.
50. BLACK, J. F., *J. Am. Chem. Soc.*, **100** (1978) 527.
51. VINK, P., *J. Polym. Sci. Polym. Symp.*, **40** (1973) 169.
52. ADAMS, J. H., *J. Polym. Sci.*, A-1, **8** (1970) 1279.
53. GUILLORY, J. P. and COOK, C. F., *J. Polym. Sci.*, A-1, **9** (1971) 1529.
54. ZANNUCCI, J. S. and LAPPIN, G., *Macromol.*, **7** (1974) 393.
55. VINK, P. and VAN VEEN, T. J., *Europ. Polym. J.*, **14** (1978) 533.
56. HELLER, H. J. and BLATTMAN, H. R., *Pure Appl. Chem.*, **30** (1972) 145.
57. DECKER, C. and MAYO, F. R., *J. Polym. Sci., Polym. Chem. Edn.* **11** (1973) 2847.
58. GRASSIE, N. and LEEMING, W. B. H., *Europ. Polym. J.*, **11** (1975) 819.
59. CHEW, C. H., GAN, L. M. and SCOTT, G., *Europ. Polym. J.*, **13** (1977) 361.
60. HOWARD, J. A. and CHENIER, J. H. B., *Can. J. Chem.*, **54** (1976) 390.

61. HOWARD, J. A., OHKATSU, Y., CHENIER, J. H. B. and INGOLD, K. U., *Can. J. Chem.* **51** (1973) 1543.
61a. HUSBANDS, M. S. and SCOTT, G., *Europ. Polym. J.* **15** (1979) 249.
62. Liston, T. V., Ingersoll, H. G. Jr. and Adams, J. Q., *ACS, Div. Pet. Chem. Prepr.*, **14** (1969) A 83.
63. Ranaweera, R. P. R. and Scott, G., *J. Polym. Sci., Polym. Lett.*, **13** (1975) 71.
64. RANAWEERA, R. P. R. and SCOTT, G., *Europ. Polym. J.*, **12** (1976) 591.
65. CHAKRABORTY, K. B. and SCOTT, G., *Polym. Deg. Stab.*, **1** (1979) 37.
65a. AL MALAIKA, S. and SCOTT, G., *Europ. Polym. J.*, **16** (1980) 503.
66. CARLSSON, D. J. and WILES, D. M., *J. Polym. Sci., Polym. Chem. Edn.*, **12** (1974) 2217.
67. IVANOV, V. B., SHLIAPINTOHK, V. IA., HVOSTACH, O. M., SHAPIRO, A. B. and ROZANTSEV, E. G., *J. Photochem.*, **4** (1975) 313.
68. ANISIMOVA, O. M., IVANOV, V. B., ANISIMOV, V. M., SHAPIRO, A. B. and ROZANTSEV, E. G., *Izv. Akad. Nauk. Ser. Khim. SSSR*, (1978) 2263.
69. ALLEN, N. S., *J. Appl. Polym. Sci.*, **22** (1978) 3277.
70. NEIMAN, M. N. and ROZANTSEV, E. G., *Izv. Akad. Nauk SSSR, Ser. Khim.*, (1964) 1178.
71. MELCHORE, J. A., *Ind. Eng. Chem. Prod. Res. Dev.*, **1** (1962) 232.
72. SAVIDES, C., *SPE Journal*, **29** (1973) 38.
73. CHAKRABORTY, K. B. and SCOTT, G., *Europ. Polym. J.*, **13** (1977) 1007.
74. PAC, J., SEDLAR, J., PETRUJ, J. and PARIZEK, M., *Plaste. Kaut.*, **22** (1975) 724.
75. CARLSSON, D. J., MENDENHALL, G. D., SUPRUNCHUK, T. and WILES, D. M., *J. Am. Chem. Soc.*, **94** (1972) 8960.
76. USILTON, J. J. and PATEL, A. R., *Polymer Prepr. (New Orleans Meeting, 1977)*, **18**(1) (1977) 393.
77. HODGEMAN, D. K. C., *J. Polym. Sci., Polym. Lett.*, **16** (1978) 161.
78. FLOOD, J., RUSSELL, K. E. and WAN, J. K. S., *Macromol.*, **6** (1973) 669.
79. GUILLORY, J. P. and BECKER, R. S., *J. Polym. Sci., Polym. Chem. Edn.*, **12** (1974) 993.
80. SEDLAR, J., PETRUJ, J., PAC, J. and NAVRATIL, M., *Polymer*, **21** (1980) 5.
81. GRATTAN, D. W., CARLSSON, D. J. and WILES, D. M., *Polym. Deg. Stab.*, **1** (1979) 69.
82. DUBINSKAYA, A. M., BUMIAGIN, P. JU., OBINTSOVA, R. R. and BERLIN, A. A., *Vysokomol. Soed.*, **A10** (1968) 410.
83. BEVINGTON, J. C. and GHAMEN, N. A., *J. Chem. Soc.*, (1956) 3506.
84. TÜDŐS, F., BEREZSNICH, T. F. and AZORI, M., *Acta Chim. Hung.*, **24** (1960) 91.

Chapter 5

NEW DEVELOPMENTS IN THE DEGRADATION AND STABILISATION OF POLYVINYL CHLORIDE

K. S. MINSKER, M. I. ABDULLIN, S. V. KOLESOV

Bashkir State University, Ufa, USSR

and

G. E. ZAIKOV

Institute of Chemical Physics, Academy of Sciences of the USSR, Moscow, USSR

SUMMARY

This chapter deals with the main concepts of the modern theory of polyvinyl chloride (PVC) degradation and stabilisation. A number of key problems are discussed; in particular, the structure of the real PVC macromolecule and the reasons for the low thermal stability of PVC; the kinetics and mechanism of HCl elimination and macrochain cross-linking; the formation peculiarities of the UV/visible spectrum of degraded PVC and the mechanism of action of the main types of thermal stabilisers. In contrast to the universally accepted theory of PVC degradation activated by β-chloroallyl groups, new concepts on the primary role of oxygen-containing chloroallyl groups (∿∿C(O)—CH=CH—CHCl∿∿) have been developed. The appropriate experimental evidence is discussed and the new concepts have been used to construct well-grounded PVC degradation and stabilisation schemes on the basis of soundly based kinetic equations.

1. INTRODUCTION

Polyvinyl chloride (PVC) has been a marketable product for nearly 50 years. Nowadays it is one of the large-tonnage polymers. PVC production worldwide comprises 15–30% of the overall production of synthetic resins amd plastics, i.e. over ten million tons.[1]

However, despite the remarkable success achieved in the development of formulations that make it possible to produce about 4000 industrial and domestic products demanding a wide range of manufacturing conditions the selection of effective stabilisers is carried out empirically in most cases and involves a considerable element of fortuity. This is due to the fact that so far almost no theories have been developed to explain the main types of PVC ageing. No solution has yet been found to general theoretical and experimental problems, which include: (a) the identification of the active sites responsible for the low stability of macromolecules, long the main enigma of PVC degradations; (b) the determination of the principal elementary processes that take place during PVC degradation and stabilisation; (c) the determination of the kinetic parameters of the above processes; and (d) the establishment of a connection between the structure of stabilisers and active additives, on the one hand, and their reactivity towards PVC, on the other.

Only an understanding of these processes makes it possible to devise an effective theory of PVC degradation and stabilisation that would permit prediction of practical performance. The main problem lies in finding the reason for the abnormally low stability of PVC.

2. CHEMICAL STRUCTURE OF PVC

The decomposition temperature $(T_{dec.})$ of commercial PVC is about 365–400 K. However, $T_{dec.}$ of low-molecular compounds that simulate the structure of vinyl chloride units in PVC macromolecules is considerably higher (575–600 K).[2-4] This is usually accounted for by the presence of defects (i.e. abnormal groups) in PVC macromolecules.

The following groups are normally considered to be the main defects forming part of the main polymer chain, which comprises essentially vinyl chloride (VC) units.

(1) Unsaturated ($CH_2{=}CH\sim$, $CH_2{=}CCl\sim$, $CHCl{=}CH\sim$) and

saturated $(CH_3—\overset{\xi}{C}HCl)^{7-9}$ end-groups (UEG and SEG, respectively). These result from chain transfer to monomer and chain termination by disproportionation. The ends of some macromolecules also contain fragments of the initiators.

(2) Vicinal (1,2) chlorine atoms[9-11] that form as a result of macro-radical recombinations.

(3) Chain branches of various lengths[9,12-14] that form as a result of chain transfer to polymer (normally above 70–80% mass conversion of the monomer), isomerisation of macroradicals or 'head-to-head' addition to another monomer molecule with a simultaneous isomerisation.

(4) Internal $>C=C<$ bonds[6,9,10,14-16] formed in the process of the polymerisation. These may result from partial HCl elimination in the moieties with a chlorine atom at the tertiary carbon atom and in groups with vicinal chlorine atoms, or as a result of vinyl chloride copolymerisation during the initial stages of the process with acetylenic impurities (divinyl, acetylene, vinyl-acetylene, etc.).

Without going into experimental details which are fully described in the scientific literature, there is good reason for believing that:

(a) In all probability, groups with vicinal chlorine atoms are practically absent from PVC macromolecules; if they are formed, one of the 1, 2 chlorine atoms eliminates in the form of HCl in the course of synthesis, drying, and storage, leading to the formation of internal isolated $>C=C<$ bonds.

(b) PVC may contain from 0 to 40 branchings per 1000 monomer units, depending on the manufacturing process. The number of branches normally increases with an increase in the polymerisation temperature,[17] with the degree of the monomer conversion,[18,19] with a decrease in the intensity of mixing[19] and with reduction of the water:monomer volume ratio to less than 2:1. The number of long branches amounts to several per cent (0·02–1 per 1000 monomer units) of their total number along with short (CH_3), and medium (C_4H_9), branches.[12,20] So far as the alternative between tertiary $\geqslant C—H$ and $\geqslant C—Cl$ groups is concerned, attention has been drawn to the possibility of a rather high content of $\geqslant C—Cl$ defect groups.[17,19] However, they have not yet been identified in PVC,[21] suggesting the probability of HCl elimination with the formation of internal

$>$C$=$C$<$ bonds due to the mobility of tertiary chlorine atoms.[22]

(c) In practice each macromolecule contains one EUG (approximately 10^{-3} mol/mol PVC)[23,24] with the structure $CH_2=CH\text{ⱴ}$.[24]

(d) About 10^{-5}–10^{-4} mol/mol PVC of internal $>$C$=$C$<$ bonds ($\bar{\gamma}_0$) have been identified in PVC.[23,24]

(e) Normally there are no polyene sequences $(>$C$=$C$<)_{n>1}$ (PS) in the initial PVC samples[23-25] except in coloured emulsion products, where they appear as a result of the drying of PVC under severe conditions (e.g. 430–445 K, 1–2 h).[26]

Thus, unsaturated isolated terminal and internal $>$C$=$C$<$ bonds as well as branches of various lengths are the defective structures actually present in PVC.

3. REASONS FOR THE LOW STABILITY OF PVC

3.1. β-Chloroallyl Groups (CAG)

Not long ago an inverse relationship between the PVC molecular weight (the number of end-groups) and its dehydrochlorination rate seemed to be generally accepted.[27-29] However, there is reliable experimental evidence,[15,23-25,30-32] including that obtained during the investigation of the thermal degradation of fractionated PVC, to indicate that EUG do not affect the polymer thermal stability. This possibility as well as the possible activation of the process of HCl elimination from PVC by branchings of the type $\text{ⱴ}CH_c\!-\!\underset{\displaystyle \overset{|}{\underset{\xi}{CH_2}}}{CH}\!-\!CH<$ may therefore be ignored.

A body of opinion in the literature suggests that the main reason for the low stability of PVC lies in the presence of internal unsaturated β-chloroallyl groups $\text{ⱴ}CH_2\!-\!CH=CH\!-\!CHCl\text{ⱴ}$ (CAG) in the polymer.[3,4,6,14,16,33-35] This view is quite natural, since, in the first place, during PVC degradation β-chloroallyl groups form as part of the macromolecule by random HCl elimination:

$$-CHCl\!-\!CH_2\!-\!CHCl\!-\!CH_2\!-\!CHCl\!-\!CH_2\text{ⱴ} \xrightarrow[-HCl]{k_r}$$

$$\text{ⱴ}CHCl\!-\!CH_2\!-\!CH=CH\!-\!CHCl\!-\!CH_2\text{ⱴ} \quad (1)$$

In the second place, both experimental data on the degradation of

TABLE 1

KINETIC PARAMETERS FOR THE THERMAL DEGRADATION OF SOME LOW-MOLECULAR WEIGHT CHLORINE-CONTAINING COMPOUNDS IN THE LIQUID PHASE[3]

Compound	Decomposition temperature range (K)	E_{act} $(kcal/mol)$	$\log A$	$k \times 10^5$ at 448 K (s^{-1})
8-Chlorohexadecane	509–557	36·0	10·6	0·01
2,4-Dichloropenta-decane	514–565	16·3	2·2	0·17
8-Chlorotridec-6-ene	438–469	25·0	8·1	8·7
4-Chlorodec-2-ene	438–469	22·2	6·5	5·0
4-Chlorododec-2-ene	430–453	22·7	6·9	6·8
7-Chloronona-3,5-diene	343–369	19·4	8·0	3400
6-Chloroocta-2,4-diene	360–386	17·9	7·1	2600

low-molecular compounds simulating irregular PVC structures[2-4,36-38] and theoretical calculations[33,39] show that unsaturated $>C=C<$ bonds adjacent to the VC units in the PVC macromolecules sharply increase (by several orders) the rate of HCl evolution (Table 1).

$$\text{\wedge\wedge\wedge CHCl—CH}_2\text{—CH—CH—CHCl—CH}_2\text{\wedge\wedge} \xrightarrow[-HCl]{k_{p1}}$$

$$\text{\wedge\wedge\wedge CHCl—CH}_2\text{—CH=CH—CH=CH\wedge\wedge} \quad (2)$$

This has given rise to the view that PVC dehydrochlorination, initiated by CAG, is a 'zipper' reaction.[40,41]

Finally, a clear linear dependence has been established between the rate of HCl elimination from PVC and the number of internal $>C=C<$ bonds, $\bar{\gamma}_0$ (they were assumed to be β-chloroallyl groups $\text{\wedge\wedge\wedge CH}_2\text{—CH=CH—CHCl\wedge\wedge}$) in the initial samples of polymer products: $W_{HCl} = k\bar{\gamma}_0$[6,10,16,24,25,30] (see, for example, Fig. 1, curve 1).

There seems to be no other experimental evidence of activation of PVC dehydrochlorination by β-chloroallyl groups.

Nevertheless, this concept has become so deep-rooted that other experimental data which differ basically from the idea of a β-chloroallyl activation of HCl elimination are ignored. For instance, the highest values of the rate constants for PVC dehydrochlorination reactions activated by β-chloroallyl groups, k_{p1}, determined in the

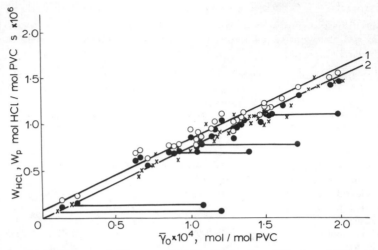

FIG. 1. Dependence of (1) rate of HCl elimination (W_{HCl}) and (2) rate of polyene formation (W_p) on $\bar{\gamma}_0$ during thermal degradation (448 K, 10^{-2} Pa) of various PVC samples: ●, ○, by ozonolysis; ×, by alkaline hydrolysis.

study of thermal degradation of some low-molecular CAG models (calculated for 448 K) are 5×10^{-4}–5×10^{-5} s^{-1} (Table 1).[3,4] This means that 0·5–5 h are required to evolve one mole of HCl at 448 K; essentially a slow reaction. It is well known that only 30 min is required for the formation of rather long PS (($\supset$C=C$\subset$)$_{n>5-10}$) in the

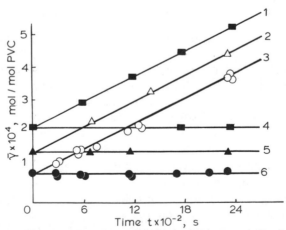

FIG. 2. Variation of the concentration of internal unsaturated groups in various PVC samples during dehydrochlorination (448 K; 10^{-2} Pa): 1, 2, 3 by ozonolysis; 4, 5, 6 by alkaline hydrolysis.

polymer.[10,42–44] The rate constant of HCl elimination from PVC activated by internal single $>C=C<$ bonds is $7 \cdot 5 \times 10^{-3} \, s^{-1}$, i.e. in this case HCl evolves faster by an order of magnitude.[24,25,30]

During the course of PVC degradation, due to random HCl elimination, the content of β-chloroallyl groups in the macromolecules increases steadily (Fig. 2, curve 1). In the case of chloroallyl activation of PVC degradation, an increase in the rate of the overall polymer dehydrochlorination is to be expected in accordance with the data in Fig. 3. It can be seen from the observations, however, that in a corrected experiment the rate of HCl elimination remains constant up to fairly large conversions.[27,45] The proposed compensation for the accumulation of internal unsaturation $(>C=C<)_{n>1}$ in the polymer, which would result from termination of the polyene growth,[24,30] lacks experimental support. In the case of the experimentally observed linear dependence of HCl yield with time during PVC dehydrochlorination, the calculated values of the termination constants (k_t) differ substantially for various samples; they are a function of the initial content of isolated internal $>C=C<$ bonds $(\bar{\gamma}_0)$ in macrochains (Fig. 3).

It should be noted that any deviation from the linear dependence of HCl evolution upon time, which is often observed experimentally, is

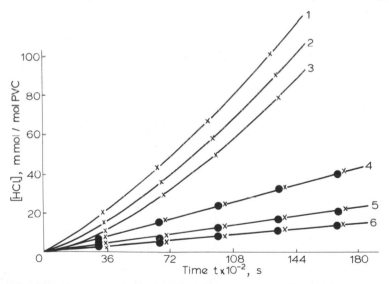

FIG. 3. Kinetic curves of PVC dehydrochlorination (1, 2, 3) without regard and (4, 5, 6) with due regard for termination of polyene growth kinetics: ●, experimental curves; ×, calculated curves. $\bar{\gamma}_0 \times 10^4$ mol/mol PVC: 1, 4—1·90; 2, 5—1·37; 3, 6—0·68; $k_t \times 10^4 \, s^{-1}$; 4, —4·3; 5, —6·2; 6, —9·6.

accounted for either by non-isothermality of the PVC degradation process, especially at the initial stages,[45,46] or by HCl catalysis which can be observed not only in a closed system,[47-49] but also under dynamic conditions.[50,51] The HCl concentration in a closed system, $U(t)$, is given by eqn. (i),

$$U(t) = \frac{W_0}{k\sigma} \left(\exp \frac{k\sigma V_1}{V} t - 1 \right)$$ (i)

where W_0 is the rate of a non-catalytic PVC dehydrochlorination, k is the effective rate constant of the catalytic HCl elimination from PVC, σ denotes HCl solubility in the polymer, V_1 is the polymer sample volume and V is the closed system volume. According to eqn. (i) the autocorrelation factor for PVC degradation in a closed volume is equal to $k\sigma V_1/V$ and is determined not only by the rate constant of the catalytic reaction but by the ratio of the polymer and the system volume as well.

Under dynamic conditions, the achievement of a stationary state is limited by the PVC sample dimensions which should be several times less than the critical size expressed by the relation (ii)[50,51]

$$l_{cr} < \pi \sqrt{\frac{D}{k}}$$ (ii)

where l = film thickness, D = coefficient of HCl diffusion, k = effective rate constant of catalytic reaction, W_0 = rate of initial (non-catalytic) reaction, W = rate of catalytic reaction, which is derived from eqn. (iii).

$$W = \frac{8 W_0 D}{l^2} \sum_{n=1}^{\infty} \frac{1 - \exp\left[k - \frac{\pi^2 D}{l^2} (2n-1)^2 \right] t}{(2n-1)^2 \frac{\pi^2 D}{l^2} - k}$$ (iii)

From (iii) it follows that the stationary rate of HCl evolution, which is equal to the rate of the primary process W_0 in thin films, increases with increasing film thickness, and when the film is very thick, the rate of HCl elimination from PVC increases exponentially. That is, even if HCl has been effectively removed, there is still a possibility of an autoaccelerating PVC dehydrochlorination. It has been estimated that the PVC sample critical dimensions are $l_{cr} = 2 \cdot 4 \times 10^{-2}$ cm at 463 K, $l_{cr} = 1 \cdot 8 \times 10^{-2}$ cm at 473 K, and $l_{cr} = 1 \cdot 5 \times 10^{-2}$ cm at 483 K.[51] In other words, under the same experimental conditions, a change in temperature may cause a drastic change in the kinetic mode of the process (i.e. from the stationary to non-stationary and vice versa).

Finally attention must be paid to the fact that the rate of HCl evolution from PVC synthesised by any of the known polymerisation methods in the absence of O_2, appears to be essentially low even if the content of internal $>C=C<$ bonds ($\bar{\gamma}_0$) is relatively high. However, during the course of sample storage in the presence of atmospheric oxygen and with the initial value of $\bar{\gamma}_0$ unchanged, the rate of HCl elimination was observed to increase up to a maximum value of W_{HCl} which was typical of a specific value of $\bar{\gamma}_0$ (Fig. 1).

Similarly, mild oxidation of predegraded PVC samples (440 K, 10^{-3} Pa), with the overall number of internal $>C=C<$ bonds retained, led to an increase in W_{HCl} (from $1{\cdot}5\times10^{-6}$ to $2{\cdot}0\times10^{-6}$ mol s) with $\bar{\gamma}_0 = 2{\cdot}62\times10^{-4}$ mol/mol PVC). This effect is not observed in PVC synthesised by special methods; for example, low-temperature (233 K) polymerisation with C_4H_9Li as the initiator. This polymer always contains an essentially low quantity of internal $>C=C<$ bonds (*ca* $(1-2)\times10^{-5}$ mol/mol PVC) and is remarkable for its high stability ($W_{HCl} = (1-2)\times 10^{-7}$ HCl mol/mol PVC s) (Fig. 1).

3.2. Oxidation Products

A consideration of the role of oxidative processes in increasing the PVC dehydrochlorination rate (with the values of $\bar{\gamma}_0$ unchanged) and the linear dependence $W_{HCl} = f(\bar{\gamma}_0)$ (Fig. 1) leads to the conclusion that the structures responsible for the decrease in PVC stability should necessarily form with the participation of oxygen atoms. At the same time it should be taken into account that the experimentally observed rate constant of the PVC dehydrochlorination activated by internal $>C=C<$ bonds ($k_p \approx 10^{-2}$ s^{-1}) is in a good agreement with those of HCl elimination from low-molecular model compounds containing conjugated double bonds ($\sim$CH$=$CH$-$CH$=$CH$-$CHCl$-$CH$_2\sim$) (Table 1). However, as noted above, conjugated bonds ($>C=C<)_{n>1}$ are not present in PVC.[23-25]

The experimental facts presented, taken as a whole, lead to the suggestion that real PVC samples should comprise conjugated oxygen-containing unsaturated groups, e.g. $\sim$C(O)$-$CR$=$CH$-$CHCl$-$CH$_2\sim$ (where R stands for H or branches of any length). These should form· easily in the course of polymer formation and on storage, with the oxidation by atmospheric oxygen of hydrogen atoms β to solitary

$>$C$=$C$<$ bonds, involving the formation of unsaturated ketone fragments (KAG)[25,32,52] (reaction 3).

$$\text{mCH}_2\text{—CH}=\text{CH—CHClm} \xrightarrow{\text{O}_2} \text{mCH}\overset{\displaystyle \text{OOH}}{\underset{\displaystyle |}{-}}\text{CH}=\text{CH—CHClm} \longrightarrow$$

$$\text{mC}\overset{\displaystyle \text{O}}{\underset{\displaystyle ||}{-}}\text{CH}=\text{CH—CHClm} \quad (3)$$

In principle other oxygen-containing groups, such as hydroperoxides or alcohols, including the product of keto-enol arrangement of the keto-allyl group, may be present in PVC macromolecules and may lead to activation of PVC decomposition. Elucidation of their exact nature and reactivity is the problem to be solved in the near future.

A systematic study of numerous commercial and laboratory PVC samples has unequivocally shown that normally all macromolecules contain a noticeable quantity of internal $>$C$=$C$<$ bonds, most of them being KAG (CAG is almost absent) (Table 2).

4. EXPERIMENTAL EVIDENCE FOR THE EXISTENCE OF KETO-ALLYL CHLORIDE GROUPS (KAG) IN PVC

The presence of keto-allyl chloride groups of the structure
$$\text{mC}\overset{\displaystyle }{\underset{\displaystyle ||}{-}}\text{CH}=\text{CH—CH}\overset{\displaystyle }{\underset{\displaystyle |}{-}}\text{m} \quad \text{(KAG) within PVC macromolecules is}$$
$$\text{O}=\text{CH—}\text{Cl}$$

supported by the following experimental facts, including the reactions that are specific to KAG:

(a) In the IR spectra of rather thick films (0·05–0·1 cm) weak absorption bands are observed at 1600 and 1675 cm^{-1} due to the stretching vibrations of conjugated carbonyl groups.[53]

(b) The PVC molecular weight diminishes after alkaline hydrolysis, a reaction (4) which is characteristic of unsaturated ketones.[23–25,30,54,55]

$$\text{mCH}=\text{CH—}\overset{\displaystyle \text{O}}{\underset{\displaystyle ||}{\text{C}}}\text{—CHClm} \xrightarrow[\text{KOH}]{\text{H}_2\text{O}} \text{mC}\overset{\displaystyle O}{\diagdown}_{\text{H}} + \text{CH}_3\text{—}\overset{\displaystyle }{\underset{\displaystyle ||}{\text{C}}}\text{—CHClm} \quad (4)$$

TABLE 2
CONTENTS OF C–C DOUBLE BONDS IN PVC MACROMOLECULES

Sample	$\bar{M}_{\eta}^{\circ} \times 10^{-3}$	$>\!C\!=\!C\!<$ bond content (10^4 mol/mol PVC)					Remarks
		$\Sigma[>\!C\!=\!C\!<]$	Internal $>\!C\!=\!C\!<$ bonds, $\bar{\gamma}_0$		UEG^a	PCS	
			CAG	KAG			
Block (mass) PVC	61.5	53.0	0	2.0	51.0	0	
	83	47.0	0	0.2	46.8	0	
	100	34.0	0	1.0	33.0	0	
	108	28.0	0	1.2	26.8	0	
	108	30.0	1.0	0.3	28.7	0	Polymerised in the absence of O_2
	120	35.0	1.2	0.1	33.7	0	
	290	50.0	0	0.2	49.8	0	Polymerised in the absence of O_2
Emulsion PVC	68	21.0	0	2.3	18.3	0	
	96	18.0	0	0.8	17.3	0	
	106	20.0	0	1.2	18.8	0	
	106	21.9	0	0	18.8	3.1	Coloured product
Suspension PVC	112	18.0	0	0.7	17.3	0	
	125	20.0	0	0.5	19.5	0	
	132	20.0	0	1.7	18.7	0	
	61.5	48.0	0	2.0	46.0	0	
	79.5	26.0	0	1.6	24.4	0	
	106	44.0	0	1.0	43.0	0	
	132	22.0	0	1.7	20.3	0	
	140	33.0	0	0.8	32.2	0	
	140	40.0	1.0	0	32.2	6.8	$TT^b = 20$
	140	48.0	1.9	0	32.2	13.9	$TT^b = 40$
	140	53.0	2.9	0	32.2	17.9	$TT^b = 60$
	140	60.0	3.8	0	32.2	24.0	$TT^b = 80$

[a] Abbreviations: UEG, unsaturated end-groups; PCS, polyconjugated systems.
[b] TT, thermodegradation time (min) at 448 K, 10^{-2} Pa.

This is accompanied by the appearance of a strong absorption band in the region of 1715 cm^{-1} in the IR spectra (stretching vibrations of the end $>$C$=$O group).[53] It is important to note that β-chloroallyl groups, $\sim\sim$CH$_2$—CH$=$CH—CHCl—CH$_2\sim\sim$, do not degrade during alkaline hydrolysis. For example, in the course of the PVC thermal dehydrochlorination in the absence of oxygen (10^{-2}–10^{-3} Pa) a continuous increase in the content of internal $>$C$=$C$<$ bonds (as determined by a change in the PVC molecular weight after oxidative ozonolysis[23-25,28] is always observed. It occurs at the expense of a random dehydrochlorination of PVC macromolecules with the formation of β-chloroallyl groups via reaction 1. During alkaline hydrolysis, however, the number of macromolecular ruptures does not change in the course of degradation[23] and remains equal to the content of internal $>$C$=$C$<$ bonds in the initial PVC samples ($\bar{\gamma}_0$) (Fig. 2).

(c) Another typical reaction of unsaturated ketones is the nucleophilic addition of organic phosphites in the presence of proton donors at temperatures above 300–330 K. It involves the attack of phosphite on the β-carbon atom of a conjugated system of KAG with subsequent protonation of the bipolar ion and formation of a stable ketophosphonate[56,57] (reaction 5).

$$\tag{5}$$

The formation of ketophosphonate structures within PVC macromolecules via reaction 5 leads to the removal of internal unsaturated $>$C$=$C$<$ bonds. The consequence is that oxidative ozonolysis (as well as alkaline hydrolysis) of the polymer product no longer results in macromolecular degradation and the respective decrease in the PVC molecular weight (Table 3).

It is significant that reaction 5 does not occur with CAG ($\sim\sim$CH$_2$—CH$=$CH—CHCl—CH$_2\sim\sim$). The model reaction of 4-chloropent-2-ene with tributylphosphite (350–450 K) serves as an example.[56] Similarly,

TABLE 3

VARIATIONS IN THE NUMBER OF INTERNAL DOUBLE BONDS IN PVC DURING INTERACTION WITH ORGANIC PHOSPHITES AT 10^{-2} Pa

Composition	Molar ratio $\dfrac{OPh}{PVC^a}$	Time (h)	Tempera- ture (K)	Intrinsic vis- cosity, $[\eta]$ (dl/g)		$\bar{\gamma}_0 \times 10^4$ (mol/mol PVC)
				Before ozonolysis	After ozonlysis	
PVC-1	—	—	—	1·23	1·90	1·0
PVC-1+OPh-1	0·02	2	377	1·23	1·23	0
PVC-1+OPh-1	0·02	8	373	1·23	1·22	0
PVC-1+OPh-2	0·02	5	353	1·23	1·23	0
PVC-2	—	—	—	1·03	0·92	1·1
PVC-2+OPh-1	0·015	11	353	1·03	1·03	0

^{a} Abbreviations: PVC-1, $\bar{M}_n^\circ = 127\,500$;   PVC-2, $\bar{M}_n^\circ = 99\,000$;   OPh-1,

$$H \left[O \!-\!\!\left\langle\!\bigcirc\!\right\rangle\!\!-\! C(CH_3)_2 \!-\!\!\left\langle\!\bigcirc\!\right\rangle\!\!-\! O \!-\! P(OC_8H_{17}\text{-i}) \right]_n OC_6H_5; \; OPh\text{-}2, \; C_6H_{13}OP(OC_{10}H_7\text{-}\alpha)_2.$$

the method of competing reactions of a phosphite (350–360 K) with a 1 : 1 (mol) mixture of methyl vinyl ketone (a KAG model) and 4-chloropent-2-ene (a CAG model) has demonstrated that organic phosphites react readily and selectively with methyl vinyl ketone, while 4-chloropent-2-ene remains unchanged after the reaction, apart from a certain amount (<5–10%) of dehydrochlorination.[56] Dialkyl-3-oxo-butylphosphonate, $CH_3\!-\!\underset{O}{\overset{\parallel}{C}}\!-\!CH_2\!-\!\underset{O}{\overset{\parallel}{P}}(OR)_2$, has been identified as the principal product (65–70 mass %) whose structure was confirmed by elemental analysis as well as by IR spectroscopy (absorption bands at 1280 cm^{-1}, ($\geqslant$P=O) and 1730 cm^{-1} ($>$C=O)). Trialkyl and alkylaryl phosphites are more active in this reaction than triaryl phosphites.

(d) Independent evidence for the presence of KAG ($\sim\!\!\sim$C(O)—CH=CH—CHCl—CH$_2\sim\!\!\sim$) groups in PVC is the ability of the polymer to react with conjugated dienes by the Diels–Alder reaction, unknown in PVC until quite recently. Electron-accepting groups in CAG act as dienophiles in reaction 6:

$$\underset{\sim\!\!\sim}{\overset{O}{\overset{\parallel}{C}}}\!-\!CH\!=\!CH\!-\!CHCl\sim\!\!\sim + \underset{\overset{\parallel}{CH}}{\overset{HCR}{}}\!-\!\underset{\overset{\parallel}{CH}}{\overset{HCR'}{}} \longrightarrow \underset{\sim\!\!\sim}{\overset{O}{\overset{\parallel}{C}}}\!-\!CH\!-\!CH\!-\!CHCl\sim\!\!\sim$$

$$\overset{HCR \qquad HCR'}{\underset{CH=CH}{\diagdown\qquad\diagup}} \qquad (6)$$

TABLE 4

CHANGE IN NUMBER OF $\sim\!\!\sim\!\!C(O)$—CH=CH— GROUPS IN PVC DURING IN-
TERACTION WITH CONJUGATED DIENES (353 K, 10^{-2} Pa)

Compound	Molar ratio diene PVC	Time (h)	Intrinsic viscosity, $[\eta](dl/g)$			$\bar{\gamma}_0 \times 10^4$ (mol/mol PVC)
			Before ozon- olysis	After ozon- olysis	After hydro- lysis	
PVC-1 ($\bar{M}^\circ_n = 129,600$)	—	—	1·24	1·16	1·17	1·03
PVC-1 + 5-methyl- hepta-1,3,6-triene	0·01	2	1·24	1·24	1·23	0·09
PVC-1 + 3-methylene-6, 10-dimethylunde- ca-1,5,10-triene	0·01	2	1·24	1·23	1·23	0·09
PVC-1 + 2,5,10,15-tetra- methylhexade- ca-1,5,7,9,15-pentene	0·01	2	1·24	1·23	1·24	0
PVC-1 + cyclopentadiene	0·01	2	1·24	1·24	1·24	0
PVC-1^a + cyclopentadiene	0·02	2	1·24	1·10	1·24	0
PVC-2 ($\bar{M}^\circ_n = 105,000$)	—	—	1·07	1·02	1·02	0·76
PVC-2 + piperylene	0·02	2	1·07	—	1·07	0
PVC-2 + cyclopentadiene	0·01	1	1·07	1·07	1·07	0
PVC-2 + isoprene	0·01	2	1·07	1·07	1·06	0·13

^{a} PVC predegraded in vacuo (448 K, 15 min).

In particular, the interaction of PVC with cyclopentadiene, piperylene,
isoprene, 5-methylhepta-1,3,6-triene and related compounds containing conjugated $>$C=C$<$ bonds (350 K, 0·5–1 h) results, as in the case
of organic phosphites, in the elimination of unsaturated groups within
the PVC macromolecules, so it is easy to monitor the process by the
method of oxidative ozonolysis and alkaline hydrolysis of the treated
polymer (Table 4).

5. KINETIC ASPECTS OF THE THERMAL DEHYDROCHLORINATION OF PVC

The overall PVC dehydrochlorination process is a multistage sequence
of parallel and successive reactions. Two processes are easily identified
on an experimental basis: random dehydrochlorination (reaction 1)
proceeding at the rate W_r, and the formation of polyene sequences
proceeding at the rate W_p. In the general case (Table 5) the overall

rate is given by eqn. (iv).

$$W_{HCl} = W_r + W_p \tag{iv}$$

In accordance with experimental facts the rate W_r is practically independent of the origin of the polymer (bulk, emulsion, suspension, etc.),[24,25,28,30,59–61] and of its molecular weight, and is therefore constant. This is a fundamental property of PVC which shows that all vinyl chloride macromolecular units take an equal part in the process of random HCl elimination, while the rate of formation of the polyene system, W_p, may vary considerably.

TABLE 5

KINETIC PARAMETERS OF PVC DEHYDROCHLORINATION (448 K, 10^{-2}–10^{-3} Pa)

$\bar{\gamma}_0 \times 10^4$ (mol/mol PVC)	$W_{HCl} \times 10^6$ (mol HCl/ mol PVC s)	$W_r \times 10^7$ (mol HCl/ mol PVC s)	$W_p \times 10^6$ (mol HCl/ mol PVC s)	$\bar{M}_n^\circ \times 10^{-3}$
0·2	0·22	0·77	0·14	83
0·2	0·22	0·88	0·13	293
0·7	0·53	0·73	0·47	215·5
0·8	0·71	0·83	0·63	137·5
0·8	0·70	0·78	0·62	109
0·9	0·72	0·78	0·64	104
1·0	0·82	0·76	0·74	138
1·1	0·88	0·85	0·80	70
1·2	0·92	0·70	0·85	106
1·2	1·04	0·86	0·95	138·5
1·3	1·03	0·83	0·95	96
1·3	1·04	0·73	0·97	122
1·3	1·05	0·78	0·97	109
1·4	1·12	0·81	1·04	115
1·4	1·12	0·85	1·04	94·5
1·4	1·15	0·88	1·06	116
1·5	1·15	0·79	1·07	114·5
1·5	1·20	0·83	1·12	118
1·5	1·16	0·77	1·08	108
1·5	1·16	0·79	1·08	105
1·6	1·20	0·78	1·12	117·5
1·7	1·26	0·80	1·18	102·5
1·8	1·45	0·83	1·37	116
1·9	1·48	0·87	1·39	124·5
2·0	1·52	0·87	1·43	61·5
2·0	1·57	0·81	1·49	109

TABLE 6
KINETIC PARAMETERS OF PVC DEHYDROCHLORINATION REACTIONS (10^{-2}– 10^{-3} Pa)

Tempera-ture (K)	$k_r \times 10^8$ (s^{-1})	$k_p \times 10^4$ (s^{-1})	E^r_{act} (kJ/mol)	$log A_r$	E^p_{act} (kJ/mol)	$log A_p$
448	8·0	75				
438	3·7	30	$90 \pm 6·5$	$3·4 \pm 0·7$	147 ± 7	$14·9 \pm 0·8$
428	2·2	10				
418	1·3	4·8				

The kinetic parameters of PVC dehydrochlorination reactions proceeding at the rates of W_r and W_p are summarised in Table 6. It seems very pertinent that the study of hundreds of commercial and laboratory samples, irrespective of their methods of formation, has revealed a strict correspondence between the dehydrochlorination rate W_p (W_r being a constant value) and the values of $\bar{\gamma}_0$ calculated from the variation in the PVC molecular weight during alkaline hydrolysis, which measures the content of internal unsaturated oxygen-containing groups, KAG[24,25,28,30,61,62] (Fig. 1, straight line 2).

As can be seen, the straight line extrapolates to zero. This is an indication of the fact that the end-groups, including the unsaturated ones (UEG) whose content in PVC is an order higher than that of KAG (see Table 2), are not so active with respect to the specific reaction of the polyene sequence formation as the in-chain unsaturated groups ∿C(O)—CH=CH—CHCl—CH$_2$∿.

On the basis of the above facts the process of PVC dehydrochlorination can generally be described by four parallel–successive reactions.[24,25,28,30,62]

(1) Random HCl elimination from normal PVC units, involving the formation of β-chloroallyl groups with the rate constant k_r (reaction 1).

$$k_r = 0·8 \times 10^{-7} \text{ s}^{-1} \text{ (448 K)} \qquad \text{(v)}$$

(2) HCl elimination involving the formation of polyconjugated systems of double bonds ($\rangle C=C \langle)_n$ initiated by internal KAG groups (reaction 7) with rate constant k_p given by eqn (vi).

∿C(O)—CH=CH—CHCl—CH$_2$Cl—CH$_2$—CHCl∿ →

∿C(O)—CH=CH—CH=CH—CHCl∿ + HCl   (7)

$$k_p = 0·75 \times 10^{-2} \text{ s}^{-1} \text{ (448 K)} \qquad \text{(vi)}$$

(3) Slow HCl elimination, involving the formation of polyconjugated bonds $(>C=C<)_n$, and activated by β-chloroallyl
$\sim\!CH_2-CH=CH-CHCl\sim$ groups, with the rate constant k_{p1}
(reaction 2).

$$k_{p1} \simeq 10^{-5}-10^{-4}\,s^{-1}\,(448\,K)^{3,4} \qquad \text{(vii)}$$

It has been shown recently that HCl elimination from a
halide-containing polymer product of the structure $\sim\!CH_2-$
$C(CH_3)=CH-CHCl-C(CH_3)_2\sim$ proceeds with the rate constant $\approx\!10^{-5}\,s^{-1}\,(448\,K)$.

(4) HCl elimination involving the formation of $(>C=C<)_n$ and
activated by in-chain $(>C=C<)_n$ bonds, with the rate constant k_{p2}, eqn. (viii), which is in agreement with the linear
dependences $[HCl] = f(t)$, $(W_p = \text{constant})$.

$$\sim\!CH_2-(CH=CH)_2-CHCl-CH_2\sim \rightarrow$$
$$\sim\!CH_2-(CH=CH)_3-\sim + HCl \qquad (8)$$
$$k_{p2} \simeq 10^{-2}\,s^{-1}\,(448\,K)^{3,4} \qquad \text{(viii)}$$

In accordance with this scheme PVC dehydrochlorination can be
represented as follows:

$$\text{PVC} \xrightarrow{k_r} A_1 \xrightarrow{k_{p1}} A_2 \Big\langle \begin{matrix} \xrightarrow{k_t} \text{non-active products} \\ \xrightarrow{k_{p2}} \text{active products } (A_2) \end{matrix} \qquad (9)$$

where k_t is the growth termination constant for $(>C=C<)_n$.
If in the initial PVC, the content of internal isolated β-chloroallyl
groups (A_1) and that of conjugated $\sim\!\underset{\underset{O}{\|}}{C}-CH=CH\sim$ and

$\sim\!(CH=CH)_{n>1}\sim$ bonds (A_2) are A_1^0 and $A_2^0 = \gamma_0$, respectively, a change
in their concentration in the course of PVC dehydrochlorination, with
reference to the termination of the polyene growth kinetic chains, can
be represented by eqns. (ix) to (xi).

$$\frac{dA_1}{dt} = k_r a_0 - k_{p1} A_1 \qquad \text{(ix)}$$

$$\frac{dA_2}{dt} = k_{p1} A_1 - k_t A_2 \qquad \text{(x)}$$

Hence

$$A_1 = \frac{k_r a_0}{k_{p1}}[1 - \exp(k_{p1}t)] + A_1^0 \exp(-k_{p1}t) \qquad \text{(xi)}$$

where a_0 denotes HCl content in the initial PVC. Substituting the values of A_1 from (x) in (xi) we arrive at

$$\frac{dA_2}{dt} = k_r a_0[1 - \exp(-k_{p1}t)] + k_{p1}A_1^0 \exp(-k_{p1}t) - k_t A_2 \qquad \text{(xii)}$$

Hence

$$A_2 = \frac{k_r a_0 - k_{p1}A_1}{k_{p1} - k_t}\exp(-k_{p1}t)$$

$$+ \left(\bar{\gamma}_0 - \frac{k_r a_0 - k_{p1}A_1^0}{k_{p1} - k_t} - \frac{k_r a_0}{k_t}\right)\exp(-k_t t) + \frac{k_r a_0}{k_t} \qquad \text{(xiii)}$$

Assuming, as stated above, that the rate constants for the formation of $(>\!C\!=\!C\!<)_n$ initiated by KAG (k_p) and PS (k_{p2}) are approximately equal, the overall PVC dehydrochlorination rate can be described by eqn. (xiv)

$$\frac{d[\text{HCl}]}{dt} = k_r a_0 + k_{p1}A_1 + k_{p2}A_2 \qquad \text{(xiv)}$$

Substitution of A_1 and A_2 values from (xi) and (xii) into (xiv) gives eqn. (xv).

$$[\text{HCl}] = \left(2k_r a_0 + \frac{k_{p2}k_r a_0}{k_t}\right)t - \left(\frac{k_r a_0 - k_{p1}A_1^0}{k_{p1}}\right)\left(\frac{k_t + k_{p2} - k_{p1}}{k_t - k_{p1}}\right)$$

$$\times [1 - \exp(-k_{p1}t)] + \frac{k_{p2}}{k_t}\left(\bar{\gamma}_0 - \frac{k_r a_0}{k_t} - \frac{k_r a_0 - k_{p1}A_1^0}{k_{p1} - k_t}\right)$$

$$\times [1 - \exp(-k_t t)] \qquad \text{(xv)}$$

Since normally $A_1^0 = 0$, and $A_2^0 = \bar{\gamma}_{0>}$ at least at the initial stages, $k_t t \ll 1$, and $k_{p1}t \ll 1$, eqn. (xv) is reduced to (xvi).

$$[\text{HCl}] = k_r a_0 t + k_p \bar{\gamma}_0 t \qquad \text{(xvi)}$$

It is significant that eqn. (xvi) defines the experimentally observed linear dependence of W_{HCl} and W_p on the KAG content in macromolecules, while the set of parameters k_r, k_p and $\bar{\gamma}_0$ yields a satisfactory description of the overall PVC dehydrochlorination kinetics irrespective of the conditions and method of its synthesis.

The total combination of experimental facts, together with the kinetic mechanisms in an admissible approximation, shows that unsaturated oxygen-containing groups of the KAG type are practically the only labile groups of importance within the PVC macromolecules. This is also responsible for the considerable progress made in the solution of the principal enigma of PVC, namely the causes of its own abnormally low stability, as well as providing the basis for understanding the mechanism and kinetics of PVC ageing.

6. CROSS-LINKING OF PVC MACROMOLECULES

Apart from macromolecular dehydrochlorination, the formation of cross-linked structures is observed during PVC degradation. The formation of a cross-linked insoluble polymer product (gel) is preceded by an induction period τ_g. It is significant that if the catalytic influence of active sites on the macrochain cross-linking process is left out, the induction period before gelation is also determined by the content of keto-allyl groups in the polymer (Fig. 4).[24,28,29,63]

These results indicate that the formation of cross-links between macromolecules occurs through interaction between blocks of poly-conjugated systems (PCS) of different lengths, since each keto-allyl group yields a block of $(>C=C<)_n$ bonds during τ_g. The process of intermolecular cross-link (C) accumulation is described by eqn. (xvii)

$$\frac{dC}{dt} = k_{cross}\,[\text{PCS block}]^n = k_{cross}\bar{\gamma}_0^n \qquad \text{(xvii)}$$

where n denotes the reaction order according to the concentration of the polyene blocks and equals 2 (as determined via logarithmic transformation of the dependence $\tau_g = f(\bar{\gamma}_0)$).

At time τ_g, omitting the consumption of PCS blocks,

$$C_g = k_{cross}\bar{\gamma}_0^2\tau_g \qquad \text{(xviii)}$$

where C_g is the critical content of cross-linkages between PVC macrochains that gives rise to the appearance of gel in the system.

According to Flory's statistical theory[64] the necessary and sufficient condition for the formation of an infinite network from a system of macromolecules with any initial molecular mass distribution (MMD) lies in satisfying the relation

$$X_{cr} = 1/(\bar{P}_w - 1) \approx 1/\bar{P}_w \qquad \text{(xix)}$$

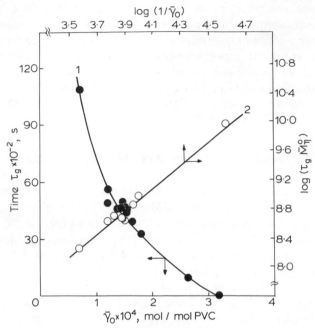

FIG. 4.   Dependence of gelation time on KAG content during PVC thermal degradation (448 K, 10^{-2} Pa) (1), and logarithmic transformation of this dependence (2).

where X_{cr} denotes the critical probability that any unit of the primary polymeric chain has formed a cross-bond.

Moving on to average molecular weights and bearing in mind that the values of the PVC average-viscosity molecular weight, $\bar{M}_\eta^0$, are easily determined experimentally, C_g can be expressed by eqn. (xx):

$$C_g = \frac{mZ'}{Z\bar{M}_\eta^\circ} \qquad (\text{xx})$$

where m represents the molecular weight of the $\sim\!\!CHCl\!-\!CH_2\!\sim$ unit, $Z = \bar{M}_w^\circ/\bar{M}_n^\circ = 2$, and $Z' = \bar{M}_\eta^\circ/\bar{M}_n^\circ = 1\cdot86$.

Combining eqns. (xviii) and (xx) yields the expression (xxi) that makes it possible to calculate the rate constant for macromolecular cross-linking, k_{cross}, for PVC fractions with any initial MMD:

$$k_{cross} = \frac{mZ'}{Z\bar{M}_\eta^\circ\bar{\gamma}_0\tau_g} \qquad (\text{xxi})$$

The calculated values of k_{cross} are listed in Table 7.

TABLE 7

KINETIC PARAMETERS FOR CROSS-LINK FORMATION DURING THE THERMAL DE-
GRADATION OF PVC (*in vacuo*, 10^{-2} Pa)

$\bar{M}_n^{\circ} \times 10^{-3}$	$\bar{\gamma}_0 \times 10^4$ (mol/mol PVC)	Tempera-ture (K)	$\tau_g \times 10^{-3}$ (s)	k_{cross} (mol PVC $mol^{-1} s^{-1}$)
140	0·7	448	10·74	7·1
152	1·0	448	5·16	7·4
92	1·4	448	3·96	7·2
115	1·4	448	4·32	6·7
95	1·4	448	4·56	6·9
103	1·6	448	3·36	6·8
116	1·8	438	4·20	3·7
116	1·8	448	2·70	5·7
116	1·8	458	2·40	6·5
116	1·8	468	1·68	9·2
152	2·7	448	0·90	5·9

$E_{act}^{cross} = 50 \pm 8$ kJ/mol; $\log A^{cross} - 6·3 \pm 0·5$.

The content of cross-links accumulating during PVC thermodegradation, as calculated by eqn. (xvii), coincides satisfactorily with
their measurement by an independent method, equilibrium swelling
(O) of gel in cyclohexanone at 298 K (Table 8). This is evidence for
the adequacy of the kinetic scheme suggested and for the real process
of PVC macromolecular cross-linking. From the point of view of
mechanism of cross-linkage formation, the following experimental
evidence is in favour of the Diels–Alder reaction.[6,65–67]

(1) Increasing τ_g in the presence of active dienophiles (e.g. the
    derivatives of maleic acid) that compete with polyenes in the
    diene condensation reaction.[65,66]

(2) The presence of dienophilic keto-allyl groups in PVC capable
    of reacting with conjugated dienes.[58]

(3) The similarity of the kinetic parameters for the PVC cross-
    linking reaction during thermal degradation *in vacuo* to the
    values known for the Diels–Alder reaction ($E_{act} = 40$–
    80 kJ/mol; $\log A = 5$–7). In all probability each block of con-
    jugated $>C=C<$ bonds partakes at least twice in the cross-
    linking reaction: first as a diene, then as a dienophile (due to
    the presence of KAG), which ensures the required condition
    for the formation of a three-dimensional network, i.e. two
    cross-links per primary polymer chain.

TABLE 8

CONTENT OF CROSS-LINKAGES IN THERMALLY DEGRADED PVC at 448 K, 10^{-2} Pa ($\bar{M}_n^\circ = 116\,000$; $\bar{\gamma}_0 = 1\cdot8 \times 10^{-4}$ mol/mol PVC)

Network parameters of cross-linked PVC[a]	Degradation conditions, $t \times 10^{-3}$ $(s)/T(K)$					
	1·8/468	3·6/468	4·8/458	5·4/458	9·0/438	18/438
$\bar{M}_c \times 10^{-5}$	0·59	0·25	0·99	0·46	1·12	0·42
$n_c \times 10^3$	0·54	1·07	1·01	1·14	1·08	2·16
$\nu_c \times 10^{-20}$	6·5	12·8	12·12	13·68	12·96	25·92
γ_c	1·05	2·48	0·63	1·35	0·55	1·48

[a] Abbreviations: n_c denotes the content of cross-linkages (mol/mol PVC); ν_c is the number of cross-links per unit volume; γ_c is the cross-linking index, i.e. the number of cross-linkages per one macromolecule; $\bar{M}_c$ stands for the molecular weight of the polymer chain section contained between cross-links, to be found from the Flory-Rehner relation:

$$\bar{M}_c = 2\rho V_1 / [V_2^{\frac{1}{3}}(1-\mu)]$$

where $V_2 = 1/(1 + Q\rho/d)$; V_1 and d stand for the solvent molar volume and density, respectively; μ is the thermodynamic parameter of the polymer–solvent interaction ($\mu = 0\cdot38$–$0\cdot40$); ρ is the polymer density.

Thus the process of PVC degradation, as well as the kinetics of the polymer dehydrochlorination, are determined by the concentration of keto-allyl groups in the macromolecules. Consequently the problem of inhibiting the macrochain cross-linking reactions is directly associated with elimination of labile abnormal groups present in the polymer. It should be noted that 8–10% of the macromolecules contain labile keto-allyl groups. However, this proportion cannot alone be responsible for the experimentally observed gel yield and, ultimately, for the transformation of PVC into a completely cross-linked product. The relatively high rate of gel accumulation is evidently accounted for by the increasing contribution of other functional macromolecular groups to PVC structurisation as the polymer dehydrochlorination process intensifies. They include the polyconjugated sequences (PCS) whose formation is initiated by CAG. Although the formation of PCS blocks, activated by the above groups, proceeds at a much lower rate than under the influence of KAG, their role in the increase of polyene concentration becomes very important at sufficiently long times of PVC thermodegradation.

7. FORMATION OF POLYENE SEQUENCES DURING THE DEGRADATION OF PVC

PVC degradation is always accompanied by loss of initial colour resulting from the formation of chromophoric groups in the polymer macromolecule. Discolouration is one of the most important factors that may account for the reduced service life of many polymeric products.

In analysing the kinetics of polyene formation and distribution according to their length, as well as in describing the processes of PVC dehydrochlorination and cross-linking, an investigator should proceed from the presence of keto-allyl groups in PVC. While activating HCl elimination during PVC thermodegradation, KAGs become incorporated into the forming polyenes, affecting their specific properties; in particular their ability to absorb light selectively in the near-UV and visible spectral ranges.

Ultraviolet/visible spectrophotometry makes it possible to study experimentally the kinetics of polyene formation during PVC degradation.[10,14,42,68–70] It is accepted[69] that a poorly resolved absorption spectrum of a degraded PVC sample is a superposition of the spectra of individual polyenes and is difficult to interpret. A rational method of obtaining information from poorly resolved spectra lies in simulating the form of the spectral line on a computer, as well as in selecting the 'best' parameters for the model to reach a minimum difference between the experimental and simulated spectra, and in identifying the parameters of the 'best' model with those of the simulated poorly resolved spectrum.[69–71].

Simulation of spectra requires data, obtained under comparable conditions, on the wavelengths, λ, and the extinction coefficients, ε, of the main absorption maximums for all the polyenes that contribute to the formation of the PVC absorption spectrum. Within the range 300–520 nm, where the electronic spectrum of PVC is located, the light is absorbed by $\sim CH_2-(CH=CH)_n\sim$ polyenes with $n = 3–14$ and polyenes of the type $O=C-(CH=CH)_n\sim$ with $n = 2–11$. A reduction

in the length of oxygen-containing polyenes absorbing light within a definite range results from the bathochromic effect of the $>C=O$ group, its value depending on the number of conjugated double bonds, n. It should be noted that the literature contains rather limited information on the spectral characteristics suitable for identification of

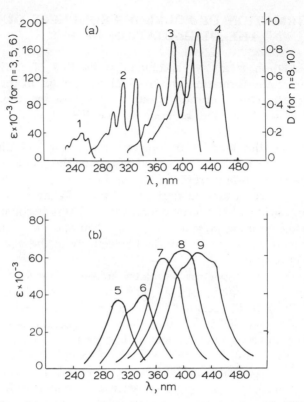

FIG. 5.   Ultraviolet spectra of model polyenes of the type H–(CH=CH)ₙ–H (a) and CH₃–(CH=CH)ₙ–CHO (b). Peaks 1 and 5, $n = 3$; peak 6, $n = 4$; peaks 2 and 7, $n = 5$; peak 8, $n = 6$; peak 9, $n = 7$; peak 3, $n = 8$; peak 4, $n = 10$.

individual polyene compounds,[73-77] and the missing data required for the construction of model spectra (Fig. 5) are obtained by calculations on the basis of the Lewis–Calvin equation ($\lambda^2 = kn$),[78] as well as by using the linear dependence $\varepsilon = f(n)$. According to ref. 69, for example,

$$\varepsilon_{max} = 32\ 821 \cdot 43n - 52\ 599 \cdot 88 \qquad \text{(xxii)}$$

For the optical density D at point λ_i of a spectral curve the relation (xxiii) holds:

$$D(\lambda_i) = l \sum_{n=1}^{N} c_n \varepsilon_n(\lambda_i) \qquad \text{(xxiii)}$$

where $\varepsilon_n(\lambda_i)$ is the extinction coefficient for polyenes of length n at point λ_i, c_n denotes the concentration of individual polyenes containing n double bonds, N is the number of polyenes, and l is the solution layer thickness.

If an unknown concentration of an individual polyene sequence, c_n, is selected as the variable parameter of the simulated spectral function and the sum of the deviation squares at all the spectral-function points investigated is assumed as the formal criterion for comparing the experimental and model spectra, the problem of simulating is reduced to the search for certain parameters c_n to reach the best coincidence possible of the simulated $\sum\limits_{i=1}^{k} D^{\mathrm{mod}}(\lambda_i)$ and experimental $\sum\limits_{i=1}^{k} D^{\mathrm{exp}}(\lambda_i)$ spectra (k is the number of points in a tabulated spectrum), i.e. it amounts to minimisation of the value

$$\sum_{t=1}^{k} [D^{\mathrm{mod}}(\lambda_t) - D^{\mathrm{exp}}(\lambda_i)]^2 \qquad \text{(xxiv)}$$

The calculated results show (Fig. 6) that a satisfactory coincidence of the experimental and calculated spectral curves within a wide enough range of wavelengths is achieved only by superposition of the simu-

lated spectra for $\mathrm{H\text{-}(CH\!\!=\!\!CH)_n\text{-}H}$ and $\mathrm{CH_3\text{-}(CH\!\!=\!\!CH)_n\text{-}C}\!\!\overset{\displaystyle O}{\underset{\displaystyle H}{\diagdown}}$

polyene types. An independent use of both models does not make it possible to reproduce satisfactorily the experimental UV/visible spectrum of a degraded PVC.

Thus it can be asserted with some certainty that polyenes of at least two types, $\mathrm{\text{-}(CH\!\!=\!\!CH)_n\text{-}}$ and $\overset{\displaystyle O}{\diagup}\mathrm{C\text{-}(CH\!\!=\!\!CH)_n\text{-}}$, contribute to the UV/visible spectrum of a degraded PVC, their spectral characteristics being essentially different.

According to the above experimental evidence, during the initial stages of PVC thermodegradation ($(3\cdot6\text{-}5\cdot4)\times10^3$ s at 448 K) no termination of the HCl elimination kinetic chains is observed, which suggests the formation of rather long polyene sequences of structure

$\mathrm{O\!\!=\!\!\overset{\displaystyle \xi}{C}\text{-}(CH\!\!=\!\!CH)_n\text{-}}$ ($n \approx 25\text{-}30$ in $3\cdot6\times10^3$ s at 448 K, with reference to the established value of $k_p = 0\cdot75\times10^{-2}\,\mathrm{s}^{-1}$). Consequently it seems quite probable that the polyenes of various lengths and different

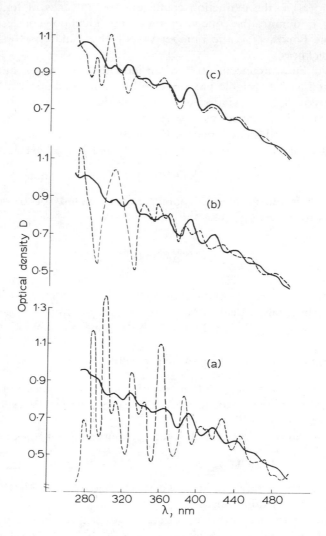

FIG. 6. Experimental UV/visible spectra of degraded (448 K, 10^{-2} Pa, $3 \times$ 10^3 s) PVC and spectra obtained via computer simulation on the basis of the spectra of individual polyenes of the type H–(CH=CH)$_n$–H (a), CH$_3$–(CH=CH)$_n$–CHO (b), and their superpositions (c). —— Experimental curves; – – – – calculated curves.

$$\sim\!\!C(O)\!-\!CH\!=\!CH\!-\!CH\!=\!CH\!-\!CH\!=\!CH\!-\!CH\!=\!CH\!-\!CH\!=\!CH\!-\!(CH\!=\!CH)_n\!-\!\sim$$

$$+$$

$$\sim\!\!(CH\!=\!CH)_m\!-\!CH\!=\!CH\!-\!CH\!=\!CH\!-\!CH\!=\!CH\!-\!C(O)\!\sim$$

$$\longrightarrow$$

$$\tag{10}$$

P_1 P_2 P_3

$\sim\!\!C(O)\!-\!CH\!=\!CH\!-\!CH\!=\!CH\!-\!CH\!=\!CH\!-\!CH$

$\quad CH\!=\!CH \qquad CH\!-\!CH\!=\!CH\!-\!(CH\!=\!CH)_n\!-\!\sim$

$\quad CH\!-\!CH \qquad CH\!-\!C(O)\!\sim$

$-\!(CH\!=\!CH)_m\!-\!CH\!=\!CH\!-\!CH\!=\!CH\!-\!CH\!=\!CH$

chemical structures result from secondary reactions that disrupt the conjugation chains in polyene sequences of the expected length. The nature of these reactions can at present only be conjectured. In the first place macromolecular cross-linking reaction between blocks of conjugated bonds,[24,30,63,68] reaction 10, should be considered (see page 199) where P_1, P_2 and P_3 denote polyenes of different length and different chemical structures. It has been shown[68] that the formation of cross-links starts practically from the beginning of the dehydrochlorination process: that is, it can fully account for the disrupted polyene sequences even at low conversions of HCl.

Other secondary processes can also occur, for example HCl re-addition to polyenes.[44,79]

8.   CHEMICAL STABILISATION OF PVC

8.1.   Reactions of KAG

The abnormally high rate of PVC dehydrochlorination is attributed to the presence of keto-allyl chloride groups (KAG) of the type $\sim\sim$C(O)—CH=CH—CHCl$\sim\sim$ in the macromolecules. An irreversible decomposition of KAG rather than chemical modification of any of the three active sites $>$C=O, $>$C=C$<$ or $>$CH—CH$\sim\sim$ leads to a substantial
                                                                   |
                                                                   Cl
increase in the PVC stability. The reaction rate of the random HCl elimination from similar sequences of vinyl chloride units W_r is responsible for the lower values of polymer dehydrochlorination rate observed when the KAG is fully and effectively decomposed ($W_p = 0$).

A KAG is known to be rather reactive.

$$\overset{\overset{\displaystyle O \atop \|}{\delta^-}}{\sim\sim C \underset{\pi}{\leftarrow} CH} \overset{\delta^+}{=} CH - \overset{\delta^+}{CH}\sim\sim \atop | \atop Cl \tag{11}$$

The carbon atom β to the $>$C=O group possesses a considerable positive charge. This accounts for its electrophilic nature and its ability to interact with the reagents containing atoms with immobile electron pairs. At the same time the γ-carbon atom also lacks electrons. This facilitates an anionoid departure of the chlorine atom in substitution reactions.

A number of possibilities are therefore to be expected during PVC stabilisation in accordance with the KAG reactivity.

(1) An irreversible interaction of KAG with the stabiliser at $>\!C\!=\!O$ and $>\!C\!=\!C\!<$ bonds, involving the disruption of conjugation.

(2) A reversible reaction of the stabiliser at the $>\!C\!=\!C\!<$ bond, involving the disruption of conjugation.

(3) An irreversible substitution of labile chlorine atoms.

(4) A reversible substitution of labile chlorine atoms in KAG or CAG.

8.2. Irreversible Disruption of Conjugation in KAG

8.2.1. Saturation of Carbon–Carbon Double Bonds

If polyene sequences have already formed in the PVC macro-molecules the reaction of the stabiliser only at the initial keto-allyl group of the polyenyl fragment, $\sim\!\!\sim\!C(O)\!-\!CH\!=\!CH\!-\!(CH\!=\!CH)_{n>1}\!-\!CHCl\!\sim\!\!\sim$, will not lead to stabilisation since in this case the PVC dehydrochlorination process is activated by the sequence of conjugated $(>\!C\!=\!C\!<)_{n>1}$ bonds.

In the general case of double bond saturation in KAG during reaction with a stabiliser, the succession of reactions can be represented in the following way:

$$
\begin{array}{c}
\underset{(x_1)}{\overset{\displaystyle\overset{O}{\underset{\|}{}}}{-C-CH=CH-CHCl\!\sim\!\!\sim}}
\end{array}
\left\{
\begin{array}{l}
\xrightarrow[\;(+S)\;]{k} \quad \underset{(y_1)}{\overset{\displaystyle\overset{O}{\underset{\|}{}}}{-C-CH-CH-CHCl\!\sim\!\!\sim}}\\[2em]
\xrightarrow[-HCl]{k_P} \quad \underset{(x_2)}{\overset{\displaystyle\overset{O}{\underset{\|}{}}}{-C-(CH=CH)_n-CHCl-}}
\end{array}
\right. \qquad (12)
$$

$$
k \Big\downarrow{\scriptstyle +S} \qquad {\scriptstyle -HCl}\searrow{\scriptstyle k_P}
$$

$$
\underset{(y_2)}{\overset{\displaystyle\overset{O}{\underset{\|}{}}}{-C-CH-CH-(CH=CH)_{n-1}CHCl-}}
$$

If we confine ourselves to analysing the stabilisation processes during the initial stages of PVC degradation, then the accelerating effect of β-chloroallyl groups on the polyene growth (rate constant k_{p1}) can be neglected, as actually occurs in most cases, and the stabiliser (S) consumption can also be neglected, since in real PVC compositions the concentration c_0 is great compared with $\bar{\gamma}_0$ ($c_0/\bar{\gamma}_0 > 10$). The reaction scheme is then described by eqns. (xxv)–(xxviii).

$$\frac{d[HCl]}{dt} = k_r a_0 + k_p(x_1 + x_2 + y_2) \tag{xxv}$$

$$\frac{dx_1}{dt} = -kx_1 c_0 - k_p x_1 \tag{xxvi}$$

$$\frac{dx_2}{dt} = k_p x_1 - kx_2 c_0 \tag{xxvii}$$

$$\frac{dy_2}{dt} = kx_2 c_0 \tag{xxviii}$$

Since at the very beginning $x_1 = \bar{\gamma}_0$ and $x_2 = y_2 = 0$, the HCl yield versus time can be represented by eqn. (xxix).

$$[HCl] = \left(k_r a_0 + \frac{k_p^2 \bar{\gamma}_0}{k_p + kc_0}\right)t + \frac{k_p kc_0 \bar{\gamma}_0}{(k_p + kc_0)^2}[1 - \exp(k_p + kc_0)t] \tag{xxix}$$

The kinetic curves $[HCl] = f(t)$ calculated according to eqn. (xxix) for different samples $(k = 10^{-2}\text{–}10^2 \,(\text{mol/mol PVC})^{-1}\,\text{s}^{-1})$, as well as the dependence of the PVC dehydrochlorination rate on the initial concentration, c_0, of the stabilising reagent, are shown in Fig. 7a. A typical linear dependence of HCl yield on time is observed over the whole range of values considered for k and c_0. The effectiveness of the compounds reacting with labile groups will evidently be dependent on the relation between k and k_p. Calculations show that with c_0 approximately 10^{-3} mol/mol PVC (which is close to the concentrations actually used) an almost complete inhibition of the polyene growth will be experimentally observed with $kc_0/k_p > 10^2$.

8.2.2. Removal of Carbonyl Groups

Such an interaction is associated with the reduction of carbonyl or the addition of a chemical compound to a $>C=O$ bond.

The reduction of $>C=O$ groups accounts for the decrease in the PVC dehydrochlorination rate in the presence of organo-Si and organo-Ge hydrides having the structure R_3MH $(R = Et, Bu)$.[62,80–82] No reduction of $>C=C<$ bonds occurs in KAG.[62] However, due to the disruption of conjugation in $-C(O)-CH=CH-$ groups the polymer loses its ability to undergo hydrolysis with the rupture of macromolecules. As a result the experimentally found value of $\bar{\gamma}_0$ diminishes. The HCl elimination curves in this case are linear which is

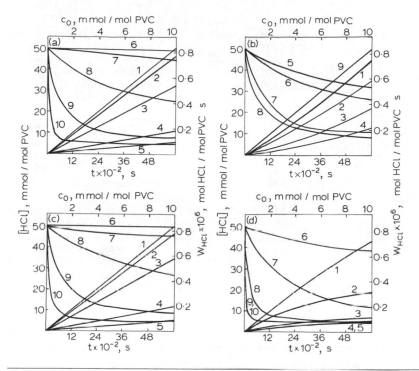

Variable	Curve									
	1	2	3	4	5	6	7	8	9	10
(a) k, (mol/mol PVC s)$^{-1}$	10^{-2}	10^{-1}	1	10	10^2	10^{-2}	10^{-1}	1	10	10^2
(b) k, (mol/mol PVC s)$^{-1}$	1	1	10	10	1	1	10	10	—	—
k_{deg} (s^{-1})	10^{-3}	10^{-4}	10^{-3}	10^{-4}	10^{-3}	10^{-4}	10^{-3}	10^{-4}	—	—
(c) k, (mol/mol PVC s)$^{-1}$	10^{-2}	10^{-1}	1	10	10^2	10^{-2}	10^{-1}	1	10	10^2
k_{deg}(s^{-1})	10^{-2}	10^{-4}	10^{-4}	10^{-4}	10^{-4}	10^{-2}	10^{-4}	10^{-4}	10^{-4}	10^{-4}
(d) k, (mol/mol PVC s)$^{-1}$	10^{-2}	10^{-1}	1	10	10^2	10^{-2}	10^{-1}	1	10	10^2

FIG. 7.   Solution graphs for sets of differential equations: (a) (xxv)–(xxviii); (b) (xxx)–(xxxii); (c) (xlii)–(xliv); (d) (xlv)–(xlvii) at 448 K, 10^{-2} Pa, $\bar{\gamma}_0 = 10^{-4}$ mol/mol PVC. Various values of the kinetic parameters k and k_{deg} are given in the table. (a) Curves 1–5, HCl evolution for $c_0 = 5$ mmol/mol PVC, curves 6–10, dependence of W_{HCl} on c_0; (b) curves 1–4, HCl evolution for $c_0 = 5$ mmol/mol PVC, curves 5–8, dependence of W_{HCl} initial value on c_0, curve 9, HCl evolution during degradation of non-stabilised PVC; (c) curves 1–5, HCl evolution for $c_0 = 5$ mmol/mol PVC, curves 6–10, dependence of W_{HCl} on c_0; (d) curves 1–5, HCl evolution for $c_0 = 5$ mmol/mol PVC, curves 6–10, dependence of W_{HCl} initial value on c_0.

TABLE 9

EFFECT OF EPOXIDE STABILISERS AND ETHYLENE GLYCOL ON PVC DEHYDRO-
CHLORINATION (448 K, 10^{-2} Pa, BARIUM STEARATE AS HCl ACCEPTOR)

Stabiliser	$c_0 \times 10^3$ (mol/mol PVC)	Dehydrochlorination rate $\times 10^6$ (mol HCl/mol PVC s)		
		W_{HCl}	W_r	W_p
None	—	0·78	0·08	0·70
Butyl epoxy-	5·0	0·52	0·08	0·44
stearate	10·0	0·46	0·08	0·38
	20·0	0·47	0·08	0·39
	50·0	0·47	0·08	0·39
Octyl epoxy-	5·0	0·56	0·08	0·48
stearate	10·0	0·54	0·08	0·46
	20·0	0·40	0·08	0·32
	50·0	0·40	0·08	0·32
Ethylene glycol	20·0	0·68	0·08	0·60
	50·0	0·64	0·07	0·57
	100·0	0·31	0·08	0·23
	200·0	0·33	0·08	0·25

in accordance with the variant under consideration. The reduction of
$>$C$=$O groups proceeds rather easily. The rate constant for the
reaction at 343 K is $3·5 \times 10^{-2}$ (mol/mol PVC)$^{-1}$ s^{-1} for Bu$_3$SiH and
$1·2 \times 10^{-3}$ (mol/mol PVC)$^{-1}$ s^{-1} for Bu$_3$GeH. The decrease in the
number of the polyene active growth sites results in a considerable
colour-stabilising effect and prevents the formation of a cross-linked
polymer product.

High reactivity towards carbonyl groups is observed with epoxides
and 1,2-glycols (reaction 13)[82,83] and their introduction in PVC leads
to a decrease in the polyene growth rate (see Table 9).

$$\begin{matrix} >\!\!C\!\!=\!\!O + R\!\!-\!\!CH\!\!-\!\!CH\!\!-\!\!R' \\ \diagdown\!\!\diagup \\ O \end{matrix} \quad\longrightarrow\quad \begin{matrix} O\!\!-\!\!CHR \\ \diagup \quad | \\ C \quad | \\ \diagdown \quad | \\ O\!\!-\!\!CHR' \end{matrix} \quad (13)$$

$$>\!\!C\!\!=\!\!O + R\!\!-\!\!CH\!\!-\!\!CH\!\!-\!\!R'$$
$$\qquad\qquad |\quad\ |$$
$$\qquad\qquad OH\ \ OH$$

The possibility of dioxolane formation is evidenced by the reaction
of epichlorohydrin with methyl vinyl ketone, the latter being the

simplest model of KAG. The reaction proceeds at 323–333 K in the presence of catalytic quantities of HCl.[82]

8.3. Modification of Active Sites by Saturation of Internal Double Bonds

The thermal stability of PVC increases with mild chlorination under conditions that rule out the interaction of chlorine with normal units of the polymer.[15,16,84] The HCl elimination curves of the mildly chlorinated PVC, as in other cases of irreversible blocking of internal unsaturated groups, have a linear character, and the values of W_{HCl} are proportional to the number of the remaining active sites (Fig. 8). It should be borne in mind, however, that under chlorination conditions[85,86] there is a possibility of an increase in the polymer dehydrochlorination rate. Disregard of the peculiarities of the thermal decomposition of chlorinated PVC may give rise to an erroneous interpretation of the experimental facts and to the conclusion that mild chlorination of PVC does not increase the polymer stability.[81]

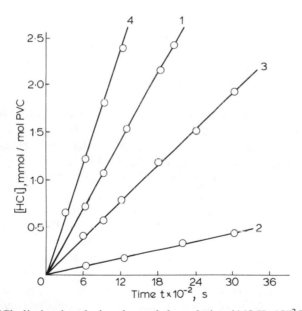

FIG. 8.   HCl elimination during thermal degradation (448 K, 10^{-2} Pa) of PVC and CPVC. 1, PVC-1; 2, PVC-1 subjected to soft chlorination to eliminate internal $>C=C<$ bonds; 3, PVC-2; 4, PVC-2 chlorinated to chlorine content of 65·2 mol %.

8.4. Reversible Disruption of Conjugation in KAG

A stabiliser can react chemically with active sites, for example by saturating double bonds.

$$
\begin{array}{c}
\xrightarrow[k_{\text{deg}}]{k} \quad -\text{C(O)}-\overset{|}{\text{CH}}-\overset{|}{\text{CH}}-\text{CHCl} \\
(y_1)
\end{array}
$$

$$-\text{C(O)}-\text{CH}=\text{CH}-\underset{\underset{(x_1)}{|}}{\text{CH}}- \qquad\qquad\qquad\qquad (14)$$

$$
\xrightarrow[-n\text{HCl}]{k_{\text{p}}} \quad -\text{C(O)}-(\text{CH}=\text{CH})_{\overline{n+1}}\text{CHCl}-\xrightarrow[-\text{HCl}]{k_{\text{p}}}
$$
$$(x_2)$$

Further reaction of the stabiliser with x_2 does not result in stabilisation of the polymer. If we neglect the change in the initial reagent concentration, c_0, during the initial stages of the PVC degradation and bear in mind that at any time $(y_1 + x_1 + x_2) = \bar{\gamma}_0$, the reaction process scheme can be described by the set of equations (xxx)–(xxxii).

$$\frac{d[\text{HCl}]}{dt} = k_r a_0 + k_p (x_1 + x_2) \qquad\qquad (\text{xxx})$$

$$\frac{dx_1}{dt} = -kx_1 c_0 - k_p x_1 + k_{\text{deg}}(\bar{\gamma}_0 - x_1 - x_2) \qquad\qquad (\text{xxxi})$$

$$\frac{dx_2}{dt} = k_p x_1 \qquad\qquad (\text{xxxii})$$

These can be solved analytically to obtain the function $[\text{HCl}] = f(t)$; however, it is more convenient to carry out a numerical solution by means of a computer. The kinetic curves of HCl yield versus time for various computed values of k and k_p, as well as the corresponding dependences of initial PVC dehydrochlorination rates on c_0, are given in Fig. 7(b).

Autocatalysis is specific to the process of HCl elimination in the presence of stabilisers that interact reversibly with KAG involving the disruption of conjugation. The PVC dehydrochlorination rate increases exponentially (the rate depending on the ratio of the three constants, k, k_{deg} and k_p) throughout the process, varying within the range from W_r to $W_{\text{HCl}} = W_r + W_p$, which is typical of a non-stabilised polymer. With $c_0 = 10^{-3}$–10^{-2} mol/mol PVC the stabilising effect depends on the numerical value of the relation $(kc_0/k_{\text{deg}})/k_p$. A decrease in the initial rate of HCl elimination, which is of great practical importance, will be observed when the relation $(kc_0/k_{\text{deg}})/k_p \geqslant 10^5$ holds. This is valid for a reversible saturation of $>C=C<$ bonds in PVC by dienes via the Diels–Alder reaction.[58,87] In this case, reaction 15, the CAG containing electron-accepting side-groups act as dienophiles.

$$—C(O)—CH=CH—CHCl— + CHR \diagup\!\!\!\diagup CHR' \underset{k_{deg}}{\overset{k}{\rightleftarrows}}$$
$$\diagdown CH—CH \diagup$$

$$\sim\!\!\sim\!\!C(O)—CH—CH—CHCl—$$
$$RCH \qquad CHR' \qquad (15)$$
$$CH=CH$$

The reaction of PVC with cyclopentadiene, piperylene and some other compounds containing $(>C=C<)_n$ (353 K) leads to a considerable decrease in or a complete elimination of internal unsaturated groups $(\bar{\gamma}_0)$ and to a corresponding decrease in the initial PVC dehydrochlorination rate, W_{HCl}, due to the decreasing rate W_p (Fig. 9). Under conditions of thermal degradation, adducts of KAG in PVC and of conjugated dienes decompose due to the reversibility of the Diels–Alder reaction. This results in the restoration inside the polymer of the initial number of dehydrochlorination-active sites $\bar{\gamma}_0$ and of the initial value of W_{HCl}.

The change in the number of active sites and HCl yield versus time during the modified PVC degradation is described by eqns. (xxxiii)–(xxxvi),

$$\frac{d[HCl]}{dt} = k_r a_0 + k_p x \qquad \text{(xxxiii)}$$

$$\frac{dx}{dt} = k_{deg}(\bar{\gamma}_0 - x) \qquad \text{(xxxiv)}$$

whereupon with $x = \bar{\gamma}_0$ initially,

$$x = \bar{\gamma}_0[1 - \exp(-k_{deg}t)] \qquad \text{(xxxv)}$$

and

$$[HCl] = (k_r a_0 + k_p \bar{\gamma}_0)t - \frac{k_p}{k_{deg}} \bar{\gamma}_0[1 - \exp(-k_{deg}t)] \qquad \text{(xxxvi)}$$

The rate of HCl elimination at the initial stage of the diene-modified PVC degradation is dependent on the thermal stability (the value of

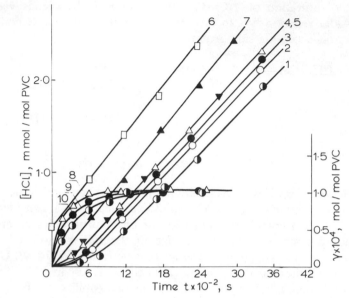

Fig. 9.   HCl evolution (1–7) and variation in KAG content (8–10) during degradation of PVC (1, 4, 5) and predegraded PVC (7) in the presence of conjugated dienes; of PVC pre-processed (353 K, 2–6 h, removal of diene residues by extraction with ether) with conjugated dienes (2, 3, 8, 9, 10) *in vacuo* (448 K, 10^{-2} Pa). 1,2,10,5-Methylhepta-1,3,6-triene; 3, 9, cyclopentadiene; 4,10,2,5,10,15-tetramethylhexadeca-1,5,7,9,15-pentene; 5,6,4-methylene-6,10-dimethylundeca-1,5,10-triene; 7, PVC without additives.

k_{deg}) of the groups

$$\text{\tiny\textasciitilde\textasciitilde}C(O)\!-\!\underset{\underset{\displaystyle CH=CH}{\overset{\displaystyle\diagup\quad\diagdown}{RCH\qquad CHR'}}}{\overset{\diagup\qquad\diagdown}{CH\!-\!CH}}\!-\!CHCl\text{\tiny\textasciitilde\textasciitilde}$$

For diene compounds the restoration of $\bar{\gamma}_0$ and HCl yield are satisfactorily described by eqns. (xxxv) and (xxxvi) with the respective values of k_{deg} (Table 11).

The case of conjugated dienes elucidates the peculiarities of PVC stabilisation with the additives that react reversibly only with $>\!C\!=\!C\!<$ bonds in KAG.

TABLE 10

EFFECT OF MACROMOLECULAR CHEMICAL STRUCTURE OF SUSPENSION-CHLORINATED
PVC (CPVC) ON ITS THERMAL STABILITY (448 K, 10^{-2} Pa)

Sample	Chlorine content (mass %)	$\bar{M}_n \times 10^{-3}$	Content of structural units (molar %)			$W_{HCl} \times 10^6$ (mol HCl/ mol PVC s)
			$-CH_2-CHCl-$	$-CHCl-CHCl-$	$-CCl_2-CHCl-$	
PVC_1	56·7	118·5	100	0	0	0·81
PVC_2	56·7	140	100	0	0	0·65
$CPVC_1^1$	56·9	116·5	96	4	0	0·76
$CPVC_1^2$	57·2	113·5	96	4	0	0·74
$CPVC_1^3$	57·7	118	92	4	4	0·83
$CPVC_2^4$	60·0	122	86	8	6	0·73
$CPVC_2^5$	64·0	122	70	16	14	0·63
$CPVC_2^6$	64·3	75·5	70	14	16	3·90
$CPVC_2^7$	65·2	78	64	20	16	2·03

(1) At every temperature an equilibrium is established between the adducts and KAG.

(2) The stabilising effect is observed only at the initial stage of degradation and disappears more or less quickly (depending on the compound selected) as the process continues.

(3) Diene compounds exhibit no stabilising action towards the degraded PVC in which polyene sequences with $n \geqslant 3$ have formed from KAG (Fig. 9). Thermal degradation in the presence of dienes (D) proceeds in accordance with reaction 16 (see page 210).

The form of the experimental curves for HCl accumulation in the presence of conjugated dienes is similar to those resulting from the solution of the differential equations (xxx)–(xxxii). Consequently the decrease in W_{HCl} achieved will be determined by the relation between the rates of the forward and reverse reactions. This can be illustrated by associating the rate of HCl evolution with the ratio of the experimentally determined equilibrium concentrations of the adducts and the active (hydrolysable) groups ($\delta = y_i/x_i$), which is proportional to the relation between the rates of the adduct formation and decomposition in the polymer. On the assumption that the accelerating effect of y_2 groups (like β-chloroallyl groups) on HCl elimination can be neglected, and that y_3 activates dehydrochlorination just like x_1, x_2 and x_3,

$$\sim\text{C(O)}\text{—CH}=\text{CH}\text{—CHCl/CH}_2\sim \underset{\substack{\downarrow\\k_p\big| -\text{HCl}}}{\overset{D}{\rightleftharpoons}} \sim\text{C(O)}\text{—CH}\text{—CH}\text{—CHCl}\text{—CH}_2\sim$$

$$(x_1) \qquad \qquad R\diagdown\!\diagup R' \quad (y_1)$$

$$\sim\text{C(O)}\text{—CH}=\text{CH}\text{—CHCl}\text{—CH}_2\sim \underset{\substack{\downarrow\\k_p\big| -\text{HCl}}}{\overset{D}{\rightleftharpoons}} \sim\text{C(O)}\text{—CH}\text{—CH}=\text{CH}\text{—CHCl}\text{—CH}_2\sim \quad (16)$$

$$(x_2) \qquad \qquad R\diagdown\!\diagup R' \quad (y_2)$$

$$\sim\text{C(O)}\text{—CH}=\text{CH}\text{—}(\text{CH}=\text{CH})_2\text{CHCl}\text{—CH}_2\sim \underset{\substack{\downarrow\\k_p\big| -\text{HCl}}}{\overset{D}{\rightleftharpoons}} \sim\text{C(O)}\text{—CH}\text{—CH}\text{—}(\text{CH}=\text{CH})_2\text{CHCl}\text{—CH}_2\sim$$

$$(x_3) \qquad \qquad R\diagdown\!\diagup R' \quad (y_3)$$

TABLE 11

EFFECT OF CONJUGATED DIENES ON PVC DEHYDROCHLORINATION (448 K; 10^{-2} Pa)

Additive concentration, c_0, 10^{-2} mol/mol PVC

PVC^a	Additive	Initial rate of PVC dehydrochlorination $\times 10^6$ (mol HCl/mol PVC s)			δ	$k_{deg} \times 10^3$ (s^{-1})	
		W_{HCl}	W_r	W_p		From eqn. (xxxvi)	From eqn. (xxxv)
PVC-1	None	0·68	0·08	0·60	—	—	—
PVC-2	None	0·83	0·08	0·75	—	—	—
PVC-3	None	1·24	0·08	1·16	—	—	—
PVC-1	Cyclopentadiene	0·43	0·08	0·35	2·8	1·3±0·3	3·4±0·5
PVC-3	Cyclopentadiene	0·40	0·08	0·32	2·7	1·0±0·3	—
PVC-2	5-Methylhepta-1,3,6-triene	0·28	0·08	0·20	2·8	1·1±0·3	2·5±0·5
PVC-2	Octa-1,3,6-triene	0·33	0·08	0·25	2·1	—	—
PVC-2	3-Methylene-6,10-di-methylundeca-1,5,10-triene	0·38	0·08	0·30	1·2	1·8±0·3	—
PVC-1	Piperylene	0·38	0·08	0·30	0·9	2·1±0·3	3·4±0·5
PVC-2	2,5,10,15-Tetramethyl-hexadeca-1,5,7,9,15-pentene	0·38	0·08	0·30	1·2	1·8±0·3	5·5±0·5
PVC-2	Methyl sorbate	0·46	0·08	0·38	1·0	2·8±0·3	—
PVC-1	Sorbic acid	0·40	0·08	0·32	0·77	1·7±0·3	—

^{a} PVC characteristics $\bar{M}_\eta^\circ/(\bar{\gamma}_0 \times 10^4)$ mol/PVC mol: PVC-1, 105,000/0·76; PVC-2, 129,600/1·03; PVC-3, 81,500/1·6.

the scheme can be described by the eqns. (xxxvii)–(xl).

$$\frac{d(x_1 + y_1)}{dt} = -k_p x_1 \qquad \text{(xxxvii)}$$

$$\frac{d(x_2 + y_2)}{dt} = k_p x_1 - k_p x_2 \qquad \text{(xxxviii)}$$

$$\frac{d(x_3 + y_3)}{dt} = k_p x_2 \qquad \text{(xxxix)}$$

$$\frac{d[HCl]}{dt} = k_r a_0 + k_p (x_1 + x_2 + x_3 + y_3) \qquad \text{(xl)}$$

If in equilibrium $y_1/x_1 = y_2/x_2 = y_3/x_3 = \delta$, and at the initial moment $x_1 + y_1 = \bar{\gamma}_0$, and $x_2 = x_3 = y_2 = y_3 = 0$, HCl yield versus time will be described by eqn. (xli).

$$[HCl] = (k_r a_0 + k_p \bar{\gamma}_0)t + \frac{k_p \bar{\gamma}_0}{1+\delta}\, t \exp\left(-\frac{k_p}{1+\delta} \cdot t\right)$$

$$-2\,\delta\bar{\gamma}_0\left[1 - \exp\left(-\frac{k_p}{1+\delta} t\right)\right] \qquad \text{(xli)}$$

A coincidence of the curves calculated by equation (xli) and the experimental curves for a number of compounds investigated is achieved with the values of δ listed in Table 11. As can be seen, the

(17)

TABLE 12
KINETIC PARAMETERS FOR THE REACTION OF TRI(2-ETHYLHEXYL)PHOSPHITE
WITH KAG IN PVC (10^{-2} Pa)
Phosphite concentration, c_0, 10^{-2} mol/mol PVC

PVC characteristics		T (K)	Rate constant $k \times 10^3$ (s^{-1})
$\bar{M}_n^\circ \times 10^{-3}$	$\bar{\gamma}_0 \times 10^4$ $(mol/mol\,PVC)$		
119	1·42	298	0·11±0·06
107·8	1·52	289	0·06±0·03
107·8	1·52	298	0·11±0·06
107·8	1·52	303	0·37±0·04
107·8	1·52	313	1·00±0·04
70·4	1·62	298	0·11±0·06

Effective Arrhenius parameters: $E_{act} = 71 \pm 8$ kJ/mol; $\log A = 10 \pm 1$.

greater the value of δ, the higher the effectiveness of the diene as an inhibitor of the PVC dehydrochlorination process.

Another example is the conversion of in-chain KAG into saturated structures during the interaction of PVC with organophos-phites.[25,30,57,87–89] Under rather mild conditions (293–313 K, 1–8 h) organic phosphites undergo 1,4-addition to CAG.[56,57,89]

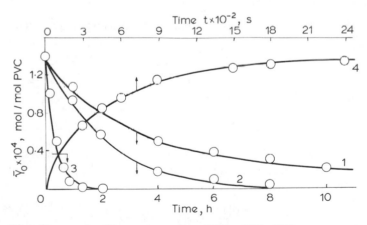

FIG. 10. Change in the content of —C(O)—CH=CH— groups in PVC during interaction with tri(iso-octyl)phosphite (1–3) and during thermal expos-ure of phosphorylated polymer (4). 1, 289 K; 2, 298 K; 3, 313 K; 4, 448 K. $c_0 = 10$ mmol/mol PVC.

The interaction of phosphites with KAG proceeds at a noticeable rate even at room temperature (Table 12) and is easily measured quantitatively by the decreasing content of internal unsaturated $>C=C<$ groups as evidenced by oxidative ozonolysis or alkaline hydrolysis of the phosphorylated polymer (Fig. 10).[57]

Ketophosphonate groups are more thermostable than KAG. Nevertheless, under polymer degradation conditions they undergo thermal decomposition with regeneration of internal unsaturated groups $\bar{\gamma}_0$ (Fig. 10, curve 4). Analysis of the KAG accumulation curve by eqn. (xxxv) shows that $k_{\text{deg}} = (0 \cdot 95 \pm 0 \cdot 05) \times 10^{-3} \, \text{s}^{-1}$ (448 K) for a ketophosphonate·group in PVC in the presence of tri-isooctyl phosphite.

8.5. Irreversible Replacement of Labile Chlorine Atoms

Unlike the above cases, the replacement of a labile chlorine atom in $\text{\textasciitilde C(O)}\text{--}(\text{CH}=\text{CH})_{\overline{n}}\text{CHCl}\text{\textasciitilde}$, whether it is the initial group ($n = 1$) or one formed in the course of degradation ($n > 1$), leads to an irreversible termination of the polyene growth active site at any stage of the PVC decomposition process:

$$
\begin{aligned}
&\text{\textasciitilde C(O)}\text{--}\text{CH}=\text{CH}\text{--}\underset{\underset{\text{Cl}}{|}}{\text{CH}}\text{\textasciitilde}
\left\{
\begin{array}{l}
\xrightarrow[\text{BX}]{k} \text{\textasciitilde C(O)}\text{--}\text{CH}=\text{CH}\text{--}\underset{\underset{\text{X}}{|}}{\text{CH}}\text{\textasciitilde} + \text{BCl} \\[2ex]
\xrightarrow[\text{HCl}]{k_p} \text{\textasciitilde C(O)}\text{--}(\text{CH}=\text{CH})_{\overline{2}}\underset{\underset{\text{Cl}}{|}}{\text{CH}}\text{\textasciitilde}
\end{array}
\right. \\[3ex]
&\quad\quad\quad \swarrow^{k}_{\text{BX}} \quad\quad\quad\quad\quad\quad\quad\quad \searrow^{k_p}_{-\text{HCl}} \\[1ex]
&\text{\textasciitilde C(O)}\text{--}(\text{CH}=\text{CH})_{\overline{2}}\underset{\underset{\text{X}}{|}}{\text{CH}}\text{\textasciitilde} \quad\quad\quad\quad \text{polyene}
\end{aligned}
\tag{18}
$$

As a result the rate W_p decreases. If variation of c_0 during the process is neglected, the reaction scheme is described by the set of differential equations (xlii)–(xliii).

$$
\frac{\text{d}[\text{HCl}]}{\text{d}t} = k_r a_0 + k_p x
\tag{xlii}
$$

$$
\frac{\text{d}x}{\text{d}t} = -kxc_0
\tag{xliii}
$$

with a simple solution:

$$[HCl] = k_r a_0 t + \frac{k_p}{k c_0} \bar{\gamma}_0 [1 - \exp(-k c_0 t)]$$ (xliv)

The kinetic curves of $[HCl] = f(t)$ calculated for various k (Fig. 7(c)) show an autoretardation of the PVC dehydrochlorination process in the presence of a stabiliser when $k \geqslant 10^{-1} (\text{mol/mol PVC})^{-1} \text{s}^{-1}$. Effective compounds will be those with the numerical values of $k \simeq 10-10^2$. Even at a concentration of $c_0 \simeq 10^{-3}$ mol/mol PVC, the stabiliser will bring about a fast retardation of the process down to the rate of a non-stabilisable random HCl elimination from the normal polymer repeat (W_r). The stabilising efficiencies of any compound towards PVC samples with different intrinsic stabilities (i.e. different $\bar{\gamma}_0$) will hardly differ.

The reduction of PVC macromolecular segments containing allyl chlorine atoms by trialkylsilanes corresponds to the latter case. Reaction 19 is catalysed by HCl.[80,81]

$$\text{\small ∿-(CH=CH)}_n\text{-CHCl∿} \xrightarrow{\text{HCl}} \text{∿-(CH=CH)}_n\text{-}\overset{\oplus}{\text{CH}}\text{∿}$$
$$\swarrow {\scriptstyle \text{Et}_3\text{SiH}} \qquad [\text{ClHCl}]^- \qquad (19)$$
$$\text{∿-(CH=CH)}_2\text{-CH}_2\text{∿} + \text{HCl} + \text{Et}_3\text{SiCl}$$

A decreased dehydrochlorination rate is characteristic of the reduced polymer.[81]

8.6. Reversible Replacement of Labile Chlorine Atoms

If a PVC stabiliser undergoes reaction with the labile chlorine atoms by substitution in ∿C(O)-(CH=CH)$_n$-CHCl∿ (where $n \geqslant 1$) and the substituted groups X are able to decompose under thermal degradation conditions, stabilisation of the polymer product will involve the sequence of reactions shown in the reaction scheme on page 216.

Both deactivation and regeneration of the active polyene sites occur throughout the process. The active site concentration will depend on the relation of the rates of the substitution and decomposition reactions. Since the stabiliser is irreversibly consumed in successive substitutions of labile chlorine atoms, a change in the initial concentration, c_0, should be taken into account in any kinetic description. As the substituent concentration is equal to the difference $(\bar{\gamma}_0 - x)$ at any

$$
\begin{array}{c}
\text{—C(O)—CH=CH—CH}\hspace{-2pt}\sim \\
\overset{\displaystyle |}{\underset{}{\text{Cl}}}
\end{array}
\left\{
\begin{array}{l}
\xrightarrow{k_{\mathrm{p}}} \text{HCl} \\[12pt]
\xrightarrow[\text{BX}]{k} \sim\hspace{-2pt}\text{C(O)—CH=CH—CH}\hspace{-2pt}\sim \\
\hspace{4.2cm} \overset{\displaystyle |}{\underset{}{\text{X}}}
\end{array}
\right.
$$

$$ \Big\downarrow {\scriptstyle k_{\mathrm{p}} \,/\, -\mathrm{HX}} \hspace{5cm} (20)$$

$$
\begin{array}{c}
\sim\hspace{-2pt}\text{C(O)}\text{—}(\text{CH=CH})_2\text{—CH}\hspace{-2pt}\sim \\
\hspace{1.6cm}\overset{\displaystyle |}{\underset{}{\text{Cl}}}
\end{array}
\left\{
\begin{array}{l}
\xrightarrow{k_{\mathrm{p}}} \text{HCl} \\[12pt]
\xrightarrow[\text{BX; }-\text{BCl}]{k} \sim\hspace{-2pt}\text{C(O)}\text{—}(\text{CH=CH})_n\text{CH}\hspace{-2pt}\sim \xrightarrow[-\mathrm{HX}]{k_{\mathrm{p}}} \cdots \\
\hspace{5.7cm}\overset{\displaystyle |}{\underset{}{\text{X}}}
\end{array}
\right.
$$

instant, the reaction scheme can be described by the set of non-linear equations (xlv)–(xlvii)

$$\frac{\mathrm{d}[\mathrm{HCl}]}{\mathrm{d}t} = k_{\mathrm{r}}a_0 + k_{\mathrm{p}}x \tag{xlv}$$

$$\frac{\mathrm{d}x}{\mathrm{d}t} = -kxc + k_{\mathrm{deg}}(\bar{\gamma}_0 - x) \tag{xlvi}$$

$$-\frac{\mathrm{d}c}{\mathrm{d}t} = kxc \tag{xlvii}$$

These can be solved by means of a computer. For $c_0 \simeq 10^{-3}$ mol/mol PVC, $k = 10^{-2}$–10^2 (mol/mol PVC)$^{-1}$ s^{-1}, and $k_{\mathrm{deg}} = 10^{-2}$ s^{-1}, typical linear dependences of the HCl evolution kinetics with time are observed (Fig. 7(d)). The greater is k, the lower are the concentrations of the active sites and hence the rate of HCl evolution, and the closer to each other are the HCl accumulation curves. With $k > 10^2$ variation of the polymer dehydrochlorination rate is a function of k_{deg} only, i.e. W_{HCl} is solely dependent on the thermal stability of the substituent group X. Since the rate of the labile chlorine atom replacement depends on the stabiliser content, an increase in c_0 diminishes the requirements for k. For instance, with $c_0 = 10^{-2}$ mol/mol PVC the PVC dehydrochlorination rate is limited by decomposition of the substituent groups even with $k \geqslant 10$ (mol/mol PVC)$^{-1}$ s^{-1}.

In the presence of carboxylates of coordinately unsaturated metals

(Zn, Cd, Pb),[28,30,32,87,90–100] organotin compounds,[17,28,30,81,87,90,96,98,101–104] a number of sulphur- and nitrogen-containing compounds and epoxides,[90,95,97,105–107] replacement of labile chlorine atoms by stable groups is associated with inhibition of PVC dehydrochlorination.

8.6.1. Carboxylates of Coordinately Unsaturated Metals

Carboxylates of coordinately unsaturated metals are most widely used for PVC stabilisation either independently or in admixture with carboxylates of alkali or alkali-earth metals. A decrease in the polyene growth rate and a retardation of the polyene colouration during thermal degradation are the most typical consequences of the reaction of Pb, Cd, and Zn salts of organic acids with labile chlorine atoms in PVC.[30,87,93,94,100,108] It should be noted that the overall elimination rate W_{HCl} in the presence of the above metal salts can vary in a complicated way—from retardation to autocatalysis of HCl elimination—depending on the nature of the metal and on the salt concentration, which is connected with the properties of accumulating electrophilic metal chlorides.[28,109–111] However, analysis of the rates of the reactions comprising the overall PVC dehydrochlorination process (W_r and W_p) always indicates a regular decrease with concentration in the polyene growth rate W_p (Table 13). For carboxylates of the metals whose chlorides do not accelerate the polymer decomposition, as in the case of lead, the kinetic curves of HCl liberation are always linear,[108] which is evidence of a constant concentration of active sites. A stationary concentration of active sites is determined by the relation between the rates of the successive exchange and degradation reactions of the replaced groups (reaction 21).

$$\text{~~C(O)—CH=CH—CH—CH}_2\text{~~} + Me(OCOR)_2$$
$$\underset{Cl}{|}$$

$$k \Big\downarrow {-MeClOCOR}$$

$$\text{~~C(O)—CH=CH—CH—CH}_2\text{~~} \xrightarrow[-HOOCR]{k_{deg}} \qquad (21)$$
$$\underset{ROCO}{|}$$

$$\text{~~C(O)—(CH=CH)}_2\text{—CH—CH}_2\text{~~} \quad - \quad \left[\begin{array}{c} \xrightarrow[-HCl]{k_p} \\ \\ \xrightarrow[Me(OCOR)_2]{k} \end{array} \right.$$
$$\underset{Cl}{|}$$

TABLE 13

EFFECT OF METAL CARBOXYLATES ON PVC DEHYDROCHLORINATION *in vacuo*
$(448 \text{ K}, 10^{-2} \text{ Pa})$

PVC characteristics $\bar{M}_n^\circ \times 10^{-3}$, $\bar{\gamma}_0 \times 10^4$ (mol/mol PVC)		Stabiliser	Content, $c_0 \times 10^3$ (mol/mol PVC)	Dehydrochlorination rate$^a \times 10^6$ (mol HCl/mol PVC s)		
				W_{HCl}	W_r	W_p
152	1·0	None	—	0·80	0·08	0·72
74·5	1·2	None	—	1·04	0·08	0·96
74·1	1·4	None	—	1·10	0·08	1·02
152	1·0	$(C_{17}H_{35}COO)Na$	1·0	0·80	0·08	0·72
152	1·0	$(C_{17}H_{35}COO)_2Ca$	1·0	0·80	0·08	0·72
152	1·0	$(C_{17}H_{35}COO)_2Ba$	1·0	0·80	0·08	0·72
152	1·0	$(CH_3COO)_2Pb$	1·0	0·44	0·08	0·36
		$(CH_3COO)_2Pb$	5·0	0·42	0·08	0·34
		$(CH_3COO)_2Pb$	10·0	0·42	0·08	0·34
74·1	1·4	$(C_3H_7COO)_2Cd$	1·0	0·37	0·08	0·29
		$(C_3H_7COO)_2Cd$	5·0	0·37	0·08	0·29
		$(C_3H_7COO)_2Cd$	10·0	0·37	0·08	0·29
74·1	1·4	$(C_{11}H_{23}COO)_2Cd$	1·0	0·51	0·21	0·30
		$(C_{11}H_{23}COO)_2Cd$	5·0	0·35	0·18	0·17
		$(C_{11}H_{23}COO)_2Cd$	10·0	0·35	0·18	0·17
94·5	1·2	$(C_{17}H_{35}COO)_2Cd$	1·0	0·83	0·53	0·30
		$(C_{17}H_{35}COO)_2Cd$	5·0	0·47	0·25	0·22
		$(C_{17}H_{35}COO)_2Cd$	10·0	0·30	0·14	0·16
152	1·0	$(C_{17}H_{35}COO)_2Pb$	1·0	0·35	0·08	0·27
		$(C_{17}H_{35}COO)_2Pb$	5·0	0·20	0·08	0·12
		$(C_{17}H_{35}COO)_2Pb$	10·0	0·20	0·08	0·12
152	1·0	None^b	—	0·07	0·02	0·05
152	1·0	$(C_{17}H_{35}COO)_2Zn^b$	1·0	0·14	0·13	0·01
		$(C_{17}H_{35}COO)_2Zn^b$	5·0	0·02	0·014	0·007

^{a} For Cd and Zn carboxylates the initial rate values are given.
^{b} 423 K.

Decomposition of the substituted groups during PVC thermal degradation is shown in independent experiments. The absorption intensity at 1735 cm^{-1} in the IR spectrum of PVC stabilised with zinc 2-ethylhexanoate passes through a maximum and this is associated with the detachment of 2-ethylhexanoic acid.[112] Decomposition of the ester groups has been demonstrated in low-molecular weight models of β-acylallyl structures and HCl has been suggested to catalyse the splitting of ester bonds.[97,107]

The carboxylate salts of Ca and Ba do not inhibit PVC decomposition. Only their combination with Cd and Zn salts (for example chlorides) is observed to produce a stabilising action. This is conditioned by regeneration of the corresponding carboxylates during the interaction between the chlorides of coordinately unsaturated metals and the Ca and Ba salts. For Ca stearate and $ZnCl_2$ such a reaction has been demonstrated by means of low-molecular weight compounds.[96,97,107]

8.6.2. Organotin Compounds

Organotin compounds with the structure R_2SnX_2, where R stands for alkyl-, aryl- and arylalkyl-substituents, and X stands for carboxy-, alkoxy-, mercapto- and some other substituents, act as highly effective stabilisers for PVC.[28,98,112] Their inhibiting effect on the PVC dehydrochlorination process, as in the case of Pb, Cd and Zn carboxylates, is connected with labile chlorine atom replacement by a fragment of the stabiliser molecule in accordance with scheme 20.[87,90,101,102,104] Dialkyltin carboxylates and mercaptides do not react with in-chain double bonds in KAG. In PVC heated in vacuo (298 K, 4–8 h) with dibutyltin derivatives, e.g. bislaurate, maleate, ethylmercaptide or thioglycollate, the content of internal groups —C(O)—CH=CH—, determined by ozonolysis and hydrolysis, does not change after extraction with diethyl ether for 8 h. Meanwhile, if stabilisers with isotope labels are used, residual radioactivity is discovered in the polymer, its value in predegraded PVC being the same as in the initial polymer. This is an independent indication of the fact that the content of active sites, $\bar{\gamma}_0$, does not change in the course of PVC degradation in vacuo.[102,113] The replacement of labile chlorine atoms by the electronegative groups of the organotin compounds results in an increase of PVC stability up to the value conditioned by the stability of the substituent (X) groups. This fact might explain the problem of the residual instability of the modified PVC raised in the scientific literature.[101,103] The dehydrochlorination kinetics of PVC modified with organotin compounds is well described by eqn. (xxxvi).

Optimal initial concentrations of organotin stabilisers are found to exist for maximum stability during PVC degradation in the presence of R_2SnX_2. They correspond to the lowest values (different for each compound) of the PVC decomposition rate, W_{HCl}.[104,114] As in the case of Zn, Cd, and Pb carboxylates, a decrease in W_{HCl} is associated with a decrease in the polyene growth rate, W_p (Table 14).

TABLE 14

EFFECT OF ORGANOTIN COMPOUNDS ON PVC DEHYDROCHLORINATION *in vacuo*
(448 K, 10^{-2} Pa)

PVC: $\bar{M}_n^o = 81\ 500$; $\bar{\gamma}_0 = 1 \cdot 69 \times 10^{-4}$ mol/mol PVC

Stabiliser	Content $c_0 \times 10^3$ (mol/mol PVC)	Initial dehydrochlorination rate × 10^6 (mol HCl/mol PVC s)		
		W_{HCl}	W_r	W_p
None	—	1·28	0·08	1·20
Bu$_2$Sn(OCOC$_3$H$_7$ / OCOC$_{17}$H$_{35}$)	1·0	1·03	0·28	0·75
	2·0	0·87	0·20	0·67
	5·0	0·58	0·12	0·46
	10·0	0·58	0·13	0·45
Bu$_2$Sn(OCOC$_{11}$H$_{23}$)$_2$	0·5	0·55	—	—
	1·0	0·45	0·18	0·27
	2·0	0·36	0·16	0·20
	5·0	0·27	0·09	0·18
	10·0	0·22	0·09	0·13
Bu$_2$Sn(OCOCH=CHCOO-iso-Bu)$_2$	1·10	0·33	0·09	0·24
	2·80	0·25	0·09	0·16
	6·0	0·25	0·09	0·16
Bu$_2$Sn(OCOCH / OCOCH)	0·50	0·40	0·10	0·30
	1·0	0·40	0·10	0·30
	3·0	0·38	0·10	0·28
	5·0	0·38	0·10	0·28
	10·0	0·38	0·10	0·28
Bu$_2$Sn(SC$_2$H$_5$)$_2$	1·0	0·45	0·10	0·35
	2·0	0·73	0·19	0·54
	3·0	0·90	0·50	0·40
	7·5	1·10	0·80	0·30
	10·0	1·10	0·80	0·30

8.6.3. *Dehydrochlorination Kinetics of PVC Stabilised with Metal Carboxylates and Organotin Compounds*

The replacement rate constants and the stability of replaced groups (k and k_{deg}) are a good indication of the stability efficiency of the compounds participating in the replacement of labile chlorine atoms. These parameters could be determined by means of a set of equations similar to (xlv)–(xlvii), but describing the accumulation of HCl (metal chlorides) by chlorine atom replacement in the chain, while the values for k and k_{deg} could be selected by the process of numerical integration on a computer. A coincidence of the calculated and experimental dependences $[HCl]_p = f(t)$ would be a formal criterion for the selection of constants. However, on the basis of a number of simplifying assumptions a simple analytical expression has been derived to describe the succession of exchange and decomposition reactions of the replaced groups. Assuming that the change in c_0 at the initial stages of the stabilised PVC degradations can be neglected, reaction 21 is described by eqns. (xlviii) and (xlix).

$$\frac{d[HCl]_p}{dt} = k_p x + kxc_0 \qquad \text{(xlviii)}$$

$$\frac{dx}{dt} = -kxc_0 + k_{deg}(\bar{\gamma}_0 - x) \qquad \text{(xlix)}$$

where $[HCl]_p$ denotes the amount of HCl evolved at the expense of polyene growth alone. In the stationary states, $dx/dt = 0$.

$$x = \frac{k_p \bar{\gamma}_0}{kc_0 + k_p} \qquad \text{(l)}$$

$$\frac{d[HCl]_p}{dt} = \frac{k_p + kc_0}{k_{deg} + kc_0} k_{deg} \bar{\gamma}_0 \qquad \text{(li)}$$

The data in Tables 13 and 14 suggest that if Pb and Cd carboxylates and organotin compounds are used, increasing c_0 leads to a decrease in the rate, W_p, down to a certain minimum value which is linearly dependent on the index of intrinsic stability, $\bar{\gamma}_0$; the dependence is similar to that of $W_p = f(\bar{\gamma}_0)$ for an unstabilised polymer (Fig. 11).

Evidently the polyene growth in this case is effected only at the expense of the substituent groups. That is why the above dependence $W_p^{lim} = f(\bar{\gamma})$ makes it possible to evaluate the stability constant k_{deg} for the substituent groups. The degradation rate constants for the substituent groups are listed in Table 15. A satisfactory coincidence of k_{deg} values for groups of similar structure, estimated for organotin compounds and Pb, Cd and Zn carboxylates, is observed. In addition the values of k which make eqn. (li) describe well the experimental dependences $W_p = f(c_0)$ have been found (Fig. 12). In this particular case it is the thermal degradation of another polymer, essentially a tercopolymer containing triads of the type

$$\text{$\sim\!\!\!\sim$(CH$=\!$CH)$_{\overline{n}}$ CHX—CH$_2$$\overline{}$CHCl—CH$_2$$\sim\!\!\!\sim$}$$
$$\text{M}_1 \qquad\qquad \text{M}_2 \qquad\qquad \text{M}_3$$

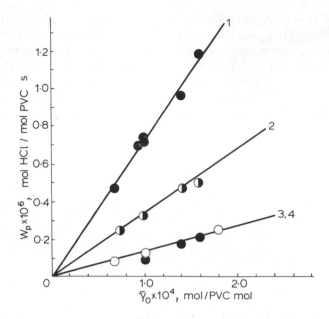

FIG. 11. Dependence of W_p on $\bar{\gamma}_0$ during degradation of various PVC samples (448 K, 10^{-2} Pa) stabilised with metal carboxylates. 1, Barium stearate; 2, lead acetate; 3, lead stearate (●); 4, cadmium stearate (○).

that is actually transformed. The structure of such a triad is continuously reproduced in the PVC chain with rapid replacement of another chlorine atom. The stability of the substituent group is determined by the retained negative influence of the adjacent conjugated $>C=C<$ bonds $(M_1)_n$. However, a possible influence of the neighbouring vinyl chloride unit M_3 should also be taken into account. For example, if the chlorine atom is replaced by an acetate group, the sequence M_1M_2 is a model of the growing end of a polyene sequence in polyvinyl acetate, for which $k_p^{PVA} = (5\cdot6 \pm 2\cdot4) \times 10^{-4}\,\text{s}^{-1}$ (448 K).[61] The rate constant of the degradation of the substituent acetate groups in PVC is $k_{\text{deg}} = (3\cdot5 \pm 0\cdot2) \times 10^{-3}\,\text{s}^{-1}$ at the same temperature (Table 15). The relation $k_p^{PVA} : k_{\text{deg}} \approx 1:10$ is evidence of a destabilising effect of the neighbouring vinyl chloride units on the decomposition of the acetate groups.

In a low-polar medium, of which the polymer is an example ($\mu = 4$), the chlorine atom detachment during replacement must be facilitated by a pre-addition to the leaving Lewis-acid moiety to reduce its

TABLE 15

REPLACEMENT RATE CONSTANTS AND STABILITY OF REPLACED GROUPS IN
$\sim\!\!C(O)\text{--}(CH\!\!=\!\!CH)_n\text{--}CHX\text{--}CH_2\sim$ (448 K, 10^{-2} Pa)

Stabiliser[a]	k $((mol/mol$ $PVC)^{-1}\,s^1)$	$k_{deg} \times 10^2\ (s^{-1})$	
		From eqn. (l)	From eqn. $(xxxvi)$
None	—	—	0·75±0·02
Pb(CH$_3$COO)$_2$	40±2	—	0·35±0·08
Cd(C$_3$H$_7$COO)$_2$	35±5	—	0·21±0·04
Bu$_2$Sn(OCOC$_3$H$_7$)(OCOC$_{17}$H$_{35}$)	27±2	—	0·27±0·04
Cd(C$_{11}$H$_{23}$COO)$_2$	—	—	0·12±0·02
Bu$_2$Sn(C$_{11}$H$_{23}$COO)$_2$	13±1	0·10±0·02	0·12+0·02
Cd(C$_{17}$H$_{35}$COO)$_2$	7±0·5	—	0·12±0·02
Pb(C$_{17}$H$_{35}$COO)$_2$	7±0 5	—	0 12±0·02
Bu$_2$Sn(OOCCH=CHCOO-iso-Bu)$_2$	15±1	0·09±0·02	0·09±0·02
Bu$_2$Sn(OOCCH=CHCOO—⟨C$_6$H$_{11}$⟩)$_2$	—	0·09±0·02	0·09±0·02
Bu$_2$Sn(OCO—CH‖OCO—CH)	40±5	0·15±0·02	0·17±0·02
Bu$_3$Sn—OCO—CH=CH—COOSnBu$_3$	—	0·15+0·02	—
Bu$_2$Sn(SC$_2$H$_5$)$_2$	27±2	0·13±0·02	0·17±0·02
Oc$_2$Sn(SC$_2$H$_5$)$_2$	27±2	0·15±0·02	0·17±0·02
Bu$_2$Sn(SCH$_2$COOCH$_2$)(SCH$_2$COOCH$_2$)	—	0·12±0·02	—

[a] Abbreviations: Bu, butyl; Oc, octyl.

basicity. This function is obviously effected by the chlorides of coordinately unsaturated metals, and the replaced chlorine evolution is
possible in the form of $[MeCl_3]^-$. Binding of metal chlorides into stable
complexes to a considerable extent suppresses the exchange reaction of
chlorine atoms for ester residues.[32,96]

Thus there is a connection between the chemical structure of metal-
containing stabilisers and their reactivity towards the replacement of

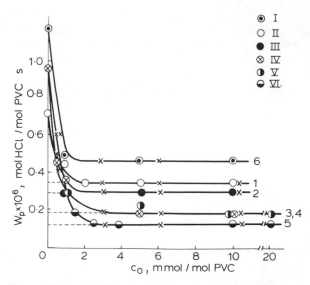

FIG. 12. Dependence of W_p on initial content of additives c_0 dibutyltin butyrate–stearate (I), Pb acetate (II), Cd butyrate (III), Cd laurate (IV), Cd stearate (V), Pb stearate (VI) (448 K, 10^{-2} Pa). $\bar{\gamma}_0 \times 10^4$ mol/mol PVC: 1, 5—1·0; 2, 3, 4—1·4; 6—1·6. ×, W_p values calculated as required by eqns. (l) and (li) with k and k_{deg} values as given in Table 15.

labile chlorine atoms in $\wedge\wedge\wedge C(O){-}(CH{=}CH)_n{-}CHCl\wedge\wedge\wedge$ groups. As one can see, with an increasing length of the radical, R, in the acid residue a concurrent decrease of both constants is observed. This is possibly connected with steric hindrance during the incorporation of carboxy-residues into the polymer chain, and with diffusion hindrances in the departure of the detached groups. Apart from the steric factors, the presence of polar (electron-donating or electron-accepting) groups in the acid residue is also expected to play a significant role. The contribution of polar factors in the stabilising efficiency of compounds participating in replacement reactions of chlorine atoms in PVC remain to be elucidated. The nature of a metal determines the possibility of the participation of a compound in the replacement reaction. Na, Ca and Ba salts do not replace chlorine atoms in labile polymeric structures. The esterification of model compounds in the presence of Ba and Ca carboxylates in solution is believed[96,115] to proceed due to the solvating effect of the solvent. No clear connection between the nature of the metal and the values of k, at least for the salts of Pb^{2+}, Cd^{2+},

Zn^{2+} or R_2Sn^{2+}, has so far been detected. Nevertheless, note should be taken of the high reactivity of organotin compounds towards labile chlorine replacement (higher values of k for R_2SnX_2 as compared with Pb and Cd salts, containing similar acid residues). This fact does not seem surprising since the most active organotin stabilisers, such as dibutyltin bis(alkylmaleates) and dibutyltin bis(thioglycollates) react slowly even with non-activated chlorine atoms in PVC.[116,117]

8.6.4. Stabilisation of PVC with Organic Compounds

Under certain conditions it is possible to involve in the replacement reaction of labile chlorine atoms in PVC, compounds that in themselves are unable to undergo this sort of interaction. Specifically, epoxy compounds and butanediol β-aminocrotonate partake in an exchange reaction in the presence of electrophilic metal chlorides as catalysts.[105-107,118] This accounts for the synergistic increase in the stabilising effect of organic compounds in combination with coordinately unsaturated metals.[28] A study of low-molecular weight model compounds has shown that epoxy compounds and compounds with labile hydrogen atoms undergo allylic chlorine replacement in the presence of $ZnCl_2$ or $CdCl_2$,[97,99,105-107] e.g. reactions 22, and 23 (see page 226):

$$R\!-\!\underset{\displaystyle \diagdown\!\!\!O\!\!\!\diagup}{CH\!-\!CH_2} + ZnCl_2 \longrightarrow R\!-\!\overset{\oplus}{C}H\!-\!CH_2\!-\!O\!-\!ZnCl_2^{\ominus}$$

$$\Big\downarrow \overset{\displaystyle -CH=CH-CH-}{\underset{\displaystyle \ \ Cl}{}}$$

$$\begin{array}{c} -CH=CH-CH- \\ | \\ RCHCl-CH_2-O \end{array} \qquad (22)$$

The dehydrochlorination rate of PVC stabilised with mixtures of Zn or Cd chlorides and butyl epoxystearate (BES) or butanediol β-aminocrotonate (β BAC) is substantially decreased, compared with the rate in the presence of organic stabilisers alone. A decrease in W_{HCl}, as in the case of metal carboxylates or organotin compounds, results in a parallel decrease in W_p (Table 16); the value of W_{HCl} depending on the nature of the metal chloride as well as on the ratio of the concentrations of the catalyst and organic compound (Fig. 13). The HCl elimination curves have a linear character.

The chemical reactions that take place during PVC stabilisation with mixtures of epoxy compounds and $MeCl_2$ can be represented as a

$$-CH{=}CH{-}\underset{\overset{|}{Cl}}{CH}\text{\small www} \;+\; [CH_3{-}\underset{\overset{|}{NH_2}}{C}{=}\overset{\overset{H}{|}}{C}{-}\underset{\overset{||}{O}}{C}{-}O{-}(CH_2)_{\overline{2}}]_2$$

$$\downarrow \text{ZnCl}_2 \;{-}\text{2HCl}$$

$$[CH_3{-}\underset{\overset{|}{NH}}{C}{=}\overset{\overset{H}{|}}{C}{-}\overset{\overset{O}{||}}{C}{-}O{-}(CH_2)_{\overline{2}}]_2$$
$$-CH{-}CH{=}CH{-}$$

$$(23)$$

$$\downarrow \text{ZnCl}_2 \;{-}CH{=}CHCHCl{-}; \;{-}\text{2HCl}$$

$$\left[\begin{array}{c} CH{-}CH{=}CH{-} \\ CH_3{-}\underset{\overset{|}{NH}}{C}{=}C{-}C(O){-}O{-}(CH_2){-} \\ \text{\small www}CH{-}CH{=}CH{-} \end{array}\right]_2$$

succession of chlorine atom replacements and ether group decompositions (reaction 24):

$$\text{\small www}C(O){-}CH{=}CH{-}\underset{\overset{|}{Cl}}{CH}\text{\small www} \xrightarrow[\underset{\underset{\text{MeCl}_2}{O}}{RCH{-}CHR';}]{k} \text{\small www}C(O){-}CH{=}CH{-}CH\text{\small www}$$
$$ClCHR{-}R'HCO$$

$$\downarrow k_{\text{deg}}$$

$$\text{\small www}C(O){-}(CH{=}CH)_{\overline{2}}\underset{\overset{|}{Cl}}{CH}\text{\small www} \xrightarrow[\underset{\overset{|}{O}}{RCH{-}CHR';\ \text{MeCl}_2}]{k} \cdots$$
$$+ClCHR{-}CHR'OH$$

$$(24)$$

The progress of the exchange reaction and subsequent decomposition of the substituent groups is evidenced by the accumulation of β-chlorohydrin in the low-molecular weight products obtained by extraction of the degraded polymer with diethyl ether, since in the IR spectra of the evolved products there is an absorption band in the region of $3400\ \text{cm}^{-1}$ typical of the OH group in β-chlorohydrin. Its formation by the interaction of HCl with an epoxy compound (EC) is ruled out in the presence of a metal-containing stabiliser–HCl acceptor. The formation of a halogen-containing ether, rather than a metal chloride (MeCl_2), in chlorine atom replacement by an ether group during CAG

TABLE 16

EFFECT OF MIXTURES OF BUTYL EPOXYSTEARATE (BES) OR BUTANEDIOL β-AMINOCROTONATE (β-BAC) WITH CHLORIDES OF COORDINATELY UNSATURATED METALS ON PVC DEHYDROCHLORINATION (448 K, 10^{-2} Pa) Ba stearate was the HCl acceptor

Organic stabiliser	Metal chloride	Concentration of components $\times 10^3$ (mol/mol PVC)		Dehydrochlorination rate $\times 10^6$ (mol HCl/mol PVC s)		
		Organic stabiliser	$MeCl_2$	W_{HCl}	W_r	W_p
None	None	—	—	0·78	0·08	0·70
BES	None	30·0	—	0·46	0·08	0·38
BES	$ZnCl_2$	5·0	1·5	0·37	0·08	0·29
BES	$ZnCl_2$	15·0	1·5	0·19	0·08	0·11
BES	$ZnCl_2$	30·0	1·5	0·11	0·08	0·03
BES	$ZnCl_2$	50·0	1·5	0·11	0·08	0·03
BES	$CdCl_2$	5·0	1·5	0·47	0·08	0·39
BES	$CdCl_2$	15·0	1·5	0·31	0·08	0·23
BES	$CdCl_2$	30·0	1·5	0·22	0·08	0·14
BES	$CdCl_2$	50·0	1·5	0·20	0·08	0·12
BES	$PbCl_2$	30·0	1·0	0·34	0·08	0·26
BES	$PbCl_2$	30·0	1·5	0·32	0·08	0·24
BES	$PbCl_2$	30·0	1·75	0·33	0·08	0·25
None	None	—	—	0·67	0·08	0·59
β-BAC	None	5·0	—	0·45	0·08	0·38
β-BAC	$ZnCl_2$	5·0	0·25	0·35	0·08	0·27
β-BAC	$ZnCl_2$	5·0	0·50	0·38	0·08	0·30
β-BAC	$ZnCl_2$	5·0	1·00	0·53	0·08	0·45
β-BAC	$CdCl_2$	5·0	0·25	0·40	0·08	0·32
β-BAC	$CdCl_2$	5·0	0·75	0·41	0·08	0·33
β-BAC	$CdCl_2$	5·0	1·50	0·37	0·08	0·29

interaction with an epoxy compound, distinguishes the process from PVC stabilisation with metal carboxylates. If the replacement of labile chlorine atoms proceeds effectively, the polyene growth rate will be negligible, and the total rate W_{HCl} during PVC degradation approximates to W_r.

The kinetics are described by eqns. (lii) and (liii).

$$\frac{d[HCl]}{dt} = k_r a_0 + k_p x \tag{lii}$$

$$\frac{dx}{dt} = -kx[c_0^{EC}]^n[c_0^{MeCl_2}]^m + k_{deg}(\bar{\gamma}_0 - x) \tag{liii}$$

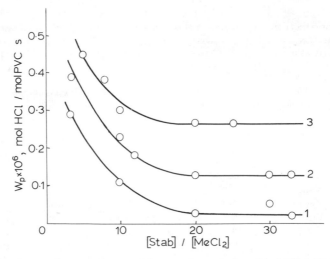

FIG. 13. Dependence of rate W_p on the relation [Stab]/[MeCl$_2$] during degradation (448 K, 10^{-2} Pa) of PVC stabilised with synergistic mixtures of additives. 1, Butyl epoxystearate–ZnCl$_2$; 2, butyl epoxystearate–CdCl$_2$; 3, butanediol β-aminocrotonate–ZnCl$_2$.

where n and m denote the reaction order for the corresponding components. The value of k_{deg} for a group of halogen-containing ethers, determined independently as required by eqn. (xxxvi) for the case of a modification with butyl epoxystearate, comes to $k_{\text{deg}} = (0.40 \pm 0.05) \times 10^{-3}\,\text{s}^{-1}$ at 448 K. It is one order higher than the value for the stability of ester groups in analogous macromolecular structures. The logarithmic transformation of equation (lii) makes it possible to estimate n, m, and k. It has been found that $n = m = 1$, and for synergistic mixtures that include butyl epoxystearate and ZnCl$_2$ $k = 86 \pm 6$ (for CdCl$_2$ $k = 42 \pm 5$) (mol/mol PVC)$^{-2}\,\text{s}^{-1}$ at 448 K.

8.7. Destruction of Polyene Sequences of Double Bonds

In the course of PVC thermal degradation the formation of polyene sequences of $(>\!C\!=\!C\!<)_{n>1}$ bonds is normally observed. Some stabilisers can react randomly with any double bond in the polyene. Termination of the active site of the polymer dehydrochlorination process will, however, occur only as a result of an addition reaction with an end-group, penultimate double bond (with reference to a β-chloroallylic group) or with any diene sequence that removes the destabilising influence of conjugated $>\!C\!=\!C\!<$ bonds in further decomposition of the adjacent vinyl chloride units. In general, such interaction can be both reversible and irreversible. It is assumed that (a) C=C bonds are converted into saturated groups via a second-order irreversible (reversible) reaction, (b) all $>$C=C$<$ bonds in the polyene have the same reactivity, and (c) the reaction rate constant k does not depend on the degree of conversion; the termination probability for the active sites of the polymer dehydrochlorination process can be defined as a function of time and by the kinetic parameters of the process.

The required parameters are easy to find by a computer simulation of the random process of interaction of the additive with $>$C=C$<$ bonds in the polyene by the Monte Carlo method. By analysing the behaviour of an ensemble of N_0 of growing polyenes, reacting reversibly with C_0 of the additive molecules with the rate constant k, it is possible to estimate the relative number of 'living' ends of polyenes $N'(t)/N_0$, as well as the number of unreacted $>$C=C$<$ bonds in the ensemble with reference to degradation of the reaction products with the constant k_{deg}. Satisfactorily reproducible results are achieved at $N_0 = 5000$ with selected model parameters. Correspondence of the simulated process with the actual one is established by comparing the experimental curves of HCl elimination and those calculated using eqn. (liv).

$$[HCl] = \int_0^t \left(k_r a_0 + k_p \bar{\gamma}_0 \frac{N'(t)}{N_0} \right) dt \qquad \text{(liv)}$$

Figure 14 shows the calculated curves for the number of 'living' active sites and the quantity of HCl evolved versus the reaction time with various values of k and k_{deg}. As in the case of a reversible interaction of the chemical additive with polyenes, the effect of a decrease in the PVC dehydrochlorination rate is expected to manifest itself when $(kc_0/k_{deg}) > k_p$.

Various reagents participate in reactions with polyenes, dienophilic compounds being the most significant among them. Interaction of dienophiles with the polyenes contained in PVC via the Diels–Alder reaction forms the basis of the stabilising action of various derivatives of maleic and fumaric acids.[28,66-68,71,112,119] In particular, a decrease in the rate of HCl elimination from PVC is observed in the presence of maleic anhydride as a result of a decrease in the number of dehydrochlorination-active sites. Diene condensation proceeds via one of the pathways in reaction 25.

(25)

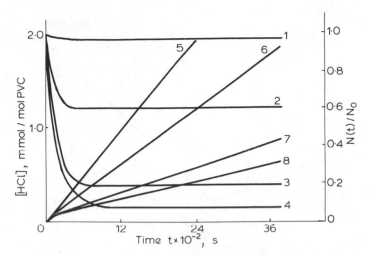

FIG. 14. Dependence of the number of 'living' (active) sites of PVC dehydrochlorination (448 K, $\bar{\gamma}_0 = 10^{-4}$ mol/mol PVC), calculated by the Monte Carlo simulation method of random additive interaction with $\searrow C{=}C\swarrow$ bonds in polyenes, versus time (1–4), and the quantity of evolved HCl calculated as required by eqn. (liv) for various k and k_{deg} and $c_0 = 20$ mmol/mol PVC. Values of k, (mol/mol PVC)$^{-1}$ s^{-1}: 1, 5—10^{-5}–10^{-4}; 2, 3, 4, 6, 7, 8—0.5×10^{-3}; k_{deg}, s^{-1}: 1, 5—10^{-4}–10^{-2}; 2, 6—10^{-4}; 3, 7—10^{-3}; 4, 8—10^{-2}.

Establishment of a stationary concentration of the polyene growth-active sites, $\bar{\gamma}_0 \cdot N'(t)/N_0$, has been found to yield kinetic curves of a linear character for $[\mathrm{HCl}] = f(t)$. Coincidence of the number of 'living' active sites, experimentally determined by the rate of PVC dehydrochlorination in the presence of maleic anhydride, with the calculated value is reached when the ratio $k/k_{deg} = 1.43$ (Fig. 15).

9.   CONCLUSIONS

The experimental data presented in this chapter make it possible to look at the reason for the abnormally low stability of polyvinyl chloride and its stabilisation in a basically new way.

The high rate of HCl elimination from PVC is accounted for by the formation of oxygen-containing unsaturated groups of the keto-allyl chloride (KAG) type; $\sim\!\!C(O){-}CH{=}CH{-}CHCl\!\!\sim$ within the macromolecule during the manufacture and storage of a polymeric

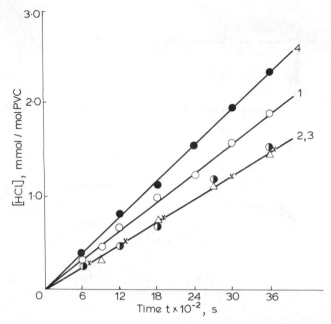

FIG. 15. HCl elimination during degradation of PVC stabilised with maleic anhydride (448 K, 10^{-2} Pa, Ba stearate used as HCl acceptor). Concentration of maleic anhydride, mmol/mol PVC: 1, 5·0 (○); 2, 20·0 (◑); 3, 40·0 (△); 4, PVC without additives; ×, points calculated via eqn. (liv) for $k/k_{deg} = 1·43$.

product; the 'KAG effect'. The observed high rate of HCl elimination from PVC, under various external influences, is due to the influence of neighbouring KAG groups with long-range order. The same effect is also operative in the formation of $\text{m—(CH=CH)}_{n>1}\text{—CHCl—CH}_2\text{m}$ polyene sequences in the course of PVC dehydrochlorination. β-Chloroallyl (CAG) activation of HCl elimination from PVC (i.e. the effect of neighbouring groups with short-range order), in the sense accepted in the literature, does not make any noticeable contribution to the kinetics of HCl elimination from PVC since groups with the structure $\text{mCH}_2\text{—CH=CH—CHCl—CH}_2\text{m}$ are rather stable.

In addition, KAG plays a major role in the process of PVC cross-linking and discolouration. Participation of the polyenes in secondary reactions results in the formation in PVC of a set of polyene sequences of various lengths and of different chemical structure. As a consequence the UV/visible spectrum of the degraded PVC is found to be a superposition of the spectra of individual polyenes of the type $O{=}\overset{\shortmid}{C}{-}(CH{=}CH)_n{-}$ and ${-}(CH{=}CH)_n{-}$.

In accordance with the above facts, the reactions that can be used for the chemical stabilisation of PVC should be selected so as to result in a reduction or complete elimination of the accelerating effects of abnormally structured groups: $\sim\sim C(O){-}(CH{=}CH)_n{-}$ in the first place, formed by decomposition of the neighbouring vinyl chloride units, followed by the units directly affected by the destabilising influence of abnormal conjugated structures.

An experimental study of stabilising additives, with reference to the significance of KAG in the process of PVC thermal degradation, has made it possible to define concretely the analogous chemical reactions in PVC responsible for their stabilising effect. In a number of instances (for example, in the case of organic phosphites, epoxy compounds and glycols) these have been found to be different reactions from those considered to be involved in the literature. In other cases the mechanism of a stabilising action follows the concepts already developed (for example, the replacement of labile chlorine atoms by more thermally stable groups by means of carboxylates of coordinately unsaturated metals or organotin compounds), although with a different kinetic interpretation. The discovery of the labile group, KAG, in PVC necessitated a detailed investigation into the mechanism and kinetics of the reaction of model KAG compounds with known PVC stabilisers. There may also be other possibilities for chemical reactions that would result in an effective stabilisation of the polymer. However, at present the primary problems in the field of chemical stabilisation of PVC consist in finding the most effective additives and combinations of additives on the basis of the mechanisms considered, rather than in elucidating the mechanisms of action of various types of stabilisers.

REFERENCES

1. SEYMOUR, R. B., Polymer News, **6** (1980) 106.
2. CHYTRY, V., OBEREIGNER, B. and LIM, D., Europ. Polym. J., **5** (1969) 379.
3. MAYER, Z., OBEREIGNER, B. and LIM, D., J. Polym. Sci., Part C, **33** (1971) 289.
4. MAYER, Z., J. Macromol. Sci., Part C, **10** (1974) 263.
5. GEDDES, W. C., Europ. Polym. J., **3** (1967) 267.
6. ABBAS, K. B. and SÖRVIK, E. M., J. Appl. Polym. Sci., **20** (1976) 2395.

234   K. S. MINSKER ET AL.

7. RAZUVAEV, G. A., PETUKHOV, G. G. and DODONOV, V. A., Vysokomol. Soedin., 3 (1961) 1549.
8. DANUSSO, F., PAJARO, G. and SIANESI, D., Chim. Ind., 41 (1959) 1170.
9. VARMA, I. K., GROVER, S. S. and GEETHA, C. K., Indian Plast. Rev., 18 (1972) 11.
10. BRAUN, D., Pure Appl. Chem., 26 (1971) 173.
11. MITANI, K., OGATA, T., AWAYA, H. and TOMARI, Y., J. Polym. Sci., Polym. Chem. Edn., 13 (1975) 2813.
12. DE VRIES, A. I., BONNEHAT, C. and CARREGA, M., Pure Appl. Chem., 26 (1971) 209.
13. BEZDADEA, E. C., BRAUN, D., BURUIANA, E., CARACULACU, A. A. and ISTRATE-ROBILA, J., Angew. Makromol. Chem., 37 (1974) 35.
14. ABBAS, K. B., J. Macromol. Sci., Part A, 12 (1978) 479.
15. MINSKER, K. S., BERLIN, A. A., KAZACHENKO, D. V. and ABDULLINA, R. G., Doklady Akad Nauk SSSR, 203 (1972) 881.
16. BRAUN, D. and QUARG, W., Angew. Makromol. Chem., 29/30 (1973) 163.
17. GUPTA, V. P. and ST. PIERRE, L. E., J. Polym. Sci., Polym. Chem. Edn., 11 (1973) 1841.
18. ZEGELMAN, V. I., TITOVA, V. A., BORT, D. N., POPOV, V. A., PAKHOMOVA, I. K., KONKHIN, YU. A., LISITSKY, V. V. and MINSKER, K. S., Plast. Massy, 8 (1980) 8.
19. CROZER, SZ. and CZLONKOWSKA, Z., J. Appl, Polym. Sci., 8 (1964), 1275.
20. RIGO, A., PALMA, G. and TALAMINI, G., Macromol. Chem., 153 (1972), 219.
21. CARACULACU, A. A., BEZDADEA, E. C. and ISTRATE-ROBILA, G. J., J. Polym. Sci., 8 (1970) 1239.
22. ASAHINA, M. and ONOZUKA, M., J. Polym. Sci., 2 (1964) 3305, 3515.
23. LISITSKY, V. V., KOLESOV, S. V., GATAULLIN, R. F. and MINSKER, K. S., Zh. Analit. Khimii, 33 (1978) 2202.
24. MINSKER, K. S., LISITSKY, V. V. and ZAIKOV, G. E., J. Vinyl. Technol., 2 (1980) 77.
25. MINSKER, K. S., BERLIN, A. A., LISITSKY, V. V. and KOLESOV, S. V., Vysokomol. Soedin., A19 (1977) 32.
26. LISITSKY, V. V., KALASHNIKOV, V. G., BIRYUKOV, V. P., MUSIKHIN, V. A. and MINSKER, K. S., Vysokomol. Soed. A23 (1981) 1081.
27. TALAMINI, G. and PEZZIN, G., Macromol. Chem., 39 (1960) 26.
28. MINSKER, K. S. and FEDOSEEVA, G. G., Destruktsiya i Stabilizatsiya Polivinilkhlorida, 2nd edn. (1979). Khimiya, Moscow.
29. BATAILLE, P. and VAN, B. T., J. Polym. Sci., Part A-1, 10 (1972) 1097.
30. MINSKER, K. S., BERLIN, A. A., LISITSKY, V. V., KOLESOV, S. V., GATAULLIN, R. F., MUKMENEVA, N. A. and ZAIKOV, G. E., Chem. Kunstst. Act., (1978) 73.
31. MINSKER, K. S., LISITSKY, V. V., KRONMAN, A. G., GATAULLIN, R. F. and CHEKUSHINA, M. A., Vysokomol. Soedin., A22, (1980) 1117.
32. ONOZUKA, M. and ASAHINA, M., J. Macromol. Sci., Part C, 3 (1969) 235.

33. VALKO, L., TVAROŠKA, I. and KOVARIK, P., *Europ. Polym. J.*, **11** (1975) 411.
34. TROITSKY, B. B., DOZOROV, V. A., MINCHUK, F. F. and TROITSKAYA, L. S., *Europ. Polym. J.*, **11** (1975) 277.
35. TROITSKY, B. B. and TROITSKAYA, L. S., *Vysokomol. Soedin.*, **A20** (1978) 1443.
36. MEYDTMANN, H. and RICK, G., *J. Phys. Chem.*, **30** (1961) 250.
37. CHYTRY, V., OBERREIGNER, B., LIM, D. and KRIVINKOVA, D., *Europ. Polym. J.*, **9** (1973) 649.
38. SHIMANOUCHI, T., TASUMI, M. and ABE, Y., *Makromol. Chem.*, **86** (1965) 43.
39. TVAROŠKA, I., KLIMO, V. and VALKO, L., *Tetrahedron*, **30** (1974) 3275.
40. WINKLER, D. E., *J. Polym. Sci.*, **35** (1959) 3.
41. GRASSIE, N., *Chem. Ind.*, (1954) 161.
42. ABBAS, K. B. and LAURENCE, R. L., *J. Polym. Sci., Polym. Chem. Edn.*, **13** (1975) 1889.
43. BAUM, B. and WARTMAN, L. H., *J. Polym. Sci.*, **28** (1958) 537.
44. KELEN, T., BALINT, G., GALAMBOS, G. and TÜDÖS, F., *J. Polym. Sci., Part C*, **33** (1971) 211.
45. VYMAZAL, Z., CZAKA, E., MEISSNER, B. and ŠTEPEK, J., *J. Appl. Polym. Sci.*, **18** (1974) 2861.
46. VYMAZAL, Z. *et al.*, *Plasty Kaučuk*, **11** (1974) 260.
47. FRENKEL, Ya. I., *Kineticheskaya Teoriya Zhidkostey* (1975). Nauka, Leningrad.
48. TROITSKY, B. B., TROITSKAYA, L. S. and RAZUVAEV, G. A., *Vysokomol. Soedin.*, **A13** (1971) 1183.
49. MINSKEL, K. S., MALINSKAYA, V. P. and PANASENKO, A. A., *Vysokomol. Soedin.*, **A12** (1970) 1151.
50. PUDOV, V. S. and PAPKO, P. A., *Vysokomol. Soedin.*, **B12** (1970) 218.
51. PUDOV, V. S., *Plast. Massy*, (2) (1976) 18.
52. GRASSIE, N., *Khimiya Protsessov Destruktsii Polimerov* (1959). Inostrannaya Literatara, Moscow.
53. BELLAMI, L., *Novyie Dannye po IK-spektroskopii Slozhnykh Molekul* (1971). Mir, Moscow.
54. SHEMYAKIN, M. M. and RED'KIN, I. A., *Zh. Obsch. Khim.*, **11** (1941) 1142.
55. SHEMYAKIN, M. M. and SCHUKINA, A. L., *Uspekhi Khim.*, **26** (1957) 528.
56. MUKMENEVA, N. A., AGADJANYAN, C. I., KIRPICHNIKOV, P. A. and MINSKER, K. S., *Doklady Akad. Nauk SSSR*, **233** (1977) 275.
57. MINSKER, K. S., MUKMENEVA, N. A., KOLESOV, S. V., AGADJANYAN, S. I., PETROV, V. V. and KIRPICHNIKOV, P. A., *Doklady Akad. Nauk SSSR*, **244** (1979) 1134.
58. MINSKER, K. S., KOLESOV, S. V. and PETROV, V. V., *Doklady Akad. Nauk SSSR*, **252** (1980) 627.
59. MINSKER, K. S., LISITSKY, V. V., VYMAZAL, Z., KOLINSKY, M., KALAL, J., SHVAREV, E. L., KOTLYAR, I. B., GORBACHEVSKAYA, I. I. and SAMOYLOVA, I. G., *Plast. Massy*, (1) (1976) 19.

60. MINSKER, K. S., LISITSKY, V. V., KOLINSKY, M., VYMAZAL, Z., BORT, D. N., LEBEDEV, V. P., PESINA, A. N. and IL'KAYEVA, E. M., *Plast. Massy*, (9) (1977) 44.
61. MINSKER, K. S., LISITSKY, V. V., ABDULLIN, M. I. and ZAIKOV, G. E., *Kunststoffe*, **5** (1980) 13.
62. MINSKER, K. S., BERLIN, A. A., LISITSKY, V. V., KOLESOV, S. V. and KORNEVA, R. S., *Doklady Akad. Nauk SSSR*, **232** (1977) 93.
63. BERLIN, A. A., MINSKER, K. S., KOLESOV, S. V. and BALANDINA, N. A., *Vysokomol. Soedin.*, **B19** (1977) 132.
64. FLORY, P. J., *Principles of Polymer Chemistry* (1953). Cornell Univ. Press, Ithaca, New York.
65. RAZUVAEV, G. A., TROITSKAYA, L. S., MYAKOV, V. N. and TROITSKY, B. B., *Doklady Akad. Nauk SSSR*, **170** (1966) 1342.
66. RAZUVAEV, G. A., TROITSKAYA, L. S. and TROITSKY, B. B., *J. Polym. Sci.*, *Part A*, **9** (1971) 2673.
67. KELEN, T., IVAN, B., NAGY, T. T., TURCSANYI, B., TÜDÖS, F. and KENNEDY, J. P., *Polym. Bull.*, **1** (1978) 79.
68. KELEN, T., *J. Macromol. Sci.—Chem.*, **A12** (1978) 349.
69. DANIELS, W. D. and REES, N. H., *J. Polym. Sci.*, *Polym. Chem. Edn.*, **12** (1974) 2115.
70. MINSKER, K. S. and KRATS, E. O., *Vysokomol. Soedin.*, **A13** (1971) 1205.
71. PLATE, N. A., LITMANOVICH, A. D. and NOA, O., *Makromolekulyarnye Reaktsii.*
72. SHTERN, E. S. and TIMMONS, S., *Elektronnaya Spektroskopiya v Organicheskoy Khimii* (1974). Mir, Moscow.
73. WOODS, G. F. and SCHWARTZMAN, L. H., *J. Am. Chem. Soc.*, **71** (1949) 1396.
74. SONDHEIMER, F., BEN-EFRAIM, D. A. and WOLOWSKY, R., *J. Am. Chem. Soc.*, **83** (1961) 1675.
75. MEBANE, A. D., *J. Am. Chem. Soc.*, **74** (1952) 5227.
76. BLOUNT, E. R. and FIELDS, M. J., *J. Am. Chem. Soc.*, **70** (1948) 189.
77. KRAUSS, W. and CRUND, H., *Z. Electrochemie*, **59** (1955) 872.
78. LEWIS, G. N. and COLVIN, D., *Chem. Revs.*, **25** (1939) 273.
79. OLEYNIK, N. P., VASILEYSKAYA, N. S. and RAZUVAEV, G. A., *Izv. Akad. Nauk SSSR*, (1968) (3), 482.
80. MYAKOV, V. N., TROITSKY, B. B. and RAZUVAEV, G. A., *Vysokomol. Soedin*, **B11** (1969) 661.
81. TROITSKY, B. B. and TROITSKAYA, L. S., *Vysokomol. Soedin* **A20** (1978) 1443.
82. IVANOVA, S. R., ZARIPOVA, A. G. and MINSKER, K. S., *Vysokomol. Soedin*, **A20** (1978) 936.
83. ELDERFIELD, R. and SHORT, F., *Geterotsiklicheskiye Soedineniya*, Ed. R. Elderfield (1961). In. Lit., Moscow.
84. LIEBMAN, S. D., REUWER, J. F., GOLLATZ, K. A. and NAUMAN, C. D., *J. Polym. Sci.*, *A–1*, **9** (1971) 1823.
85. SWEGLIADO, J. and GRANDI, F. Z., *J. Appl. Polym. Sci.*, **13** (1969) 1113.

86. KOLINSKY, M., DOSCOČILOVA, D., DRAHORADOVA, E., SCHNEIDER, B., ŠTOKR, J. and KUŠKA, V., J. Polym. Sci., A-1, 9 (1971) 791.
87. MINSKER, K. S., KOLESOV, S. V. and ZAIKOV, G. E., J. Vinyl Technol., 2 (1980) 141.
88. MINSKER, K. S., MUKMENEVA, N. A., BERLIN, A. A., KAZACHENKO, D. V., AGADJANYAN, S. I. and KIRPICHNIKOV, P. A., Doklady Akad. Nauk SSSR, 226 (1976) 1088.
89. POBEDIMSKII, D. G., MUKMENEVA, N. A. and KIRPICHNIKOV, P. A., Developments in Polymer Stabilisation—2, Ed. G. Scott (1980), p. 125. Applied Science Publishers, London.
90. AYREY, G., HEAD, B. C. and POLLER, R. C., J. Polym. Sci., Macromol. Rev., 12 (1974) 1.
91. BRAUN, D., THALLMAIER, N. and HEPP, D., Angew Makromol. Chem., 2 (1968) 71.
92. BRAUN, D. and HEPP, D., J. Polym. Sci., Part C, 33 (1975) 307.
93. KOLESOV, S. V., BERLIN, A. A. and MINSKER, K. S., Vysokomol. Soedin, A19 (1977) 381.
94. MINSKER, K. S., KOLESOV, S. V., CZAKO, E., SAVELYEV, A. P., VYMAZAL, Z. and KISELEVA, E. M., Vysokomol. Soedin, B21 (1979) 191.
95. SUZUKI, T., Pure Appl. Chem., 49 (1977) 539.
96. KLEMCHUK, P. P., Adv. Chem. Ser., 85 (1968) 1.
97. MICHEL, A., J. Macromol. Sci.—Chem., Part A, 12 (1978) 361.
98. KLIMSCH, P., Plaste Kautschuk, 6 (1977) 365.
99. HOANG, T. V., MICHEL, A. and GUYOT, A., Europ. Polym. J., 12 (1976) 337.
100. MINSKER, K. S., Plaste Kautschuk, 24 (1977) 375.
101. STARNES, W. H., Am. Chem. Soc., Polym. Prepr., 18 (1977) 493.
102. POLLER, R. C., J. Macromol. Sci.—Chem., Part A, 12 (1978) 373.
103. STARNES, W. H. and PLITZ, J. M., Am. Chem. Soc., Polym. Prepr., 16 (1975) 500.
104. MINSKER, K. S., KOLESOV, S. V. and KOTSENKO, L. M., Vysokomol. Soedin, A22 (1980) 2253.
105. HOANG, T. V., MICHEL, A. and GUYOT, A., J. Macromol. Sci.—Chem., Part A, 12 (1978) 411.
106. ANDERSON, D. F. and MCKENZIE, D. H., J. Polym. Sci., A-1, 8 (1970) 2905.
107. GUYOT, A. and MICHEL, A., Developments in Polymer Stabilisation—2, Ed. G. Scott (1980), p. 89. Applied Science Publishers, London.
108. KOLESOV, S. V., MALINSKAYA, V. P., SAVELYEV, A. P. and MINSKER, K. S., Fisiko-khimicheskiye Osnovy Sinteza i Pererabotki Polimerov, 1(47), (1976) 62.
109. MINSKER, K. S., MALINSKAYA, V. P. and SAYAPINA, V. V., Vysokomol. Soedin, A14 (1972) 560.
110. MINSKER, K. S. and MALINSKAYA, V. P., Plast. Massy, (3) (1972) 42.
111. MINSKER, K. S. and MALINSKAYA, V. P., Vysokomol. Soedin, A15 (1973) 200.
112. FOIGT, I., Stabilizatsiya Sinteticheskikh Polimerov Protiv Deystviya Sveta i Tepla (1972). Khimiya, Leningrad.

113. ALAVI-MOGHADAM, F., AYREY, G. and POLLER, R. C., *Europ. Polym. J.*, **11** (1975) 649; *Polymer*, **16** (1975) 833.
114. MINSKER, K. S., FEDOSEEVA, G. T., ZAVAROVA, T. B. and KRATZ, E. O., *Vysokomol. Soedin*, **A13** (1971) 2265.
115. BENGOUGH, W. J. and ONOZUKA, M., *Polymer*, **6** (1965) 625.
116. AYREY, G., POLLER, R. C. and SIDDIQUI, I. H., *J. Polym. Sci.*, *A*-1, **10** (1972) 725.
117. TROITSKII, B. B., TROITSKAYA, L. S., DENISOVA, V. N., NOVIKOVA, M. A. and LUSINOVA, Z. B., *Europ. Polym. J.*, **13** (1977) 1033.
118. HOANG, T. V., MICHEL, A. and GUYOT, A., *Polym. Deg. Stab.*, **1** (1979) 237.
119. TROITSKAYA, L. S., MYAKOV, V. N., TROITSKII, B. B. and RAZUVAEV, G. A., *Vysokomol. Soedin*, **A9** (1967) 2119.

Chapter 6

DEGRADATION AND PROTECTION OF POLYMERIC MATERIALS IN OZONE

S. D. RAZUMOVSKII and G. E. ZAIKOV

Institute of Chemical Physics, Academy of Sciences of the USSR, Moscow, USSR

SUMMARY

The basic reactions occurring when ozone reacts with saturated and unsaturated polymers and the nature of functional groups and unstable intermediates are considered. The action of ozone on polymers is usually accompanied by a change in surface properties (polarity, adhesion, surface tension) and the surface of unsaturated polymers becomes covered by cracks.

The possibility of polymer modifications by ozone and the protection of articles by atmospheric ozone action are also discussed.

1. INTRODUCTION

The change in properties of polymers in the presence of ozone has been studied intensively in the last few years.[1-4] On the one hand, this problem has aroused considerable interest because of a desire to modify the properties of traditional materials in order to widen their fields of application or to improve the operating characteristics of polymer products. On the other hand, a sharp increase in concentration of aggressive impurities in the atmosphere in which polymer articles operate, as a result of intensification of technological processes, has led to accelerated ageing and break-down and has also prompted

239

research in this direction. Examples of such intensification include an increase in the power and strength of the force field in which polymer dielectrics work (insulators in condensers, cable insulation, etc.). It appears that this phenomenon is to a large extent caused by the formation of ozone from atmospheric oxygen.[5]

Ozone adversely affects elastomers containing main-chain unsaturation. The surfaces of unprotected rubber goods (tyres, shoes, driving belts, etc.) crack rapidly leading to failure.[6,7]

2. OZONE-INDUCED DEGRADATION OF SATURATED POLYMERS

Saturated polymers and artifacts are often treated with ozone for scientific and industrial purposes.[8-12] Partial ozone degradation of macromolecules is accompanied by the formation of oxygen-containing functional groups and the surface of the polymer acquires new properties. Widely used carbochain polymers, such as polyethylene and polypropylene, suffer from the disadvantage that their surface tension is low and, as a result, adhesion to metals, dyes and other materials is

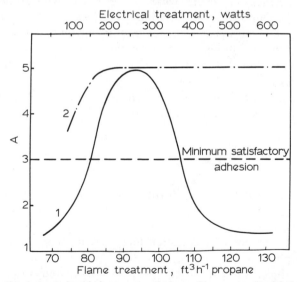

Fig. 1.   Adhesion of dye (A) to polyethylene film treated with burner flame (1) and barrier discharge (2), in relative units. Scale rating: 1, poor; 2, insufficient; 3, satisfactory; 4, good; 5, excellent.

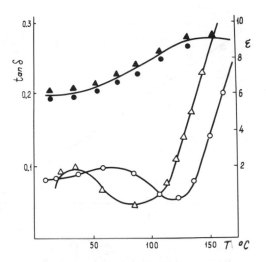

FIG. 2. Dielectric constant (ε) and dielectric loss factor (tan δ) versus temperature for polyvinyl chloride film. $\triangle$, $\blacktriangle$, control specimen; $\bigcirc$, $\bullet$, specimen exposed to ozone for 7 h.

rather poor. Treatment in a corona or barrier discharge or blowing with an ozone–air mixture greatly improves adhesion (Fig. 1).[13] Treatment of polyethylene films in a barrier discharge is widely used prior to applying a dye.[14] There are good grounds for believing that ozone is the main, although not the only, agent involved when a film is treated in a discharge.[15] The result is an increase in the surface tension and in a number of other properties of surface layers, namely hydrophilicity, the number of intermolecular bonds and crack resistance.[16] Figure 2 illustrates the change in dielectric loss tangent values (tan δ) of a polyvinyl chloride film prior to and after treatment with ozone caused by a change in a dipole-segmental mobility.[17] Ozone is also used for bleaching cellulose,[9] modification of lignin[18] and for destruction of water-soluble polymers, e.g. polyacrylamide.[19]

The action of ozone on polymer solutions is usually accompanied by a decrease in the molecular weight (Fig. 3) and accumulation of oxygen-containing functional groups (acids, ketones, peroxides, etc.).[20] Degradation proceeds readily at moderate temperature and at temperatures below 0°C due to the high reactivity of ozone. A comparison of the number of molecules of ozone reacted with the number of chain scissions shows that the number of chains broken per reaction event

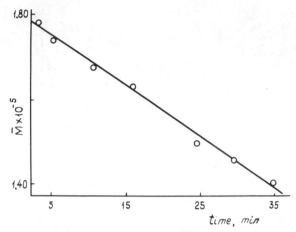

FIG. 3.   Variation in the molecular weight ($\bar{M}$) of polystyrene during its interaction with ozone in CCl_4, at 20°C.

during the initial stages of the experiment remains constant and depends on polymer structure. Table 1 gives the rate constants and the number of chain ruptures per reaction event for various polymers.[20]

Polymers containing phenyl rings in the main chain react with ozone most slowly, whereas polycyclic polymers (polynaphthalenes, polyanthracenes) or polymers with heteroatoms (polycarbonate) react much more readily. In a series of polymers with a saturated hydrocarbon chain the reaction rate increases on passing from polyisobutylene to polyvinylcyclohexane and the number of chain ruptures decreases. Polybutadiene and polyisoprene have the highest reaction rate and the smallest number of ruptures per reaction. The rate constants of the compounds given in Table 1 were determined in a bubbling reactor by following the procedure described for low-molecular compounds.[21,22]

It is convenient to consider the mechanism of the reaction of ozone with macromolecules of the polymer compounds using polystyrene as an example.[23-26] The appearance and properties of polystyrene powder are changed greatly under the action of ozone. The polymer turns yellow and the films prepared from it become brittle and opaque; dielectric properties deteriorate. The reaction is accompanied by the accumulation of functional groups typical of oxidising processes in general, namely carbonyl, peroxide and carboxyl.

Infrared spectroscopy has shown that the composition of the functional groups remains constant no matter whether ozone is reacted

TABLE 1

RATE CONSTANTS (k_1) FOR THE REACTION OF OZONE WITH DIFFERENT POLYMERS AND NUMBERS OF CHAINS BROKEN FOR EACH REACTION EVENT (φ) IN CCl_4 AT 20°C

Polymer	Structure unit	k_1 (litre/mol s)	φ
Polyphenyl		5×10^{-2}	—
Polynaphthalene		2×10^{3}	—
Polycarbonate		3	—
Polyisobutylene		$1 \cdot 2 \times 10^{-2}$	0·05
Polyethylene	$-[CH_2-CH_2]_n-$	$4 \cdot 6 \times 10^{-2}$	0·1
Ethylene–propylene copolymer		6×10^{-2}	0·06
Polypropylene		8×10^{-2}	0·1
Polystyrene		0·3	0·01
Polyvinyl-cyclohexane		0·8	—
Polyphenyl-acetylene		$1 \cdot 4 \times 10^{3}$	—

TABLE 1—*contd.*

Polymer	Structure unit	k_1 (litre/ mol s)	φ
Polybutadiene	$-[CH_2-CH=CH-CH_2]_n-$	6×10^4	0·006
Polyisoprene	$\left[CH_2-CH=\overset{\overset{\displaystyle CH_3}{\textstyle \vert}}{C}-CH_2\right]_n$	$4 \cdot 4 \times 10^5$	0·002
Cyclododeca-triene		$3 \cdot 5 \times 10^5$	—

with solid polystyrene or with its CCl_4 solutions.[27] The absorption maximum at $1740\ cm^{-1}$ is due to stretching vibrations of the $C=O$ groups (ketones) which are part of the macromolecules and are not removed upon reprecipitation (Fig. 4). The nature of these carbonyl compounds was established by their ability to react with hydrox-

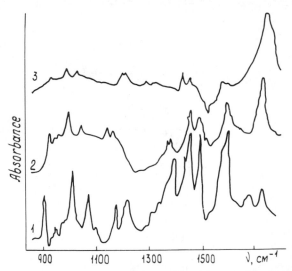

FIG. 4. Infrared spectra of polystyrene before (1) and after (2, 3) exposure to ozone: 2, for 70 min, $[O_3] = 1 \times 10^{-4}$ mol/litre; 3, for 20 min, $[O_3] = 1 \times 10^{-3}$ mol/litre.

ylamine and by relative stability towards ozone. Treatment of an ozonised polystyrene powder with an alcoholic solution of hydroxylamine chloride shifts the maximum from $1740\,cm^{-1}$ to $1680\,cm^{-1}$ (oximes). These measurements have shown[9] that the rate of aldehyde oxidation is 100 times higher than that of oxidation of ketone or $—CH_2$ groups and, therefore, their presence in the reaction products is hardly likely.

The $1000–1200\,cm^{-1}$ region of the spectrum is changed greatly during the reaction. A decrease in the number of the resolved bands and a general increase in absorption intensity indicate ozonised polystyrene molecules of reduced symmetry. This may be due to chain grafting during the process of ozonisation. Grafting is also confirmed by the fact that after ozonisation a major part of the polymer becomes insoluble and forms gels.

The primary reaction of ozone with polystyrene may follow any of the three possible routes shown in Scheme 1.

SCHEME 1

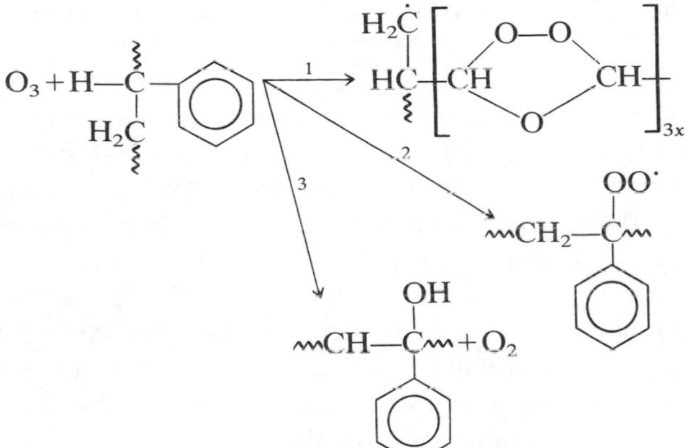

Reaction 1 would have given aromatic ozonides which would have been analysed as peroxidic compounds and should not have revealed carbonyl bands in IR spectra. The product would also have formed a large amount of CO and CO_2 upon heating to 70–100°C, and would have been decomposed with water. This, however, was not observed in the experiments.

Ozonisation of low molecular weight analogues of polystyrene, diphenylmethane[28] and isopropylbenzene leads to the formation of

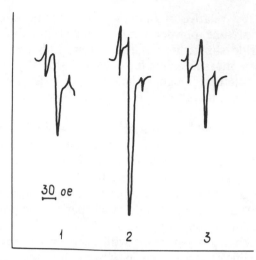

FIG. 5. Electron spin resonance spectra of polystyrene specimens interacting with ozone directly in the resonator cavity of an ESR spectrometer; 1, after 1 min; 2, after 6 min; 3, 30 min after the ozone supply is cut off. The spectra are given with reference to a Mn standard in the MgO lattice (splitting at 84 oersted).

aromatic ozonides in the form of a precipitate and analysis permits the evaluation of the share of reaction 1 in the total transformation balance. Within the temperature range from 0 to 60°C the share for both hydrocarbons varies from 1·2 to 3·6% leading to the conclusion that interaction of ozone with polystyrene follows reactions 2 and 3. When this experiment was carried out directly in the resonator of an ESR spectrometer, free radicals were observed in the reaction products (Fig. 5). A continuous delivery of ozone gives rise to the spectrum of peroxyl radicals[29] which, on cessation of the ozone stream, is transformed into a symmetrical singlet ($g = 2·0014$). The intensity of this signal is 0·3–0·5 of the initial peroxyl radical intensity. A prolonged reaction between polystyrene and ozone ($\tau \geqslant 20$ min) is accompanied by the superposition of the singlet on the signal of the peroxyl radical. Interpretation of the singlet is difficult because close analogues are absent but we can assume that it is either the ESR spectrum of the aromatic polyconjugated system of the polyphenylacetylene type[30] or that of a complex semiquinoid or phenoxyl radical.

Figure 6 shows the time dependences of the change in RO_2^- concentration for different O_3 concentrations and different sample surface areas.[27] The dependence (i)

$$\frac{[RO_2^{\cdot}]_\tau}{[RO_2^{\cdot}]_{st}} = f(\tau) \tag{i}$$

obtained by integration of eqn. (ii)

$$\frac{d[RO_2^{\cdot}]}{d\tau} = k_2[PS][O_3] - k_{dec}[RO_2^{\cdot}]^m \tag{ii}$$

has shown that under experimental conditions the stationary concentration of $RO_2^{\cdot}$ must be attained in 10^{-4}–10^{-2} s. Here $[RO_2^{\cdot}]_\tau$ and $[RO_2^{\cdot}]_{st}$ are current and stationary concentrations of peroxyl radicals, respectively, and τ is time; k_2 is the rate constant from reaction 2, k_{dec} is the rate constant of the predominant process of $RO_2^{\cdot}$ decay, $[PS]$ is the concentration of tertiary C–H bonds in polystyrene. The observed change in the signal of peroxyl radicals during the experiment may be due to interaction of ozone with the reaction products whose accumulation results in an increase of $[RO_2^{\cdot}]_{st}$. The concentration of free radicals grows linearly with an increase in the specific surface area of

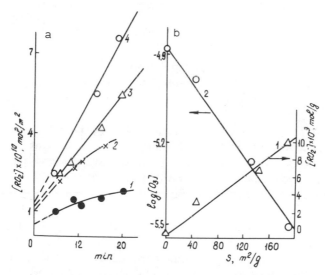

FIG. 6. Concentration of peroxyl radicals and ozone versus time of exposure (a) ozone and (b) polymer specimen surface. (a) $[O_3]$ values (mol/litre): 1, $1 \cdot 5 \times 10^{-3}$; 2, $1 \cdot 9 \times 10^{-3}$; 3, $2 \cdot 2 \times 10^{-3}$; 4, $2 \cdot 6 \times 10^{-3}$. Surface area, $S = 120\,\mathrm{m}^2$. (b) 1, Peroxyl radical concentration; 2, ozone concentration at the reactor outlet versus specimen surface area (S).

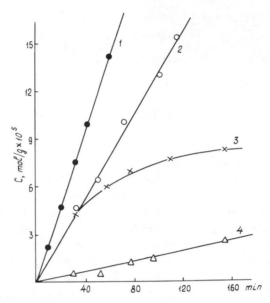

Fig. 7.   Kinetics of functional group accumulation during interaction of ozone with polystyrene. 1, Absorbed ozone; 2, ketones; 3, peroxides; 4, $[O_3]=$ 2×10^{-5} mol/litre.

the polymer(s) (see Fig. 6). Extrapolation of the experimental curves to the ordinate gives an approximate value of $[RO_2^{\cdot}]_{st}$ which rises with a change in ozone concentration at the reactor inlet from $0 \cdot 7 \times 10^{-10}$ to $1 \cdot 7 \times 10^{-10}$ mol/m^2.

The rate of ozone addition, determined from the difference between ozone concentration at the inlet and outlet, as well as the rate of accumulation of functional groups (with the exception of peroxides at the late reaction stages) remained constant with time (Fig. 7). Ozone consumption was $1 \cdot 4 \pm 0 \cdot 3$ mol per mol of functional groups. A somewhat greater than stoichiometric consumption of ozone may be due either to its participation in secondary reactions or its partial decomposition on the polymer surface.

The question of the predominant mechanism of transformation of peroxyl radicals into the reaction products is one of the main problems in the polymer destruction. It has already been mentioned that carboxyl groups accumulate in polystyrene only when intermediate radicals decompose. The dependence of the content of peroxide compounds on $[O_3]_0$ shows that at the initial reaction stages the relation-

SCHEME 2

$$O_3 + RH \rightarrow RO_2^{\cdot} + {}^{\cdot}OH \tag{2}$$

$$RO_2^{\cdot} + RH \rightarrow ROOH + R^{\cdot} \tag{4}$$

$$RO_2^{\cdot} \rightarrow products + r^{\cdot} \tag{5}$$

$$2RO_2^{\cdot} \rightarrow products \tag{6}$$

$$2r^{\cdot} \rightarrow products \tag{7}$$

ship $[ROOH] = k[O_3]\tau$ holds. From Scheme 2 it follows that the mechanism of transformation of peroxyl radicals into other products affects the kinetics of peroxide accumulation:

$$[ROOH]_6 \sim \propto \sqrt{[O_3]\tau}, \qquad [ROOH]_5 \sim \propto [O_3]\tau$$

From the above relationship it is seen that the kinetics of accumulation of peroxides agrees best with the assumption that they are formed predominantly by subsequent reactions of RO_2. The decomposition of peroxyl radicals of similar structure in the liquid phase is well known[31,32] although it is not a dominating process at low temperatures (20–100°C). However, since a polymer peroxyl radical is considerably less mobile, the relative contribution from this reaction in the total transformation can be appreciably greater. Free valences may disappear as a result of bimolecular interaction of radical decomposition products with one another or with primary peroxyl macroradicals.

The initial stages of the reactions of ozone with polystyrene can be represented by Scheme 3. The proposed scheme somewhat differs from that for the thermo-oxidative destruction of polystyrene[33,34] in that the low temperatures and high rates of radical formation involved give rise to conditions under which the major proportion of the products is formed by direct decomposition of peroxyl radicals and only 15–20% arises from chain reactions. Acids constitute a small part of the reaction products and can be formed either as a result of oxidation of phenoxyl radicals or their transformation products, or as a result of decomposition of aromatic ozonides. In both cases the reaction products have to contain the same compounds, namely formic and glyoxalic acids. Upon oxidation of solid polystyrene the acids are formed in very small amounts and attempts to identify them were unsuccessful. In the case of polystyrene solution in CCl_4, both acids were identified chromatographically (Celite 545, polyethylene glycol adipate 20%, 2 m × 6 mm column, 140°C, helium at 100 ml/min).

S. D. RAZUMOVSKII AND G. E. ZAIKOV

SCHEME 3

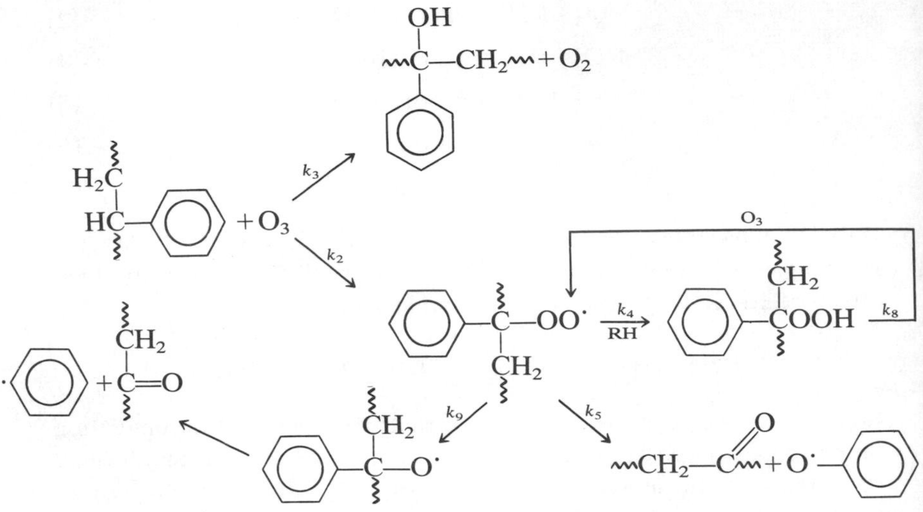

A change in the ESR spectrum during the reaction indicates that the disappearance of a free valence in the compounds whose ESR spectrum is of the singlet type (phenoxyl radicals) is due to their interaction with ozone or with the products of decomposition of peroxyl radicals since in the absence of ozone these radicals are rather stable. The assumption of the reaction with $RO_2^.$ contradicts both the relationship $[RO_2^.] = f(\tau)$ and the observed transformation of the $RO_2^.$ signal into a singlet after the ozone supply has stopped.

The appearance of intermolecular cross-links is a feature specific to the ozonisation of polystyrene in the solid phase. No such reaction is observed in dilute solution in chloroform. This cross-linking may be due to (a) the formation of polymer ozonides by attack at a phenyl ring; (b) formation of peroxide bonds between the chains during recombination of $RO_2^.$ or $RO^.$; or (c) reaction of hydroxyl, phenoxyl or hydrocarbon radicals with one another. Reactions (a) and (b) would give rise to unstable bonds, broken easily either at 100–110°C or under the action of HF. However, neither boiling of a swollen gel in solvents nor prolonged storage of a gel swollen in CCl_4 over a KI/H_2SO_4 mixture decomposes the network structure of the gel. Evidently, the structure of the cross-links is either $\diagup C{-}O{-}C\diagdown$ or $\diagup C{-}C\diagdown$.

A possible reaction of the type shown in Scheme 4 was proposed by

SCHEME 4

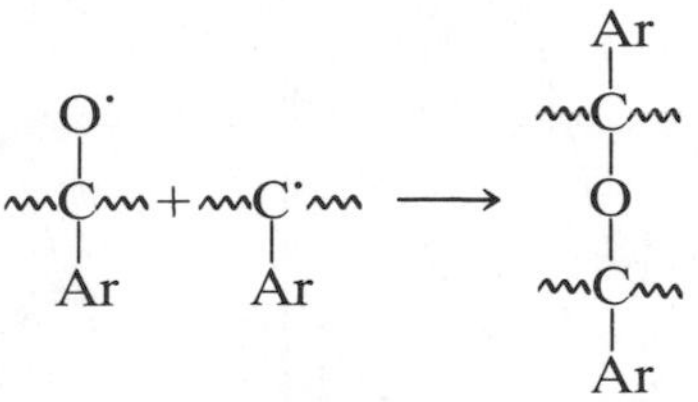

$$
\begin{array}{ccc}
\text{O}^{\bullet} & & \overset{\displaystyle \text{Ar}}{\underset{\displaystyle }{}} \\
\mid & & \sim\!\!\!\sim\!\text{C}\!\sim\!\!\!\sim \\
\sim\!\!\!\sim\!\text{C}\!\sim\!\!\!\sim + \sim\!\!\!\sim\!\text{C}^{\bullet}\!\sim\!\!\!\sim \longrightarrow & & \text{O} \\
\mid \qquad\quad \mid & & \mid \\
\text{Ar} \qquad\; \text{Ar} & & \sim\!\!\!\sim\!\text{C}\!\sim\!\!\!\sim \\
& & \mid \\
& & \text{Ar}
\end{array}
$$

Medvedev and co-workers[35] as a result of a study of the kinetics of decomposition of *tert*-butyl hydroperoxide in isopropylbenzene. The properties of articles manufactured from polymers and the possibility of their modification depends to a great extent on the site of the reaction and on the distribution of the reaction products.

Attenuated total reflection (ATR) spectroscopy has shown that the concentration of carbonyl groups in strongly ozonised polymer films decreases with distance from the surface.[36,37] This can easily be explained by considering the two processes: ozone diffusion and ozone consumption during interaction with the polymer (iii):

$$
\frac{d[O_3]}{dl} = D\,\frac{d^2[O_3]l}{dl^2} - k_v[PS][O_3]l \tag{iii}
$$

where D is the diffusion coefficient for ozone, l is distance from the surface and k_v is the rate constant of the bulk reaction. Competition between these two processes favours ozone consumption as ozone moves into the polymer depth. The depth of penetration depends on the ratio of the diffusion coefficient (D) to the rate constant of the reaction in the bulk (k_v).

The ATR method proved to be insufficiently sensitive for a detailed analysis of the topochemistry of the process under study. Much more important information was obtained by studying the kinetics of the reaction of ozone with polymer films of different thickness[38] and with polymers adsorbed on surfaces.[39]

It was found that the reaction of ozone with the polymer surface is faster by several orders of magnitude than the rate in the bulk and, consequently, the rate of formation of functional groups is higher. Figure 8 illustrates the dependence of the effective rate of ozone adsorption (W_{ef}) on the thickness of the sample. The segments cut on the ordinate axis make it possible to determine the reaction rate on the surface, the angle coefficients and the rate of the reaction in the bulk. Under the experimental conditions the rate constant of the reaction of

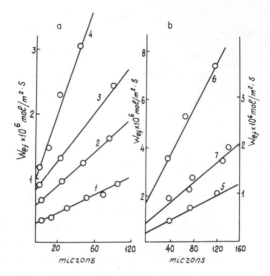

FIG. 8.   Effective ozonisation rate, W_{ef}, versus film thickness for polystyrene (a) and low-density polyethylene (b). 1, 20°C; 2 and 6, 30°C; 3, 40°C; 4, 50°C; 5, 60°C; 7, 70°C.

the adsorbed ozone with polystyrene at 18°C is $k_{ads} = 0·05$ litre/mol s whereas in the bulk $k_v = 0·22$ litre/mol s.[39] The difference in the rates is caused by a relatively larger ozone concentration on the surface than in the volume due to physical adsorption.

A study of ozone adsorption on polystyrene has shown that it is close to ideal and is well described by known laws; in particular, at temperatures close to room temperature it obeys Henry's law[40] (iv).

$$[O_3]_{ads} = \beta[O_3]_{gas} \qquad (iv)$$

where $\beta \simeq 100$. An average value for the heat of adsorption is $3·2 \pm 0·2$ kcal/mol.

At the same time solubility of ozone in near-the-surface polymer sites is close to or somewhat less than solubility in the corresponding liquids.[38]

$$[O_3]_v = \alpha[O_3]_{gas} \qquad (v)$$

where $\alpha \simeq 0·6$. Thus, the reaction of ozone with a polymer proceeds, mainly, on the surface of the polymer. This conclusion agrees well with the experimentally observed dependence of the reaction rate and

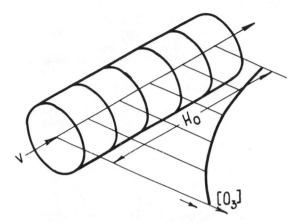

FIG. 9.   Distribution of ozone concentration over the reactor height.

concentration of free radicals on the surface area (Fig. 6) and with the absence of any dependence of the reaction rate on the mass of the sample.[38]

In a typical experiment the gas flow containing ozone was passed at a flow rate v (litre/s) through a cylindrical layer of polymer powder (for instance, polystyrene) of height H (Fig. 9). The choice of a reactor with a small cross-section ($d = 0.6$–1.2 cm) and a high H to d ratio ($\geqslant 10$) ensured a constant rate of gas flow. Ozone diffusion to the polymer surface did not limit the reaction rate. Varying particle size, type of filling (continuous or discontinuous) or H/d (from 6 to 20) did not affect the reaction rate, and the rate constant remained the same with varying v. The rate of delivery of the gas phase and the amount of the polymer were chosen so that the ozone concentration at the outlet of the reactor was appreciably lower than that at the inlet. Most of the experiments were conducted at $v = 0.9 \times 10^{-3}$ litre/s, $\alpha = 0.6$ cm and $H = 4$–10 cm (the weight of polystyrene was 0.14–0.45 g). After passing through the layer H, the concentration of ozone in the gas decreased according to eqn. (vi),

$$[O_3]_{gas} = [O_3]_0 \exp \left(-\beta k_{ads}[PS] \cdot \frac{SP}{v} \right) \qquad \text{(vi)}$$

where S is the specific surface area (m²/g); P is the weighed amount of polystyrene (g); β is the proportionality coefficient from the bulk concentration $[O_3]$ gas (mol/l) to the surface $[O_3]_{ads}$ (mol/m²).

TABLE 2
RELATIVE FUNCTIONAL GROUP CONCENTRATIONS IN POLYSTYRENE ON
TREATMENT WITH OZONE
$[O_3] = 3 \times 10^{-5}$ mol/litre; $P = 0.601$ g; treatment time 50 min

Sample Source	Relative functional group concentration, $A \times 10^5$ (mol/g)		
	$\left[-\underset{\underset{O}{\parallel}}{C}- \right]$	[—O—O—]	$\left[-C\overset{\displaystyle O}{\underset{\displaystyle OH}{}} \right]$
inlet	10·0	3·5	1·4
outlet	4·0	1·3	0·8

In accordance with this model (Fig. 9) the rate of accumulation of functional groups must vary along the reactor. This was experimentally confirmed; Table 2 gives the concentrations of functional groups in the upper and lower layers of polystyrene. From eqn. (vi) it follows that $\log [O_3]_{gas}/[O_3]_0$ must be a linear function of S. This was also confirmed experimentally (see Fig. 6).

Table 3 relates the values of $[O_3]_{gas}$ and $\beta k_{ads}[PS]$ at different temperatures. The dependence, $\log \beta k_8[PS] = f(1/T)$, is a straight line, from the slope of which the total activation energy can be determined ($E_\Sigma = 4.1$ kcal/mol). The activation energy of the chemical process is, evidently, greater than the total activation energy by the value of the heat of adsorption (ca 3 kcal/mol), i.e. $E = E_\Sigma + \rho \approx 7$ kcal/mol. The data of Figs 7 and 8 permit the calculation of the effective rate constant for the reaction of ozone with polystyrene hydroperoxide (Table 4).

The action of ozone on other polymers (polyethylene, polyvinyl-cyclohexane) is also accompanied by the formation of peroxyl radicals.[20,41] Figure 5 presents ESR spectra recorded upon the interaction

TABLE 3
DEPENDENCE OF THE RATE OF REACTION OF OZONE WITH POLYSTYRENE
ON TEMPERATURE (°C)
$S = 190$ m^2/g; $P = 0.14$ g; $[O_3]_0 = 1.4 \times 10^{-5}$ mol/litre

Parameter	Temperature (°C)					
	20	9	0	−11	−20	−42
$10^5 \times [O_3]_{gas}$	0·29	0·42	0·65	0·8	0·92	1·10
$10^5 \times \beta k_1[PS]$	5·5	3·6	2·5	1·5	0·7	0·4

TABLE 4

RATE CONSTANTS OF THE REACTIONS PROCEEDING DURING THE INTERACTION OF OZONE WITH POLYSTYRENE AT $20°C$

$$S = 190 \text{ m}^2/\text{g}; \quad [O_3] = 1\cdot5\times10^{-3} + 1\times10^{-5} \text{ mol/litre}$$

k_i	Calculation equation	Value
$\beta k_1[PS]$	$\dfrac{V}{PS}\ln\dfrac{[O_3]_0}{[O_3]_{gas}} = \dfrac{0\cdot9\times10^{-3}\times0\cdot7}{0\cdot94\times190\times0\cdot43}$	$5\cdot5\times10^{-5}$ litre/m^2 s
$k_2[PS]$	$\dfrac{V_{peroxides}}{[RO_2^{\cdot}]} = \dfrac{8\cdot4\times10^{-10}}{0\cdot8\times10^{-12}}$ or $k_4\dfrac{V_{peroxides}}{V_{ketones}+V_{acids}}$	1×10^3 s^{-1}
βk_3	$\dfrac{\beta k_1[PS]}{[ROOH]_{max}} = \dfrac{5\cdot5\times10^{-5}}{8\times10^{-7}}$	$0\cdot7\times10^2$ litre/m^2 mol s
k_4	$\dfrac{\beta k_1[PS][O_3]}{[RO_2^{\cdot}]} = \dfrac{5\cdot5\times10^{-5}\times1\cdot5\times10^{-3}}{0\cdot7\times10^{-10}}$	$1\cdot3\times10^3$ s^{-1}

of ozone with powders of polyvinylcyclohexane, polycarbonate, polyethylene and polymethylmethacrylate. The polymer, in amount 0·15–0·3 g, was introduced into an ampoule 6 mm in diameter and an ozone–oxygen mixture was fed through a capillary into the powder ($v_{O_2} = 50$ ml/min, $[O_3]_0 = 2\cdot2\times10^{-3}$ mol/litre). Within 3 minutes from the beginning of ozone supply, paramagnetic species could be measured in all the polymers. The concentrations of the radicals and their g-factors are given in Table 5.

In polyvinylcyclohexane, polystyrene and polycarbonate, the shapes of the signals were similar to those of the peroxyl radical. In polyethylene the signal had a more complicated form and appeared to

TABLE 5

THE VALUES OF g-FACTOR AND $[RO_2^{\cdot}]$ ON TREATMENT OF POLYMERS WITH OZONE

Parameter	Polymer		
	Polycarbonate	Polystyrene	Polyvinyl-cyclohexane
g-factor	2·015	2·012	2·014
$10^{-15}\times[RO_2^{\cdot}]$ spin/g	1·3	3·2	8·7

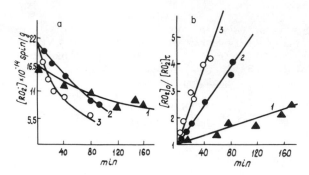

FIG. 10. Kinetics of degradation of peroxyl radical concentration $[RO_2^{\cdot}]$ (a) and time dependence of $[RO_2^{\cdot}]_0/[RO_2^{\cdot}]_\tau$ (b) in polyvinylcyclohexane at various temperatures: 1, $-10°C$; 2, $0°C$; 3, $14°C$.

be a superposition of the signals of two different radicals. After stopping the ozone stream the peroxyl radicals disappeared for several minutes. In polymers containing aromatic rings, instead of the peroxyl radicals, initially paramagnetic species were formed which give a narrow singlet absorption spectrum $(\Delta H = 8·5 \text{ Oe})$. Prolonged ozone treatment leads to the replacement of these species by peroxyl radicals. In the absence of ozone their concentration remained virtually constant for a month, but with increasing temperature they were gradually consumed.

The kinetics of the decay of peroxyl radicals were studied for polyvinylcyclohexane which had an intense signal; its disappearance was not accompanied by the appearance of the singlet. Figure 10 shows the decay of peroxyl radicals in polyvinylcyclohexane in air at different temperatures. It is seen that after stopping the ozone stream the concentration of the radicals gradually decreased and the shape of the ESR spectrum remained unchanged. To elucidate the order of the reaction, the kinetics curves were plotted in the coordinates $[RO_2^{\cdot}]_0/[RO_2^{\cdot}]_\tau = f(\tau)$. Figure 10(b) shows that the rate of the decay of peroxyl radicals is second order. The kinetic parameters of the reaction were determined by measuring the absolute initial concentration of the radicals; the activation energy of the decay of the radicals was evaluated from the temperature dependence of the rate constant described by eqn. (vii),

$$\frac{k}{\Delta l} = 4 \times 10^{-11} \exp\left(\frac{-10\,000 \pm 1000}{RT}\right) \text{cm}^3 \text{ s}^{-1} \qquad \text{(vii)}$$

where $\Delta l = V/Sl$ (V is the polymer volume, S is the specific surface area, l is the mean thickness of the surface layer where the radicals are concentrated).

Peroxyl radicals formed during ozonisation are rigidly bonded to the polymer matrix and at temperatures below the glass transition, their mobility is insufficient to explain the rate of recombination observed experimentally. On the other hand, during ozonisation of polystyrene and polyethylene[27,49] the ratio of the total amount of the oxidation products to the amount of the consumed ozone is close to 1 which means that the transport of free valence due to relay radical transfer is negligible.

A study of the accumulation of the ozonisation products in polystyrene and polyethylene[27,49] has shown that the kinetics of formation of peroxides agree well with the assumption of a predominant decomposition of the primary peroxyl radicals. Evidently, at low temperatures migration of a free valence is determined, mainly, by the reactions following the decomposition of peroxyl radicals in accord with Scheme 5 which includes a number of elementary stages:

SCHEME 5

$$RH + O_3 \rightarrow RO_2^{\cdot} + {\cdot}OH \qquad (2)$$

$$RO_2^{\cdot} + RH \rightarrow ROOH + R^{\cdot} \qquad (4)$$

$$R^{\cdot} + O_2 \rightarrow RO_2^{\cdot}$$

$$RO_2^{\cdot} \rightarrow r^{\cdot} + products \qquad (5)$$

$$r^{\cdot} + RH \rightarrow products \qquad (10)$$

$$r^{\cdot} + RO_2^{\cdot} \rightarrow products \qquad (11)$$

$$2r^{\cdot} \rightarrow products \qquad (7)$$

Hydroxyl radicals, peroxyl radicals with the free valence at the end of the chain, or low-molecular weight radicals are intermediates. They are very mobile and recombine with other radicals which results in disappearance of a free valence. When concentrations of $R^{\cdot}$ and $r^{\cdot}$ are stationary, the rate of reaction 6 is low because of the low concentration of $r^{\cdot}$; when the system contains no ozone, we obtain eqn. (viii).

$$\frac{d[RO_2^{\cdot}]}{d\tau} = \frac{2k_5 k_{11}[RO_2^{\cdot}]^2}{k_{10}[RH] + k_{11}[RO_2^{\cdot}]} \qquad \text{(viii)}$$

When $k_{10}[RH] \gg k_{11}[RO_2^{\cdot}]$, reaction of $r^{\cdot}$ with a macromolecule

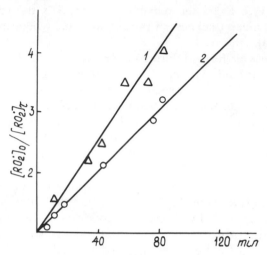

FIG. 11.   Kinetics of degradation of peroxyl radicals in polyvinylcyclohexane with various degrees of crystallinity at 0°C: 1, 45%, $[RO_2^{\cdot}]_0 = 18{\cdot}0 \times 10^{11}$ litre^{-1}; 2, 20%, $[RO_2^{\cdot}]_0 = 21{\cdot}0 \times 10^{11}$ litre^{-1}.

becomes more probable than recombination of two r$^{\cdot}$. In this case the decay of the peroxyl radicals is second order.

Literature data[43,44] show that in polymers containing amorphous and crystalline regions, recombination of peroxyl radicals, uniformly distributed over the sample volume, is neither first- nor second-order. Recombination in this case is a superposition of two second-order reactions with different rate constants for the amorphous and crystalline regions. However, no deviation from the second-order law was observed in the study of the decay of the radicals formed in polyvinylcyclohexane (Fig. 11) under the action of ozone. Evidently, due to the high lability of the C–H bond at a tertiary carbon atom in polyvinylhexane, and the small exposure time to ozone, peroxyl radicals concentrate only in a surface layer of the sample during ozonolysis. The thickness of this layer in the amorphous region is much greater than in the crystalline region due to the difference in the diffusion coefficients of ozone in the two phases, and most of the radicals are in this layer. Therefore, the radical decay curves are described by the equation of a bimolecular reaction and the value of the activation energy coincides with that found by Lebedev[43] for the decay of peroxyl radicals in the amorphous phase of Teflon.

The study of the kinetics of degradation and of the mechanism of the reactions proceeding under the action of ozone on polyethylene and polystyrene has shown that there are many features common to these two polymers.[42]

SCHEME 6

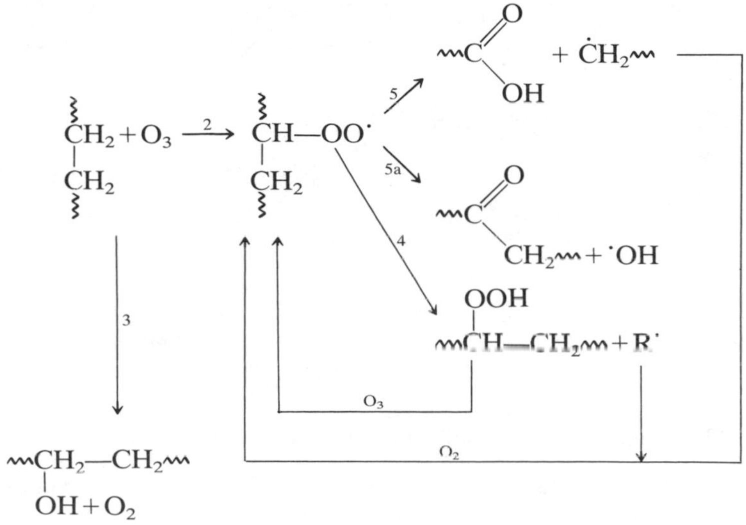

Kinetics of ozone-induced destruction of saturated polymers, when the formation of cross-links is neglected, are described by a linear law. For instance for polyethylene we can deduce eqn. (ix) by reference to Scheme 4.

$$W_g = k_\Sigma \frac{k_5}{k_5 + k_{5a} + k_4} [O_3][\text{Polymer}] \qquad \text{(ix)}$$

This agrees well with a drop of the molecular weight, accumulation of functional groups, and other parameters.

3. OZONE DEGRADATION OF UNSATURATED POLYMERS

Exposure of unsaturated polymers, primarily rubber and rubber products, to ozone results in surface cracking, stress relaxation, loss of mechanical strength, and ultimately failure. In view of the great practical importance of the detrimental effects of atmospheric ozone

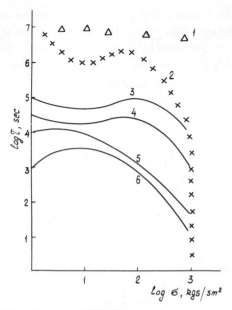

FIG. 12. Durability (τ) of polybutadiene vulcanisates in ozonised air with varying stress (σ). 1, Ozone-free air; 2, atmospheric air; 3, $[O_3]=$ $3\cdot3\times10^{-8}$ mol/litre; 4, $8\cdot5\times10^{-8}$ mol/litre; 5, 40×10^{-8} mol/litre; 6, 58×10^{-8} mol/litre.

on rubbers the process of ozone cracking has been covered in a number of reviews summarising the results of experiments concerned with ozone cracking of rubbers and with ways of preventing it.[45-54]

The degradation of rubbers under the influence of atmospheric ozone is usually observed under conditions when other processes of oxidative ageing are slow and the polymer properties remain otherwise unchanged for a long period of time. Figure 12 illustrates the results of determining the durability of polybutadiene vulcanisate in the presence of ozone at various concentrations.[55] It can be seen that in the absence of ozone the rubber is much more durable without any perceptible changes in the specimen throughout the experiment. Only mechano-chemical processes at high stresses and strains occur at a comparable rate in the presence and absence of ozone.

Serious difficulties are encountered in evaluating the results of ozone degradation of rubbers.[49,56] The existing standards in different countries are the result of efforts by many investigators to develop compatible and reproducible testing techniques which, however, are far from being perfect (see *Developments in Polymer Stabilisation—4*, Chapter 4). The dissimilar natures of the starting polymers and of other rubber compounding ingredients are responsible for wide differences in the degradation rate and make it necessary to use varying ozone concentrations to complete tests within a reasonable period of time. The evaluation of the time it takes for cracks to appear and the time it takes for the specimen to fail completely are equally important for prediction of the performance characteristics. However, in most cases only one of these parameters is determined, and the correlations between the two are not sufficiently reliable.

The lack of quantitative data in evaluating the results of studying the protective action of waxes, antiozonants and the effect of other ingredients on the ozone resistance of rubbers is particularly acute. It is well known that under the influence of ozone, cracks appear first of all on the stretched portions of test specimens. Unstrained specimens do not crack at all in most cases.[57,58] Table 6 summarises the results of testing two-layer specimens obtained by bonding together two halves of a hollow rubber cylinder with their outer layers inward (1—tension) and with their inner layers inward (2—compression) in comparison with a flat unstrained specimen (3).

TABLE 6

TIME TO CRACKING (min) AFTER EXPOSURE TO OZONE OF TWO-LAYER SPECIMENS OF DIFFERENT RUBBERS[59]

Ozone concentration[a] (mol/litre)	Polybutadiene rubber			Butadiene rubber (SKS-30)		
	1	2	3	1	2	3
1.65×10^{-6} (3.7×10^{-3})	—	—	—	85	10	20
4.46×10^{-7} (1×10^{-3})	132	11	22	40	70	—

[a] Values in parentheses, vol. %.

The time to cracking and the cracking rate are in a complex relation to the applied stress and strain.[47,60] The data presented by different authors are sometimes controversial. Some believe that there exist certain critical strains and stresses below which no cracking takes place,[57] others contend that these critical phenomena occur because of imperfect experimental procedures. Some authors find that in the low-strain region the crack propagation rate does not depend on stress, others say that there is no such region.[47]

If one assumes that in a specimen exposed to ozone, cracks start propagating after some period of time (τ_u) and that the crack propagation rate is constant,[58] it is possible to establish a relation between the time it takes for the first crack to appear (τ_c) and the time to specimen failure (τ_d).

$$\tau_d \sim \tau_u + A\tau_c \qquad (x)$$

At $\tau_u \ll A\tau_c$ this relation takes for form $\tau_c/\tau_d \simeq A$.

Some authors have observed cases where these relations are substantially constant with ozone concentration varying within broad limits (Table 7) in the presence of inert fillers and plasticisers.[61a] The ratio τ_c/τ_d is sensitive to strain and to the presence of antioxidants and antiozonants in the compound.

In most cases τ_c and τ_d are related to the ozone concentration in a similar manner.[61b,62].

$$\tau_c[O_3]^n = B \qquad (xi)$$

$$\tau_d[O_3]^n = C \qquad (xii)$$

TABLE 7
RATIO τ_c/τ_d FOR STYRENE–BUTADIENE (30% STYRENE) VULCANISATES AT DIFFERENT O_3 CONCENTRATIONS

$[O_3]10^7$ mol/litre	5·36	7·6	9·4	16·5	20·6	44·6
τ_c/τ_d	0·14	0·10	0·18	0·10	0·12	0·11

Figure 13 shows variations in τ_c for various rubbers at different ozone concentrations. The existence of such a relationship between τ_c and $[O_3]$ permits prediction of the stability of materials under operating conditions from the results of accelerated laboratory tests.[56] It can be inferred from the published data that propagation rates, V_p, of individual notches[57] and cracks[58,60] are linearly related to the ozone concentration:

$$V_p \simeq D[O_3] \qquad (xiii)$$

and are in a complex relationship with stress and strain (Fig. 14).[63]

Deformation of a specimen involves both short- and long-range order changes and may affect the macromolecular conformation, the degree of straining, and the number of strained bonds in the polymer (see *Developments in Polymer Stabilisation—4*, Chapter 3). These

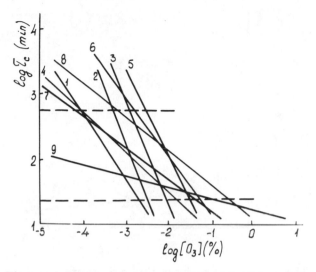

FIG. 13. Time to cracking (τ_c) for vulcanisates of various rubbers versus ozone concentration. Copolymers of butadiene with acrylonitrile (AN): 1, 18% AN; 2, 26% AN. Copolymers of butadiene with styrene (S): 3, 30% S; 7, 90% S; 9, 50% S; 4, natural rubber. Copolymer of butadiene and α-methylstyrene: 5, with 50 parts carbon black; 6, without carbon black. 8, Polychloroprene.

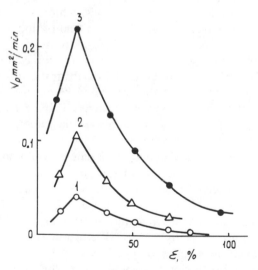

FIG. 14. Crack propagation rate (V_p) versus strain (ε) for unfilled styrene–butadiene (30% S) at various ozone concentrations, [O_3] (mol/litre): 1, $2 \cdot 2 \times 10^{-7}$; 2, $11 \cdot 0 \times 10^{-7}$; 3, $16 \cdot 5 \times 10^{-7}$.

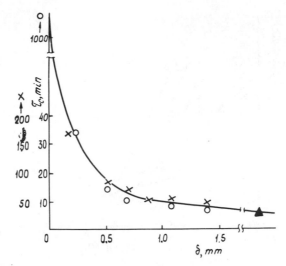

FIG. 15. Time to cracking (τ_c) versus thickness (δ) of the rubber layer above cord fabric at various ozone concentrations (O, 0·001% O_3; ×, 0·0007% O_3; ▲, 0·001% O_3, cord free).

changes affect the rate of crack formation and propagation differently and in some cases in opposite directions, which is precisely why the relation $dV_p/d\varepsilon$ changes its sign. The significant effect of redistribution of local strains on the rate of crack formation and propagation becomes evident from the results of ozone-resistance tests of rubbers with a cord-fabric layer at different depths from the surface (Fig. 15).

The time to cracking and the durability of elastomer products are also strongly influenced by polymer species. Table 8 lists the results of full-scale testing of vulcanisates of various elastomers for ozone resistance under comparable conditions.[52]

As can be seen from Table 8, elastomers containing no carbon–carbon double bonds in the main chain are more stable than unsaturated ones. Unprotected vulcanisates of general-purpose rubbers start cracking after several days of exposure, while the high-resistance elastomers based on silicone and fluorine rubbers (SKTV, SKF-26) or chlorosulphonated polyethylene (CSPE) show no visible traces of degradation even after three years of exposure.

Among the factors having to do with the nature of the elastomer two deserve special attention: the number of double bonds per unit length of the molecular chain and the nature of substituents at the C=C

TABLE 8

CRACKING RESISTANCE OF VULCANISATES BASED ON VARIOUS RUBBERS
AFTER ATMOSPHERIC AGEING

Rubber	Time to cracking (days)			
	Exposed to light		In the shade	
	10%	50%	10%	50%
SKTV	>1460	>1460	>1460	>1460
CSPE	>1460	>1460	>1460	>1460
SKF-26	>1460	>1460	>1460	>1460
Butyl rubber	>1460	800	>1460	—
SKEP	>768	752	>768	>768
Nairit	>1460	456	>1460	>1460
Nairit SKN	44	23	79	23
Natural rubber	46	11	32	32
SKMS-10	34	10	22	22
SKI-3	23	3	9	56
SKMS-30–ARKM-15	18	3	—	15
SKN-26	7	4	4	4
Natural rubber (with chalk)	18	3	— ·	—

SKTV, ·dimethylsiloxane rubber (with 1% methylvinyl siloxane); CSPE, chlorosulphonated polyethylene; SKF-26, copolymer of 1,1-difluoroethylene with hexafluoropropene; SKEP, ethylene–propylene copolymer; Nairit, polychloroprene; Nairit-SKN, a mixture of polychloroprene and butadiene–acrylonitrile copolymer; SKMS-10, copolymer of butadiene and α-methylstyrene (10%); SKI-3, synthetic polyisoprene; SKMS-30–ARKM-15, low-temperature copolymer of butadiene and α-methylstyrene (filled 15% mineral oil); SKN-26, butadiene–acrylonitrile (26%) copolymer.

bond. Comparison of the ozone resistance of isoprene rubbers with that of butyl rubbers indicates that a decrease in the double bond concentration substantially enhances the ozone resistance of polymers. This seems to be due to the hindrance involved in the movement of longer fragments of disrupted macromolecules during crack formation. In addition it becomes more probable that the surface crack, opened as a result of such movement, will not contain any reactive double bonds and that propagation of the crack will not continue.

It is more difficult to associate the high ozone resistance of poly-chloroprene, which contains a polar chlorine atom at the C=C bond, with molecular mobility but most investigators believe that this is

precisely what accounts for the ozone resistance of polychloroprene.[52,64] Studies into the temperature dependence of the time to
cracking[61c,65] have shown that as the temperature drops τ_c increases
considerably, and vice versa. At the same time, the ozone absorption
rate remains practically unchanged. The crack propagation rate differs
markedly in various polymers at a low temperature and tends to the
same limit as the temperature rises.[60] The activation energies of the
process increase from 5 kcal/mol for natural rubber to 8 kcal/mol for
butyl rubber.[61c]

Normally, the activation energies of the reaction between ozone and
C=C bonds in solution and in a gaseous phase range from 0 to
2 kcal/mol.[66,67] while those of the processes controlled by segmental
mobility are relatively high. As far as their magnitude is concerned,
they correspond to the experimentally observed activation energies
and often exceed them in absolute value.[68] Comparison of the results
of full-scale, statistical and dynamic tests[69] shows that the correlation
between them is rather poor. As a rule, cracking and degradation
occur sooner under dynamic conditions. The reasons for this accelera-
tion are not quite clear although the phenomenon itself is not unexpected.

To describe the process of ozone ageing, use is often made of
concepts borrowed from the mechanics of solids. In particular, it has
been proposed to describe the relationship between deformation and
time to cracking (τ_c) or durability by the following equation from the
theory of fatigue:[49a]

$$\tau_c = B\sigma^{-b}$$

where σ is stress; B and b are empirical coefficients. The energy
required for formation of a new surface during cracking and crack
propagation is derived by the system from the mechanical energy of
the specimen under tension or supplied by a chemical reaction.[70,71] A
major drawback of such a macromechanical approach is that it disre-
gards a basic study of the elementary stages of the complex process of
cracking and in particular, of a search for a relationship between the
chemical reactions occurring and the observed macrophysical changes
in the specimen. Most theories concerning the mechanism of the
reaction between macromolecules and ozone have been borrowed
from studies of olefins in solution.[72,73] The only phenomenon that
interested the investigators, namely the decomposition of the primary
ozonide into two fragments regarded as the main cause of cracking, has
never before been experimentally studied in polymers.

If the available data on the reactions between ozone and unsaturated macromolecules in the condensed phase and in solution are analysed, it is apparent that many important facts are lacking and whatever facts are available are controversial. Firstly, the reaction product composition data obtained by transmittance IR spectroscopy[52,74] and by the method of total internal reflectance[75] indicate that in stretched and unstretched specimens, ozonides are also present along with considerable amounts (roughly equal to the ozonide content) of carbonyl and carboxyl groups. At the same time, no cracking occurs in unstretched specimens, while calculations of the crack propagation rate involving stretched specimens suggest that only one out of 10^3 unattached ozone molecules breaks the molecular chain.[60] In the literature one can find data concerning the ozone reaction rate[76-78] which are describable by simple kinetic laws typical of ozone reactions in solution.[66] It is especially difficult to explain why the ozone reaction rate increases in the initial stage (Fig. 16). Indeed, throughout the period when the specimen is exposed to ozone the concentration of reactive groups should decrease continuously and the ozone absorption rate should also diminish monotonically.

Probably experimental errors were involved in this case since, when a flow of ozonised oxygen was passed over an individual specimen of SKI-3 rubber, ozone was readily and quantitatively absorbed, the limiting step being its diffusion from the gaseous phase to the surface.[79]

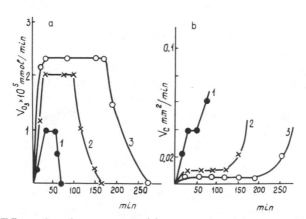

FIG. 16.   Effect of antiozonants on (a) ozone absorption rate, V_{O_3}, and (b) crack propagation rate, V_c, as a function of time when a vulcanisate of natural rubber is exposed to ozone at 26°C (20% extension). 1, Without antiozonant; 2, diphenyl-p-phenylenediamine; 3, N-phenyl-N'-isopropyl-p-phenylenediamine.

The reaction between ozone and the C=C bonds in macromolecules obeys a bimolecular law, each C=C bond acting as an independent kinetic unit. The rate constant of the reaction in solution is, in the case of high-molecular weight compounds, two to three times lower than that in the case of the corresponding low-molecular weight analogues, and in the case of rubbers, as one goes from solution to the solid phase, the value of the constant remains invariable within experimental error (Table 9).

TABLE 9

RATE   CONSTANT   OF   THE   REACTION   BETWEEN   OZONE   AND
ELASTOMERS AT $20°C^{80}$ (The polymers were supported on Aerosil)

Polymer	Rate constant $\times 10^4$ (kg/mol s)	
	Solution	Solid phase
Polybutadiene	$5\cdot1\pm2$	$6\cdot0\pm1$
cis-Polyisoprene	$10\cdot5\pm3$	14 ± 1
Polychloroprene	$1\cdot0\pm3$	$0\cdot42\pm1$
Butadiene–styrene copolymer	$4\cdot0\pm1\cdot5$	$6\cdot0\pm1$

It was found that the inductive effect of the substituent at the double bond normally influences the rate of attack of ozone at the C=C bond. Electron donors increase the rate constant, while acceptors decrease it.[81] Accordingly, a significant decrease is observed in the reaction rate as one passes from isoprene to chloroprene rubber (Table 9).

After attachment of an O_3 molecule to the double bond the sequence of transformations shown in Scheme 7 is observed.[72,73]

The first step yields an ozone complex with a double bond, which readily and rapidly transforms to a primary ozonide. The decomposition of the primary ozonide into two fragments—a carbonyl compound and an amphoteric ion—creates conditions conducive to cracking. The probability of cracking depends on the competition between the processes in which these fragments form a normal ozonide and the process of separation of the broken ends. Under normal conditions the reaction between the amphoteric ion and the carbonyl occurs in a cage. More than 50% of ozonides are derived from the fragments that were formed in the same pair.[73,82] The number of so-called cross-ozonides formed as a result of recombination of fragments from different pairs is usually smaller than that of normal ozonides. In the case of high-molecular weight compounds, the carbons at the C=C bond are less

Scheme 7

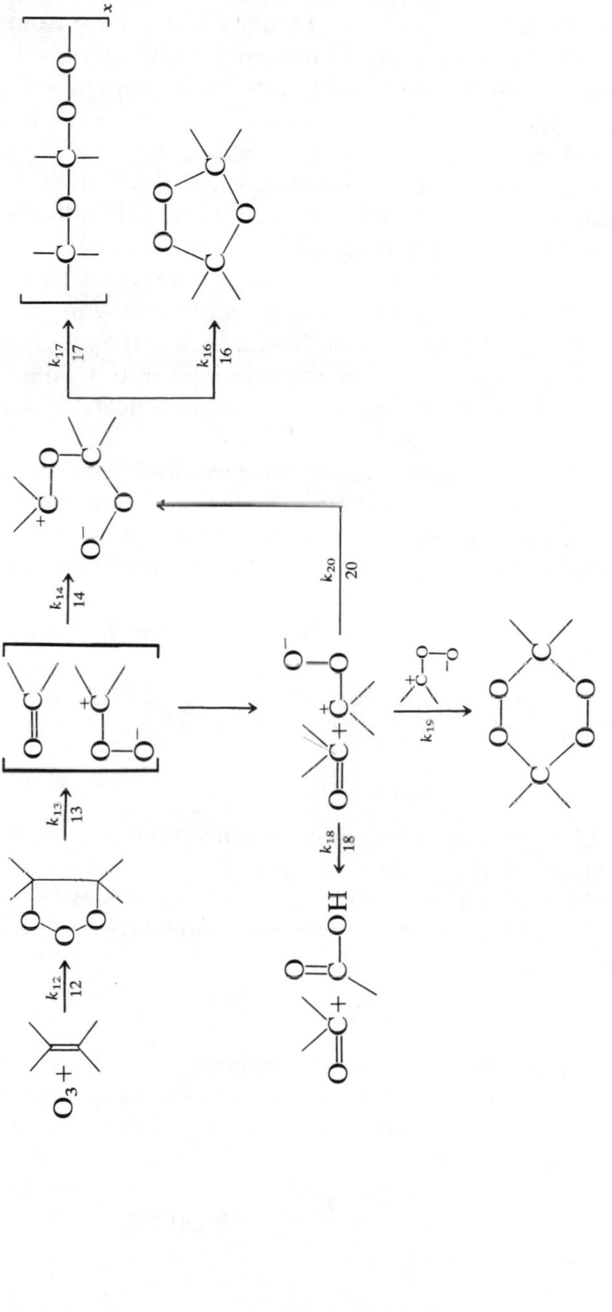

$O_3 +$

$\xrightarrow{k_{12}}$ 12

$\xrightarrow{k_{13}}$ 13

$\xrightarrow{k_{14}}$ 14

$\xrightarrow{k_{16}}$ 16

$\xrightarrow{k_{17}}$ 17

$\xrightarrow{k_{18}}$ 18

k_{19}

$\xrightarrow{k_{20}}$ 20

OH

x

mobile than in small molecules since they are associated with the atoms of the long macromolecular chain. The ratio between the processes of intra-cage recombination and emergence from the cage is evident from the decreasing molecular weight of polymer solutions with added low-molecular weight carbonyl compounds when exposed to ozone. If these are present in excess the amphoteric ions emerging from the cell are trapped. Thus, each emergence of an amphoteric ion from the cage amounts to a chain breakage. The chain scissions found experimentally are small, not exceeding several hundredths per reaction event (Table 1). The molecular weight also decreases, although at a slower rate, when rubber solutions are exposed to ozone in the absence of low molecular weight amphoteric-ion acceptors.[83-85] Isomerisation of the amphoteric ions into a carboxyl group (reaction 18) or combination of two amphoteric ions into a dimeric peroxide (reaction 19)[86] may be responsible for the decrease in molecular weight.

Ozone degradation in macromolecules may be accompanied by cross-linking processes (reaction (d) Scheme 7) between labile intermediate products in adjacent chains.[68] In the case of low-molecular weight olefins such processes lead to the formation of polymeric ozonides.

Under stationary conditions ($W[O_3]$ = constant),

$$\frac{d[A]}{d\tau} = k_{12}[C\!=\!C][O_3] - k_{14}[A] - k_{18}[A] - k_{17}[A]^2 - k_{19}[A]^2 = 0 \qquad \text{(xiv)}$$

or

$$W[O_3]_0 = k_{14}[A] + k_{18}[A] + k_{17}[A]^2 + k_{19}[A]^2 \qquad \text{(xv)}$$

where $[A]$ is the amphoteric ion concentration and W is the specific rate of supply of the gaseous mixture.

Since the reactions of chain scission and cross-linking constitute an insignificant fraction of the total process and $k[C\!=\!C][O_3] = W[O_3]_0$,

$$[A]_{stat} \approx \frac{W}{k_{14}}[O_3]_0 \qquad \text{(xvi)}$$

If we consider not rates but the number of scissions and cross-links per reaction event, we derive a simple relation associating the observed number, φ, of chain scissions (or cross-links) with the ozone concentrations:

$$\varphi = \frac{k_{18}}{k_{14}}\frac{W}{k_{14}^2}(k_{17} - k_{19})[O_3]_0 \qquad \text{(xvii)}$$

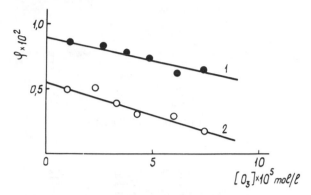

FIG. 17. Observed number of chain scissions per reaction event (φ) as a function of $[O_3]$; 1, at 25°C; 2, at −22°C.

Figure 17 represents the experimental $\varphi = f[O_3]_0$ curves for solutions of polybutadiene (SKD) at two different temperatures.[87] Intercepts on the ordinate give the following rate constant ratio: k_{18}/k_{14}. It can be seen that in the system both chain scission and cross-linking processes occur. By changing the ozone concentration one can observe both a decrease and an increase in the molecular weight. The linear relation between φ and $[O_3]_0$ is indicative of a good agreement between the experiment and the model.

When stretched specimens of vulcanisates are exposed to ozone, the conditions of recombination of fragments of broken molecules change in the stretched parts. The ends of deformed molecules become separated at the molecular relaxation rate; their pairwise distribution is disturbed with a probability approaching 100%, and the precursors of cracks appear in the form of cavities at the point of molecular scission, Scheme 8.

This may be the factor involved in the recent observation that at the early stage of cracking, cracks propagate along the boundaries of supramolecular structures[88] because this is where strains and stresses are concentrated.

It can be inferred from the above that cracking is a process determined by the specific character of the condensed phase. The interaction between adjacent chains determines the conditions for the interaction of chain-ends in the cage and the orientation of molecular movements. This interaction is difficult to simulate in solution where the interaction is weakened. At the same time, polymer solutions have

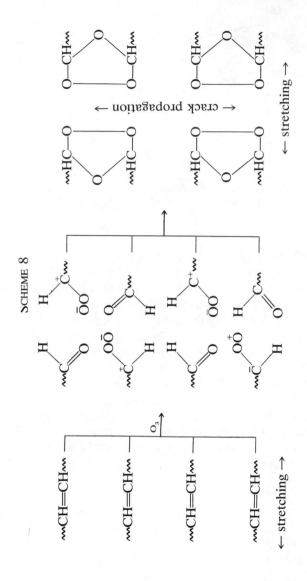

SCHEME 8

contributed to a better understanding of the mechanism of action of antiozonants in spite of the fact that the cracking process cannot be simulated in them.

4. PROTECTION OF UNSATURATED POLYMER VULCANISATES AGAINST OZONE AND THE MECHANISM OF ACTION OF ANTIOZONANTS

During World War II and immediately afterwards, various compounds were developed for the protection of rubbers against atmospheric ozone. At the same time, a search began for more effective protective techniques. Since that time large numbers of research papers and reviews have appeared in the literature,[50-54,89-95] and effective methods of protection have been found.

4.1. Incorporation of Saturated Polymers

Vulcanisates of diene elastomers can be protected against atmospheric ozone by introducing saturated polymers, wax and antiozonants into the rubber stock formula. Positive results can also be obtained by applying a layer of a saturated polymer on the surface of rubber products[96] or by modification of the surface through oxidation, treatment with mercaptans or other reagents to minimise the concentration of C=C bonds in the surface layer.[97] These methods, however, have not found wide application; more often the ozone resistance is enhanced by introduction of ethylene–propylene copolymers,[98] halogenated butyl rubbers,[99] chlorosulphonated polyethylene,[100] polyvinyl acetate and other saturated polymers[101] into the rubber stock by normal compounding procedures.

Figure 18 illustrates the relationship between the time to cracking (τ_c) and the ethylene copolymer content (EP) in compounds with different rubbers.[102] Attempts to explain what it is that gives rise to better ozone resistance include a hypothesis according to which the propagation of microcracks is slowed down when they encounter microregions occupied by saturated polymers.[101,103] Indeed, studies into the phase composition of rubber compounds indicate that after mixing, polymers form a heterogeneous system.[103-106] This hypothesis obviously does not fully represent the real picture since it does not take into account the fact that the saturated macromolecules are partially distributed inside the polymer and hence slow down the

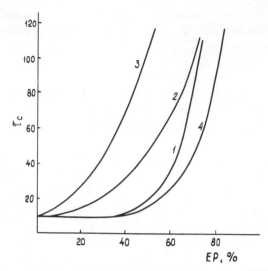

FIG. 18. Time to cracking (τ_c) versus composition of vulcanisates based on a combination of ethylene–propylene copolymer (EPK) with various rubbers, $\varepsilon = 50\%$, $[O_3] = 1 \cdot 1 \times 10^{-6}$ mol/litre; 50°C. 1, Natural rubber; 2, butadiene–styrene rubber; 3, polybutadiene rubber; 4, polyisoprene rubber.

growth of propagating microcracks by bridging their walls, thereby slowing down the exposure of fresh surface to ozone. The best results were attained when a ternary copolymer of ethylene with propylene and dienes (SKEPT) was used as an additive. It is presumed that this additive is concentrated, primarily, in the surface layer, hence its improved resistance.[105]

It is important to note that the introduction of saturated copolymers does not render rubber products absolutely ozone-resistant. Figure 19 shows time to cracking versus EPK content. It can be seen that at an ethylene–propylene diene (SKEPT) content exceeding 50%, the ozone resistance reaches a plateau and does not increase further.[107] It has been reported[98] that without any other protection, addition of 30% of SKEPT to butadiene–styrene rubbers improves the ozone resistance to the same extent as would addition of 1 part by weight of N-isopropyl-N'-phenyl-p-phenylenediamine (4010NA) and 2 parts by weight of 6-ethoxy-1,2-dihydro-2,2,4-trimethylquinoline.

The limitation of ozone resistance with concentration of saturated polymer is surprising because high concentrations give rise to phase reversal and unsaturated rubbers are then present in a dispersed

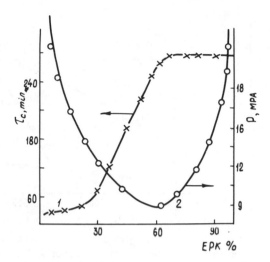

FIG. 19   Effect of isoprene (SKT-3) to ethylene propylene (EPK) ratio on the properties of vulcanisates. 1, Time to cracking (τ_c); 2, tensile strength (P), extension (ε) = 8%, [O_3] = 4·5 × 10^{-7} mol/litre.

form.[106] In this case the ozone resistance of the specimen as a whole should have been determined by the properties of the continuous phase. As can be seen from Table 8, the times to cracking are found experimentally to be much shorter than would be expected.

The main obstacle to the wide use of compositions of saturated and unsaturated polymers is the substantial variation in the mechanical properties of the rubbers made from them (Fig. 19, curve 2), as well as the complexity of the vulcanisation process.[99] Therefore, the most widely used methods of protecting rubbers against atmospheric ozone are by means of low-molecular weight additives known as antiozonants.

5.2. Antiozonants

Most of the highly effective antiozonants belong to the group of substituted N,N'-phenylenediamines (PPD) and perform a great number of diverse functions. Apart from protecting against atmospheric ozone these compounds serve as effective antioxidants and antifatigue agents.

Figure 20 summarises the results of dynamic testing of natural rubber protected by various antiozonants.[69] Plotted on the ordinate is

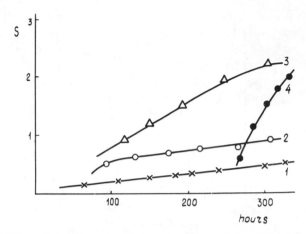

FIG. 20.   Effect of test time on the degree of degradation (S) of the surface of naturally derived rubber containing various antiozonants: 1, N,N'-di-1-methyl-heptyl-p-phenylenediamine;   2, N-phenyl-N'-isopropyl-p-phenylenediamine; 3,   6-ethoxy-1,2-dihydro-2,2,4-trimethylquinoline;   4, N,N'-dioctyl-p-phenyl-enediamine.

the mean crack area (S) per unit area of the specimen, which was used in this study as a measure of degradation:

$$S = \sum_{i=1}^{i=n} \frac{lh}{n} \qquad \text{(xviii)}$$

where n is the number of cracks, l is the length of a crack, and h is its depth.

Most of the other properties of the vulcanisates also improve perceptibly. Antiozonants increase the time to cracking,[108] improve the specimen durability,[51] minimise creep, and reduce the stress relaxation rate in the specimen.[109]

The complexity of the cracking phenomena is matched by that of antiozonant action. Several proposals have been made regarding the mechanism of this protective action in polymers; however, there is only one work in which all possible mechanisms are compared within a single experiment.[85]

The suggestion that antiozonants serve as catalysts of ozone decomposition[78] cannot be true because ozone reacts with the antiozonant,[110] and the stoichiometry and main products of the reaction are known.[84,85,111]

Nor can one accept the proposal[48,49] that, being bifunctional, antiozonants react with the products of interaction between ozone and macromolecules, binding the ends of the disrupted chains, since it is clear that the important objective is not to heal broken chains but to maintain the existing sequences of bonds. Furthermore, monofunctional antiozonants such as tributylthioureas[112] are known which cannot bind broken ends but are capable of adequately protecting rubbers against atmospheric ozone.

Another hypothesis, which states that antiozonants migrate to the surface where they react either with ozone itself or with the products of its reaction with the C=C bonds of the polymer thus forming a film that prevents ozone from penetrating inside toward the polymer, is difficult to verify. A similar effect is to be expected by analogy with waxes. However, there are two indirect arguments against any significant effect of film formation. Firstly, the products of the reaction between ozone and N-isopropyl-N'-phenyl-p-phenylenediamine (4010NA), incorporated into a specimen of cured natural rubber, do not protect it against ozone-induced degradation.[85] Secondly, calculation of the antiozonant diffusion rate indicates that migration toward the surface from the core of the specimen cannot play any significant role during cracking.[113]

Nevertheless, judging from many publications,[45-51] most researchers believe that the physical properties of antiozonants play an important role in the efficiency of their protective action, although it is not specified which properties. Probably, they should include solubility in polymers, mobility and surfactant properties. In particular, it seems that only the difference in the physical properties of tributylurea and 4010NA can explain why the times to cracking are comparable only under static conditions, while under dynamic conditions tributylurea is much less effective.[108] Comparison of the behaviour of tributylthioureas with different substituents suggests that when alkyl substituents with an open chain are replaced by cyclic ones, the correlation between the time to cracking and reactivity to ozone remains the same although the curves diverge further (see Fig. 21).[112]

The proposal[48] that antiozonants on the rubber surface compete with the C=C bonds for ozone and thereby protect vulcanisates against cracking is adequately corroborated at present. The arguments supporting this assumption have been derived using kinetic methods for studying ozone–macromolecule and ozone–antiozonant reactions in solutions.[83-85] It was shown that in the presence of antiozonants the

278   S. D. RAZUMOVSKII AND G. E. ZAIKOV

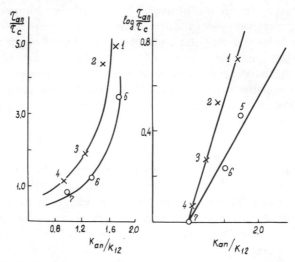

FIG. 21. Ratio of time to cracking of protected specimens (τ_{an}) to that of unprotected control specimen (τ_c) versus ratio between rate constants of the ozone–antiozonant (k_{an}) and ozone–C=C (k_{12}) reactions.

1, $C_4H_9NH{-}\underset{S}{\overset{\parallel}{C}}{-}N(C_4H_9)_2$; 2, ⬡$NH{-}\underset{S}{\overset{\parallel}{C}}{-}N(C_4H_9)_2$; ($K_{12}$)

3, ⬡$NH{-}\underset{S}{\overset{\parallel}{C}}{-}N(C_4H_9)_2$; 4, O_2N⬡$NH{-}\underset{S}{\overset{\parallel}{C}}{-}N(C_4H_9)_2$;

5, ⬡$NH{-}\underset{S}{\overset{\parallel}{C}}{-}N(C_2H_5)_2$; 6, ⬡$NH{-}\underset{S}{\overset{\parallel}{C}}{-}N(⬡)_2$;

7, ⬡$NH{-}\underset{S}{\overset{\parallel}{C}}{-}\overset{C_2H_5}{N}{-}⬡$

rate of macromolecular chain scission slowed down (Fig. 22) although the absorption of ozone continued. Products of the reaction of ozone with the antiozonant accumulated in the solution. Studies into the kinetics of this reaction have shown that it is extremely fast, its rate exceeding those of all presently known reactions involving ozone.[66] Special methods have been developed for measuring the rate constants

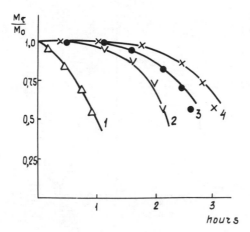

FIG. 22. Variation in the molecular weight of polybutadiene exposed to ozone in a solution at various concentrations of N-phenyl-N'-isopropyl-p-phenylenediamine (mol/litre): 1, 0; 2, $2 \cdot 7 \times 10^{-3}$; 3 and 4, $5 \cdot 3 \times 10^{-3}$. Molecular weight of polybutadiene: 1–3, 60 000; 4, 4000. CCl_4, $[O_3] = 4 \times 10^{-5}$ mol/litre, 20°C.

of the ozone–antiozonant reaction. One method is based on the competition of ozone for a reference compound with a known rate constant and the compound of interest,[114] while the other is based on comparison of the ozone dissolution rate and the rate of its reaction.[115,116]

In the first method,[114] a thiourea which reacts with ozone in two steps (Scheme 9) was selected as the reference compound:

SCHEME 9

$$C_4H_9{-}\underset{\underset{S}{\|}}{C}{-}N(C_4H_9)_2 + O_3 \xrightarrow{k_{21}} C_4H_9NH{-}\underset{\underset{O}{\|}}{C}{-}N(C_4H_9)_2 + SO_2 \uparrow$$

$$\downarrow k_{22}$$

$$C_4H_9N{=}C{=}O$$

+ unidentified
products

The first step yields sulphur dioxide. When the reaction is conducted in a bubble reactor, the presence of SO_2 can be detected by absorption in the UV region.

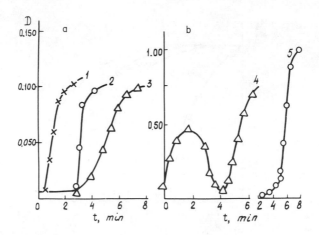

FIG. 23.   Optical density (D) of the gaseous mixture at the reactor outlet versus time.   (a) $[O_3]_0 = 6 \times 10^{-6}$ mol/litre; $\lambda = 254$ nm; $V_{O_2} = 120$ ml/min;   1, CCl_4;   2, 6×10^{-5} mol/litre of methyl oleate in CCl_4;   3, 6×10^{-5} mol/litre of tributylthiourea in CCl_4.   (b) $[O_3]_0 = 2 \times 10^{-3}$ mol/litre; $\lambda = 290$ nm; $V_{O_2} = 120$ ml/min;   4, 6×10^{-3} mol/litre of tributylthiourea in CCl_4;   5, 5×10^{-3} mol/litre of methyl oleate in CCl_4.

Figure 23 represents typical curves for the variation in the optical density (D) of the gaseous mixture at the reactor outlet for the interaction of a reference compound, methyl oleate, and tributyl-thiourea (TBTU) with ozone. Knowing the starting amounts of methyl oleate and TBTU, the gas flow rate and O_3 concentration one can easily calculate the stoichiometric coefficients of the reactions using the area above the curve before ozone appears at the reactor outlet. These were found to be 1 for methyl oleate and 2 for tributylthiourea.

Comparison of the curves representing the optical density of the gaseous mixture at the reactor outlet versus time indicates that the change in optical density at $\lambda = 280$–300 nm (Fig. 23(b), 4) differs in shape from that observed at 254 nm which is representative of the kinetics of the variation in the ozone absorption rate (Fig. 23(a), 3). The first step of the reaction between TBTU and ozone yields SO_2 whose absorption maximum can be seen at 285 nm, and this accounts for the increase in optical density during the course of the reaction.

The resulting SO_2 accumulates in the solution but is partially carried away by the gas flow (Scheme 10)

SCHEME 10

$$[O_3]_{sol} + TBTU \xrightarrow{k_{21}} [products] + [SO_2]_{sol}$$

$$[SO_2]_{sol} \overset{\alpha}{\rightleftarrows} SO_{2gas}$$

where α is the solubility coefficient. In this case the rate of the reaction between TBTU and ozone and the rate of formation of the volatile component depend on the gaseous mixture supply rate and are constant with time ($W_p = \omega[O_3]_0$), where ω is the specific rate of ozone supply (litre litre^{-1} s^{-1}), whence

$$\omega[O_3]_0 \alpha \tau = W[SO_2]_{gas} \alpha \tau + [SO_2]_{sol} \qquad \text{(xxix)}$$

or

$$\frac{\alpha[SO_2]_{sol}}{\alpha \tau} = \omega([O_3]_0 - [SO_2]_{gas}) \qquad \text{(xx)}$$

Substitution of $\alpha[SO_2]_{gas}$ for $\alpha[SO_2]_{sol}$ and integration gives eqn. (xxi):

$$\omega[SO_2]_{gas} = \omega[O_3]_0 \left[1 - \exp\left(-\frac{\omega\tau}{\alpha}\right) \right] \qquad \text{(xxi)}$$

where $[SO_2]_{gas}$ is the rate of release of volatile products, $\omega[O_3]_0$ is the rate of the reaction between ozone and TBTU, and $[1 - \exp(-\omega\tau/\alpha)]$ is the retention factor of SO_2 in the solution.

When olefin is introduced into the system, the ozone reacts with both components at the same time (eqn. (xxii)):

$$\omega[O_3]_0 = k_{21}[TBTU][O_3] + k_{12}[olefin][O_3] \qquad \text{(xxii)}$$

and the participation of TBTU in the total process is given by the expression (xxiii)

$$\frac{k_{21}[TBTU]}{k_{21}[TBTU] + k_{21}[olefin]} \qquad \text{(xxiii)}$$

then

$$\omega[SO_2]_{gas} = \omega[O_3]_0 \frac{k_{21}[TBTU]}{k_{21}[TBTU] + k_{12}[olefin]} \left[1 - \exp\left(-\frac{\omega\tau}{\alpha}\right) \right]$$

(xxiv)

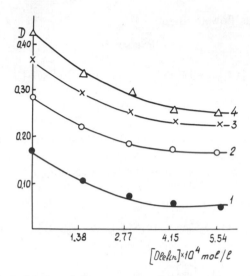

FIG. 24. Optical density of the gaseous mixture at the reactor outlet (D) versus olefin concentration [olefin] at various times: 1, 1 min; 2, 2 min; 3, 3 min; 4, 4 min. Tributylthiourea, $4 \cdot 1 \times 10^{-4}$ mol/litre; [olefin]$_0$ = $1 \cdot 5 \times 10^{-3}$ mol/litre; V_{O_2} = 120 ml/min; λ = 290 nm.

The SO_2 concentration at the reactor outlet will diminish accordingly. The dependence of $[SO_2]_{gas}$ on [olefin] is shown in Fig. 24. If one uses the ratio $[SO_2]_{gas}/[SO_2]''$ (D' and D'' are the corresponding optical densities) at different olefin concentrations [olefin]' and [olefin]'' within the same period of time, one can easily determine k_{21} if k_{12}, the rate constant of the reaction between ozone and olefin, is known.

$$k_{21} = \frac{k_{12}[\text{olefin}]'' - \dfrac{D'}{D''}[\text{olefin}]'}{[\text{TBTU}]\left(\dfrac{D'}{D''} - 1\right)} \qquad \text{(xxv)}$$

In the case of tributylthiourea the rate constant of the reaction with ozone (k_{21}) was found to be $(2 \pm 0 \cdot 2) \times 10^6$ litre/mol s.

The rate constants of the reaction of ozone with reference olefin ($k_{12} = 1 \times 10^6$ litre/mol s) and with the product of the first step of the reaction between ozone and TBTU isocyanate ($k_{22} = (3 \cdot 5 \pm 0 \cdot 5) \times 10^3$ litre/mol s) were calculated from the slopes of curves 2 and 3 in

Fig. 23(a) using the formula

$$k_{22} = \frac{\omega[O_3]_0 - [O_3]_{gas}}{\alpha[X]_\tau [O_3]_{gas}} = \frac{(D_0 - D_\tau)}{\alpha[X]_\tau D_\tau} \qquad \text{(xxvi)}$$

where $[X]_\tau$ is the current concentration of isocyanate (or methyl oleate with ozone), while D_0 and D_τ are the optical densities of the gaseous mixture at the system inlet and outlet initially.

These data suggest that the rate constant of the reaction between ozone and TBTU is much greater than those of the reactions of ozone with the C=C bonds in polymers ($k_3 = 1 \cdot 4 \times 10^5$ litre/mol s for natural rubber[80]). It should be noted that the second step of the reaction is relatively slow. This seems to be a common feature of all antiozonants. Despite the fact that they are capable of reacting with several ozone molecules, the products of the subsequent reaction steps are ineffective because of the slower rate of the reaction with ozone.

The introduction into the system of a variety of antiozonants instead of the reference olefin has made it possible to determine, from the depression of SO_2 release, the rate constants of their reactions with ozone. These values together with their rate constants with respect to the ozone dissolution rate are given in Table 10.

The tabulated data indicate that the protective action is exerted only by those compounds which react with ozone at a rate exceeding the rate of the reaction between ozone and the C=C bonds in macromolecules. The greater the reaction rate constant, the more effective the protection by the compounds tested. Figure 25 illustrates the relative efficiency τ_{an}/τ_c, expressed as the ratio between the time to cracking in stabilised rubber and the time to cracking in the control specimen,[108] versus the ratio between the rate constants of the ozone–antiozonant and ozone–>C=C< reactions (k_{an}/k_{12}). It can be seen that the resulting curve is linear in semilogarithmic coordinates (eqn. (xxvii)).

$$\log \tau_{an}/\tau_c = f(k_{an}/k_{12}) \qquad \text{(xxvii)}$$

It was established in these experiments that only one class of compounds, namely the thioureas, is excluded from the series. Their experimentally measured activity by this method turned out to be abnormally high. During tests of tyres under normal operating conditions or during dynamic tests this anomaly was not observed. It was natural to assume that the observed deviation is due not to the chemical properties of thiourea but to its specific physical behaviour in

TABLE 10

RATE CONSTANTS OF THE REACTION BETWEEN ANTIOZONANTS AND OZONE AND THEIR PROTECTIVE ACTION IN RUBBERS[114]

Compound	Formula	Time to cracking (min)	$k_{21} \times 10^{-6}$ (litre/mol s)
N,N'-Di-n-octyl-p-phenylenediamine	C_8H_{17}—NH—⟨C$_6$H$_4$⟩—NH—C_8H_{17}	840	7
N,N'-Diisopentyl-p-phenylenediamine	C_5H_{11}—NH—⟨C$_6$H$_4$⟩—NH—C_5H_{11}	870	8
N-Phenyl-N'-isopropyl-p-phenylenediamine	⟨C$_6$H$_5$⟩—NH—⟨C$_6$H$_4$⟩—NH—CH(CH$_3$)$_2$	500	7
N,N'-Di-α-methylbenzyl-p-phenylenediamine	⟨C$_6$H$_5$⟩—CH(CH$_3$)—NH—⟨C$_6$H$_4$⟩—NH—CH(CH$_3$)—⟨C$_6$H$_5$⟩	250	5
N-α-Methylbenzylanisidine	⟨C$_6$H$_5$⟩—CH(CH$_3$)—NH—⟨C$_6$H$_4$⟩—OCH$_3$	90	4
N-Butyl-N,N'-dibutylthiourea	C_4H_9NH—C(=S)—N(C$_4$H$_9$)$_2$	790	2
Methyl oleate	CH_3O—C(=O)—(CH$_2$)$_7$—CH=CH—(CH$_2$)$_7$—CH$_3$	80	1

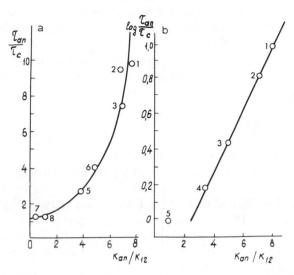

FIG. 25. Relations $\tau_{an}/\tau_c = f(k_{an}/k_{12})$ (a) and $\log \tau_{an}/\tau_c = f(k_{an}/k_{12})$ (b). 1, N,N'-dioctyl-p-phenylenediamine; 2, N,N'-di-isopentyl-p-phenylenediamine; 3, N-phenyl-N'-isopropyl-p-phenylenediamine; 4, N,N'-di-α-methylbenzyl-p-phenylenediamine; 5, N-α-methylbenzylanisidine; 6, N,N,N'-tributylthiourea; 7, 2,2-thio-bis-(6-$tert$-butyl-4-methylphenol); 8, methyl oleate.

rubbers,[47,49,117] in particular its tendency to accumulate on the rubber surface to a greater extent than other substances examined. This is also corroborated by published data concerning the relative content of various antiozonants on the surface when they were initially incorporated in equal amounts.[95]

The relation $\log \tau_{an}/\tau_c = f(k_{an}/k_{12})$ is interesting in two respects. Firstly, it can be used as a basis for the development of quantitative methods of evaluating the efficiency of antiozonants. At present no such methods are available; there are only qualitative estimates: good, satisfactory, poor, and so on,[117] which make it impossible to compare compounds of different classes or compounds of the same class taken in different amounts. Secondly, it can be an instrument in approximate calculations of the maximum antiozonant efficiency. If it is assumed that this relation persists when the rate constant k_{an} increases to a reasonable value (ca 10^8 litre/mol s), even rough estimates show that the efficiency of such an antiozonant will exceed that of the currently available ones by several orders of magnitude.

The mechanism of the physical protection of vulcanisates of unsaturated polymers against ozone is much less well understood and theoretical interpretations have not yet been elaborated. It is known that the introduction of various waxes into the original stock enhances the ozone resistance of vulcanisates, particularly under static conditions.[118–120] Waxes may be used individually and in combination with antiozonants[49,121] and it is generally considered that waxes form a protective film on the rubber surface, which slows down the reaction between ozone and the C=C bonds of the polymer. The permeability of the film is associated with the temperature of the wax film, the temperature at which it starts to soften, and its plasticity.[122]

Figure 26 shows the results of determining the time to cracking (τ_c) in the case of unfilled natural rubber containing paraffins with different softening points. The efficiency of waxes is higher at low strains; at high strains, the number of cracks decreases in the presence of waxes but then overstrains ahead of the crack and crack propagation rate may even increase.

The chemical composition of waxes influences the protection efficiency. The presence of C=C bonds, amino groups, metal salts and other additives in waxes increases the time to cracking and the time to degradation of the specimen.[123]

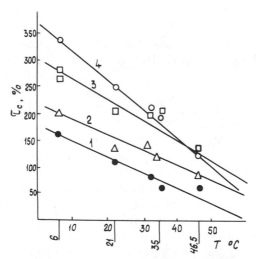

FIG. 26. τ_c (expressed as percentage of τ_c without wax) for unfilled vulcanisates of natural rubber, containing five parts by weight of paraffin wax, at various test temperatures (°C): 1, 25; 2, 35; 3, 45; 4, 57. $\varepsilon = 10\%$; $[O_3] = 4 \cdot 5 \times 10^{-7}$ mol/litre.

5. MACROMOLECULAR CHAIN SCISSION AS A METHOD OF SYNTHESIS OF OLIGOMERS WITH TERMINAL FUNCTIONAL GROUPS

The attachment of ozone to double bonds with subsequent transformation of ozonides leads to formation of two oxygen-containing functional groups (hydroxy, carboxyl or carbonyl) at the point of incorporation of ozone into the macromolecule.

The wide application of polymers and the increasingly stringent performance requirements have aroused a great deal of interest in oligomers with terminal functional groups.[124] They are used in the production of thickeners,[125] heavy-duty greases exhibiting high resistance to thermal and mechanical degradation.[126] Oligoisobutylenes with terminal phosphorus-containing groups possess good antioxidant and antiscuff properties.[125] These oligomers form the basis of multifunctional additives to greases.[127] Their introduction into rubber compounds improves processability and tackiness. Oligomers with terminal functional groups are gaining in importance as far as the synthesis of polymers with predetermined properties (e.g. highly filled polymers, modified and vulcanised rubbers) are concerned.[128,129] The synthesis of such oligomers with terminal functional groups is easy to achieve using methods of oxidative degradation, for example, by exposing polyisobutylene[130] or polystyrene[131] containing dienes to ozone (Schemes 11 and 12).

SCHEME 11

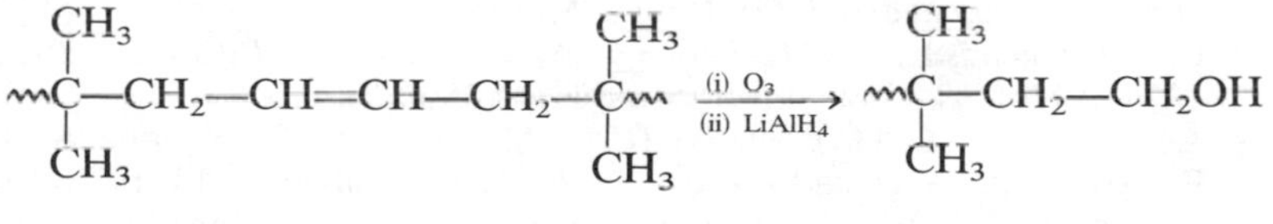

SCHEME 12

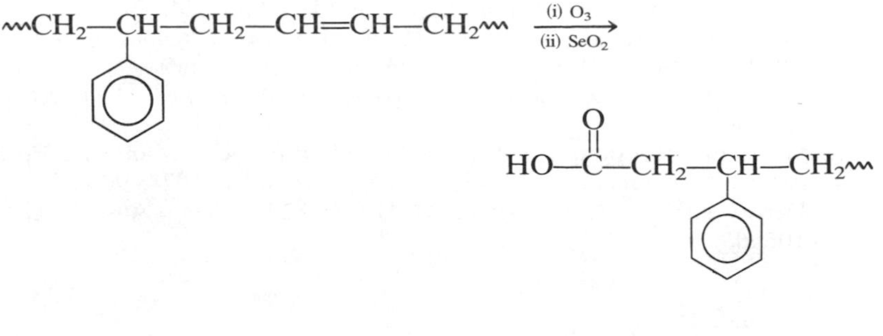

The reaction is usually conducted in the presence of pyridine,[124] methanol[132] or other additives to prevent formation of polymeric ozonides.

It is known that the properties of polymers derived from oligomers are to a great extent determined by the structure of the molecular chains arranged between the functional groups.[129] At present, the most promising path for deriving oligomers of a regular structure is considered to be partial ozonisation of stereoregular polydienes:

SCHEME 13

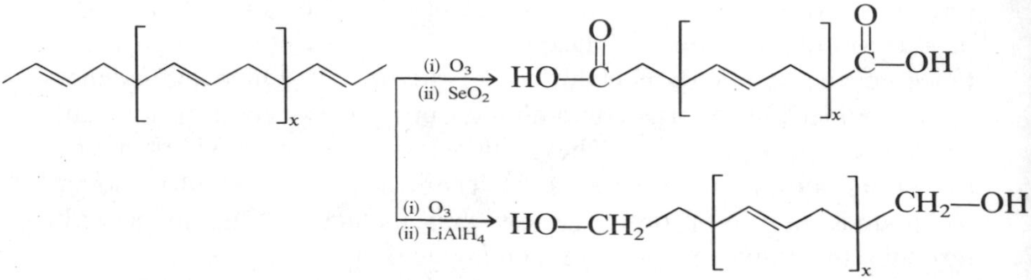

The reactions of ozone with rubbers can be exemplified in the newly developed methods for synthesising cis-1,5-polyprenols[133] and cis-1,5-polyenes.[134]

REFERENCES

1. RAZUMOVSKII, S. D. and ZAIKOV, G. E., *Ozone and its Reactions with Organic Compounds* (1974). Nauka, Moscow.
2. *Ozone. Chemistry and Technology. A Review of the Literature, 1961–1974* (1976). The Franklin Institute Press, Philadelphia.
3. ABDULLIN, M. I., GATAULLIN, R. F., MINSKER, K. S., KEFELI, A. A., RAZUMOVSKII, S. D. and ZAIKOV, G. E., *Europ. Polym. J.*, **14** (1978) 811.
4. THORSEN, W. J., WARD, W. H. and MILLARD, M. M., *J. Appl. Polym. Sci.*, **24** (1979) 523.
5. FOMIN, V. F., KOIKOV, S. N. and TSIKIN, A. N., *Elektrotekhnika*, **8** (1967) 101, 108; *Vysokomol. Soedin.*, **B11** (1969) 224.
6. MURRAY, R. W. and STORY, P. R., in *Chemical Reactions of Polymers*, Ed. E. M. Fettes (1964). Wiley-Interscience, New York.
7. ZUEV, YU. S., *Destruction of Polymers in Aggressive Media* (1972). Khimiya, Moscow.
8. MATISOVA-RYCHLA, L., RUCHLY, J., MICHEL, A., AMBROVIC, P. and PASKA, I., *J. Lumin*, **17** (1978) 13; *Chem. Abs.* **89** (1978) 90420b.
9. FRITZVOLD, B., *Nor Skogind*, **31** (1977) 811; *Chem. Abs.* **89** (1978) 106968.

10. SIROTA, A. G., *Modification of Structure and Properties of Polyolefins* (1974). Khimiya, Leningrad.
11. PALMER, R. P. and COBOLD, A. J., *Macromol. Chem.*, **74** (1964) 174.
12. KELLER, A. and UDAGAAWA, I., *J. Polym. Sci.*, A2, **8** (1970) 19; **9** (1971) 1973.
13. LEEDS, SH., *Tappi*, **44** (1961) 244.
14. TAKAKHASI, T. *Films from Polymers* (1971) p. 98. Khimiya, Moscow.
15. OWENS, D. K., *J. Appl. Polym. Sci.*, **19** (1975) 265.
16. LANLER, I. and LEBEL, P., *Chemistry and Technology of Polymers*, **1** (1960) 100.
17. BAGIROV, M. A., MALIN, V. P., GAGARYAN, YU. N. and VASILEVSKII, E. E., *Plastics*, No. 12 (1971) 37.
18. ALISSON, R. W., *APPITA*, **32** (1979) 279; *Chem. Abs.* **90** (1979) 206129g.
19. SUZUKI, J., SIZUKA, SH. and SUZUKI, SH., *J. Appl. Polym. Sci.*, **22** (1978) 2109.
20. RAZUMOVSKII, S. D., KEFELI, A. A. and ZAIKOV, G. E., in *IUPAC International Symposium on Macromolecular Chemistry*, Preprint V, 11/28; *Europ. Polym. J.*, **7** (1971) 275.
21. RAZUMOVSKII, S. D., RAKOVSKII, S. K. and ZAIKOV, G. E., *Izv. Akad. Nauk SSSR, Ser. Khim.*, (1975) 1963.
22. LISITSIN, D. M., POZNYAK, T. I. and RAZUMOVSKII, S. D., *Kinetika i kataliz*, **17** (1976) 1049.
23. STAUDINGER, H., FREY, K., GARBSCH, P. and WEHRLI, S., *Chem. Ber.*, **62** (1929) 2912.
24. CAMERON, G. G. and GRASSIE, N., *Macromol. Chem.*, **53** (1962) 72.
25. LEBEL, P., *J. Polym. Sci.*, **34** (1959) 697.
26. GAPONOVA, I. S., GOL'DBERG, V. M., ZAIKOV, G. E., KEFELI, A. A., PARIISKII, G. B., RAZUMOVSKII, S. D. and TOPTIGIN, D. YA., *Vysokomol. Soedin.*, **A20** (1978) 2038.
27. RAZUMOVSKII, S. D., KARPUKHIN, O. N., KEFELI, A. A., POKHOLOK, T V. and ZAIKOV, G. E., *Vysokomol. Soedin.*, **A13** (1971) 782.
28. SHLYAPINTOKH, V. YA., KEFELI, A. A., GOL'DENBERG, V. I. and RAZUMOVSKII, S. D., *Dokl. Akad. Nauk SSSR*, **196** (1969) 1132.
29. POKHOLOK, T. V., VIKHLYAEV, R. M., KARPUKHIN, O. N. and RAZUMOVSKII, S. D., *Vysokomol. Soedin.*, **B11** (1969) 692.
30. BERLIN, A. A., BLUMENFEL'D, L. A., CHERKASHIN, M. I., KALMANSON, A. E. and SEL'SKAYA, O. G., *Vysokomol. Soedin.*, **1** (1959) 1361.
31. RAZUMOVSKII, S. D., *Zh. Organ. Khim.*, **3** (1967) 789.
32. RAZUMOVSKII, S. D. and ZAIKOV, G. E., *Neftekhimiya*, **13** (1973) 101.
33. GRASSIE, N., *Chemistry of Polymer Degradation* (1959) p. 184. Inostrannaya Literatura, Moscow.
34. GOLDBERG, V. M., BELITSKII, M. M., KRASOTKINA, I. A. and TOPTIGIN, D. YA, *Vysokomol. Soedin.*, **A17** (1975) 2243.
35. ALEKSANDROVA, YU. A., CHUAN, Yui-Li, PRAVEDNIKOV, A. N. and MEDVEDEV, S. S., *Dokl. Akad. Nauk SSSR*, **123** (1958) 1029.
36. CARLSSON, D. J. and WILES, D. M., *Macromolecules*, **4** (1971) 174.
37. TARASEVICH, B. N., ATYAKSHEVA, L. F., PENTIN, YU. A. and ELTSE-FON, B. S., *Karbotsepnye Polimery*, (1977) Nauka, p. 200.

38. KEFELI, A. A., RAZUMOVSKII, S. D., MARKIN, V. S. and ZAIKOV, G. E., *Vysokomol. Soedin.*, **A14** (1972) 2413.
39. KEFELI, A. A., RAZUMOVSKII, S. D. and ZAIKOV, G. E., *Vysokomol. Soedin.*, **A18** (1976) 609.
40. KEFELI, A. A., RAZUMOVSKII, S. D. and ZAIKOV, G. E., *Kolloid. Zhurn.*, **38** (1976) 787.
41. KEFELI, A. A., RAZUMOVSKII, S. D. and ZAIKOV, G. E., *Vysokomol. Soedin.*, **B14** (1972) 837.
42. KEFELI, A. A., RAZUMOVSKII, S. D. and ZAIKOV, G. E., *Vysokomol. Soedin.*, **A13** (1971) 803.
43. TSVETKOV, D. D., LEBEDEV, YA. S. and VOEVODSKII, V. V., *Vysokomol. Soedin.*, **A3** (1961) 882.
44. AUERBACH, A. D. and SANDERS, L., *Polymer*, **10** (1969) 579.
45. AMBELANG, J. C., KLINE, R. H., LORENZ, O. M. and PARKS, C. R., *Rubber Chem. Technol.*, **36** (1963) 1497.
46. TUCKER, H., *Rubber Chem. Technol.*, **32** (1959) 269.
47. ZUEV, YU. S., *Zhurn. Vses. Khim. Obschchestvo Mendeleeva*, **11** (1965) 288.
48. MURRAY, R. W. and STORY, P. R., in *Chemical Reactions of Polymers*, Vol. 2, Ed. E. M. Fettes, (1964), Wiley-Interscience, New York.
49. ZUEV, YU. S., *Razrushenie Polymerov pod Deistviem Agressivnykh sred* (1972). Khimiya, Moscow. (a) *Ibid.* p. 103.
50. *Ozone. Chemistry and Technology. A Review of the Literature, 1961–1974* (1976). Philadelphia.
51. LAKE, G. I., *Rubber Chem. Technol.*, **43** (1970) 1230.
52. ANGERT, L. G., *Dostizheniya Nauki i Tekhonologii v Oblasti Reziny* (1969) p. 111. Khimiya, Moscow.
53. RAKOVSKI, S., SHOPOV, D., RAKOVSKI, K. and DOBREVA, P., *Khim. Ind.* (1980) 33.
54. BIGGS, B. S., *Rubber Chem. Technol.*, **31** (1958) 1015.
55. ZUEV, YU. S., IVANOVA, S. A. and NOVIKOVA, N. L., *Kauch. Rezina* **1** (1969) 10.
56. SKOTT, DZH. R., *Phizicheskie Ispytaniya Kauchuka i Reziny* (1968) p. 283. Khimiya, Moscow.
57. BRADEN, M. and GENT, A. N., *J. Appl. Polym. Sci.*, **3** (1960) 90, 100.
58. LAKE, G. C. and LINDLAY, P. B., *J. Appl. Polym. Sci.*, **9** (1965) 2031.
59. ZUEV, YU. S., *Khimich. Promyshl.*, (9) (1953) 21.
60. GENT, A. N. and HIRAKAWA, H., *J. Polym. Sci.*, A2, **5** (1967) 157.
61. BARTENEV, G. M. and ZUEV, YU. S., *Prochnost i Razrushenie Visokoelasticheskikh Materialov* (1964) (a) p. 316; (b) p. 334; (c) p. 335. Khimiya, Moscow.
62. NEWTON, R. C., *Rubber Chem. Technol.*, **18** (1945) 504.
63. ZUEV, YU. S. and PRAVEDNIKOVA, S. I., *Starenie i Zashchita Rezin* (1960), p. 3. Goskhimizdat, Moscow.
64. LOAN, L. D. and MURRAY, R. W., *J. Inst. Rubber Ind.*, **2** (2) (1968) 73.
65. GENT, A. N. and GRATH, J. E., *J. Polym. Sci.*, **A3** (1965) 1473.
66. RAZUMOVSKII, S. D. and ZAIKOV, G. E., *Ozone and its Reactions with Organic Compounds* (1974). Nauka, Moscow.

67. HUIE, R. E. and HERRON, J. T., *Int. J. Chem. Kin.*, *Symposium*, **1** (1975) 165.
68. RAZUMOVSKII, S. D., NIAZASHVILI, G. A., TUTORSKII, I. A. and YUREV, YU. N., *Vysokomol. Soedin.*, **A13** (1971) 195.
69. VEITH, A. G., *Rubber Chem. Technol.*, **45** (1972) 293.
70. ANDREWS, E. H., *J. Appl. Polym. Sci.*, **10** (1966) 47.
71. DE VIRES, K. L., MOORE, N. B. and WILLIAMS, M. L., *J. Appl. Polym. Sci.*, **16** (1972) 1377.
72. BAILEY, PH. S., *Ozonation in Organic Chemistry*, Vol. 1, *Olefinic Compounds* (1978). Academic Press, New York, San Francisco, London.
73. RAZUMOVSKII, S. D., *Izv. Akad. Nauk SSSR, Ser. Khim.*, (1970) 335.
74. KENDALL, F. H. and MANN, J., *J. Polym. Sci.*, **19** (1956) 503.
75. ANDRIES, J. C. and DIEM, H. E. A., *J. Polym. Sci., Polym. Lett. Edn.*, **12** (1974) 281.
76. ERICKSON, F. R., BERNTSEN, R. A., HILL, E. L. and KUSY, P., *Rubber Chem. Technol.*, **32** (1959) 1062.
77. TUCKER, H., *Rubber Chem. Technol.*, **32** (1959) 269.
78. ZUEV, YU. S., KOSHELEV, F. F., OTOPKOVA, M. A. and MIKHALEVA, S. B., *Kauch. Rezina*, **8** (1965) 12.
79. RAZUMOVSKII, S. D., KEFELI, A. A., VINITSKAYA, E. A. and ZAIKOV, G. E., *Dokl. Akad. Nauk SSSR*, **253** (1980) 1173.
80. KEFELI, A. A., VINITSKAYA, E. A., MARKIN, V. S., RAZUMOVSKII, S. D., GURVICH, YA. A., LIPKIN, A. M. and NEVEROV, A. N., *Vysokomol. Soedin.*, **A19** (1977) 2633.
81. RAZUMOVSKII, S. D. and ZAIKOV, G. E., *Zhurn. Org. Khim.*, **8** (1972) 464.
82. MURRAY, R. W., GOSSEFUCH, R. D. and STORY, P. R., *J. Am. Chem. Soc.*, **89** (1967) 2429.
83. DELMAN, A. D., SIMS, B. B. and ALLISON, A. R., *J. Anal. Chem.*, **26** (1954) 1589.
84. LAYER, R. W., *Rubber Chem. Technol.*, **39** (1966) 1584.
85. RAZUMOVSKII, S. D. and BATASHOVA, L. S., *Vysokomol. Soedin.*, **A9** (1969) 588.
86. RAZUMOVSKII, S. D., ANACHKOV, M. P. and ZAIKOV, G. E., *Tezisy V Konferentsii po problem Starenie i Stabilizatsiya Polimerov*, (1980) p. 34. Vilnus.
87. RAZUMOVSKII, S. D., ANACHKOV, M. P., ZAIKOV, G. E. and KEFELI, A. A., *Vysokomol. Soedin.*, **B24** (1981) 94.
88. KIRYUKHIN, N. N. and KABLOV, F. I., *Funktsionalnye Organicheskie Soedineniya i Polimery*, (1975) p. 236. Polytechnitscheskii Institute, Volgograd.
89. TYLEY, W. F., *Ind. Eng. Chem.*, **31** (1939) 714.
90. PINAZZI, C. P. and BILLUART, M., *Rubber Chem. Technol.*, **28** (1955) 438.
91. CREED, K. E., HILL, R. B. and BREED, J. W., *Anal. Chem.*, **25** (1953) 241.
92. THOMPSON, D. C., BAKER, R. H. and BROWNLOW, R. W., *Ind. Eng. Chem.*, **44** (1952) 850.
93. SHAW, R. F., OSSERFORT, Z. T. and TOUHEY, W. J., *Rubber World*, **130** (1954) 636.

94. BRADEN, M. and GENT, A. N., *J. Appl. Polym. Sci.*, **6** (1962) 449.
95. HOGKINSON, G. T. and KENDALL, C. E., in *Proc. 5th Rubber Technol. Conf.*, Ed. T. H. Messenger, (1962). Institution of the Rubber Industry, London.
96. British Patent, 975106 (1964); S A (1965), 3946.
97. ZUEV, YU. S. and BORSHCHEVSKAYA, A. Z., *Metody Zashchity rezin ot Ozonnogo Rastreskivaniya* (1957). Khimiya, Moscow.
98. OSSERFORT, L. T. and BERGSTROM, E. W., *Rubber Age*, **101**(9) (1969) 47.
99. OGNEVSKAYA, T. E., BOGUSLOVSKAYA, K. V., KOLOBENIN, V. N. and BOGUSLOVSKII, D. B., *Povyshenie Atmosferostoikosty Rezin za Schet Vvedeniya Ozonostoikikh Polimerov*, (1972). TsNIITEneftekhim, Moscow.
100. KAMIYA, T., *Japan Plast. Age.*, **3**(10) (1965) 11.
101. WILSON, W. A., *Rubber Age* (*NY*), **90** (1961) 85.
102. PRESTON, K., *Canad. Chem. Proc.*, **45**(9) (1961) 65.
103. ANDREWS, E. H., *Rubber Chem. Technol.*, **40** (1967) 635.
104. KHANIN, S. E., ANGERT, L. G., KULEZNEV, V. N. and MALOSHUK, YU. S., *Kolloid. Zhurn.*, **37** (1975) 99.
105. KHANIN, S. E., ANGERT, L. G., CHECHETKIN, L. N. and LAPSHOVA, A. A., *Kauch. Rezina*, **9** (1978) 22.
106. AMBELANG, J. C., PORTER, L. E. and TURK, D. L., *Rubber Chem. Technol.*, **42** (1969) 1186.
107. LEVIT, E. Z., OGNEVSKAYA, T. E., DEDUSENKO, B. N. and BOGUSLOVSKII, D. B., *Kauch. Rezina*, **5** (1979) 14.
108. LIPKIN, A. M., GRINBERG, A. E., GURVICH, YA. A., ZOLOTAREVSKAYA, L. K., RAZUMOVSKII, S. D. and ZAIKOV, G. E., *Vysokomol. Soedin.*, **A14** (1972) 78.
109. ZUEV, YU. S. and PRAVEDNIKOV, S. I., *Kauch. Rezina*, **1** (1961) 30.
110. LORENZ, O. and PARKS, C. R., *Rubber Chem. Technol.*, **36** (1963) 194.
111. RAZUMOVSKII, S. D., BUCHACHENKO, A. L., SHAPIRO, A. B., ROZANTSEV, E. G. and ZAIKOV, G. E., *Dokl. Akad. Nauk SSSR*, **183** (1968) 1106.
112. LIPKIN, A. M., ZOLOTAREVSKAYA, L. K., GRINBERG, A. E., GURVICH, YA. A., RAZUMOVSKII, S. D. and ZAIKOV, G. E., *Vysokomol. Soedin.*, **A14** (1972) 680.
113. BRADEN, M., *J. Appl. Polym. Sci.*, **6** (1962) 86.
114. LIPKIN, A. M., RAZUMOVSKII, S. D., GURVICH, YA. A., GRINBERG, A. E. and ZAIKOV, G. E., *Dokl. Akad. Nauk SSSR*, **192** (1970) 127.
115. PARFENOV, V. M., RAKOVSKI, S. K., SHOPOV, D. M., POPOV, A. A. and ZAIKOV, G. E., *Izvest. Khim. Bolg. Akad. Nauk.*, **11** (1978) 180.
116. RAKOVSKI, S. K., CHERNEVA, D. R., SHOPOV, D. M. and RAZUMOVSKII, S. D., *Izvest. Khim. Bol. Akad. Nauk*, **9** (1976) 711.
117. AMBELANG, I. C. and LORENZ, O., *Rubber Chem. Technol.*, **36** (1963) 1533.
118. BENNETT, H., *Commercial Waxes*, 2nd edn. (1956). Chemical Publishing Co., New York.
119. BACKER, D. E., *Rubber Age* (*NY*), **77** (1955) 58.

120. BUSWELL, A. G. and WATTS, J. T., *Rubber Chem. Technol.*, **35** (1962) 421.
121. ZINCHENKO, N. P. and VINOGRADOVA, T. N., *Zashchita Shinnykh Rezin ot Vozdeistviya Ozona i Utomleniya* (1969). TsNIIEneftekhim, Moscow.
122. ZUEV, YU. S. and KARANDASHEV, B. P., *Plastifikatory i Zashchitnye Agenty iz Neftyanogo Syrya*, Ed. I. P. Lukashevich, (1970) p. 161. Khimiya, Moscow.
123. ZUEV, YU. S. and POSTOVSKAYA, A. F., *ibid.*, p. 136.
124. ALIMOGLU, A. K., BEMFORT, K. G., LEDVIS, A. and MALLIK, S. U., *Vysokomol. Soedin.*, **A21** (1979) 2403.
125. KIRPICHNIKOV, P. A., *Vysokomol. Soedin.*, **A21** (1979) 2457.
126. SEROBYAN, A. K., BERESNEV, V. V. and KIRPICHNIKOV, P. A., USSR Patent 554268 (1975); *Bull. Tsobretenii*, **14** (1977) 73.
127. SEROBYAN, A. K., BERESNEV, V. V., KIRPICHNIKOV, P. A., KOMMISSAROVA, T. A. and KHURUNNOVA, A. F., USSR Patent 594162 (1975); *Bull. Tsobretenii*, **7** (1978) 110.
128. GORDIENKO, V. I., VALUEV, V. I., SHLYAKHTER, R. A. and PETROV, G. N., *Kauch. Rezina*, **2** (1978) 5.
129. KUZMINSKII, A. S., BERLIN, A. A. and ARKINA, S. N., *Uspekhi Khim. Fiz. Polimerov* (1974), p. 231. Mir, Moscow.
130. GRISBAUM, K., *Brenstoff-Chemie*, **50** (1969) 212.
131. ODINOKOV, V. N., TOLSTIKOV, G. A., MONAKOV, YU. B. and BERG, A. A., *Khimiya i Fiziko-khimiya Vysokomolekulyarnykh Soedinenii* (1975) p. 165. BF AN SSSR, Ufa.
132. ODINOKOV, V. N., IGNATYUK, V. K., TOLSTIKOV, G. A., MONAKOV, YU. B., BERG, A. A., SHARIPOVA, A. M., RAFIKOV, S. S. and BERLIN, A. A., *Izv. Akad. Nauk SSSR, Ser. Khim.*, (1976) 1552.
133. TOLSTIKOV, G. A., ODINOKOV, V. N., IGNATYUK, V. K., SEMENOVSKII, A. V. and MOISEENKOV, A. M., *Dokl. Akad. Nauk SSSR*, **234** (1977) 901.
134. ODINOKOV, V. N., IGNATYUK, V. K. and TOLSTIKOV, G. A., *Mezhdunarodnyi Simpozium po Makromolekulyarnoi Khimii, Tezisy Kratkikh Soobshchenii*, Vol. 4 (1978) p. 68. Nauka, Tashkent.

Chapter 7

POLYMER STABILISATION AT HIGH TEMPERATURES

G. P. GLADYSHEV and O. A. VASNETSOVA

Institute of Chemical Physics, Academy of Sciences of the USSR, Moscow, USSR

SUMMARY

General approaches to the problem of stabilisation of polymers and polymer-based compositions are discussed, most attention being given to the chemistry and kinetics of the processes. Stabilisation can be achieved using both 'chain' and 'non-chain' inhibitors. At high temperatures the non-chain inhibition methods often have advantages over the conventional methods based on the inhibition of chain processes. From the practical point of view, the stabilisation of heat-resistant polymers against thermal oxidative degradation may be most effectively achieved by the use of oxygen acceptors and radicals generated in the polymer matrix either during preparation and processing or during service. It is noted that in many stabilised systems the polymer lifetimes are largely controlled by diffusion processes and are often practically independent of the inhibition mechanism. The theoretical results reported in this chapter permit the prediction of service lifetimes of organic materials and compositions.

1. INTRODUCTION

Investigations into polymer stabilisation at high temperatures constitute a special branch of polymer stabilisation science, since the techniques developed for normal temperatures are generally inapplicable at high temperatures.[1-18] Most high temperature stabilisation studies have been concerned only with the effect of individual agents on the ageing of polymeric materials, or with the problem of using conventional antioxidants and inhibitors whose activities considerably decrease at temperatures of 200–250°C. The available data on the mechanisms of stabilisation of heat-resistant polymers,[11-26] i.e. polymers intended for use at temperatures above 200–250°C, have been contradictory. The reasons for this have often been the empirical nature of investigations of stabilisation of individual materials and too narrow a treatment of individual problems. The overly practical approach to the problem has been the reason why many authors have not taken proper account of the specific experimental conditions. Thus some experimenters have studied the processes under conditions in which they are limited by the diffusion of initiators or degradation products but fail to account for this fact in establishing the mechanism.

Polymer degradation at high temperatures is usually due to a combination of ageing mechanisms but, unlike ageing of many polymeric materials at low temperatures, degradation at high temperature may occur either via chain or non-chain processes. The multiplicity of ageing mechanisms is, therefore, the factor hampering the development of general approaches to stabilisation at high temperatures.

In recent years, some general concepts have been proposed to explain stabilisation at high temperatures[13,14,26] and they have already led to the solution of a number of important practical problems. In view of the great amount of work devoted to degradation and stabilisation of polymers at high temperatures, attention in this paper is centred on the chemical aspects of the processes and the mechanisms of stabilisation that have been thoroughly investigated both experimentally and theoretically; a few empirical results are quoted only as examples. A more comprehensive review of the current state of the art can be found elsewhere.[13,14]

2. GENERAL APPROACHES TO POLYMER STABILISATION

The problem of the stabilisation of high temperature resistant polymers involves essentially the retardation (inhibition) of certain processes taking place at high temperatures.[1-5,12,16] Available stabilisation methods may be subdivided into two groups. The first includes those involving addition of chemical agents, referred to as stabilisers or inhibitors. The second includes methods of chemical modification of polymers not tangibly affecting the initial material properties.

The use of inhibiting agents is the most popular. The stabilisers may take part in one or more of the following reactions.[13]

(1) Free radical reactions resulting in deactivation of active sites and chain termination.

(2) Reactions with the molecular degradation products participating in reactions which destroy the polymer.

(3) Reactions with active chemical agents penetrating into the polymer from the environment.

(4) Deactivation of reactive sites in macromolecules (e.g. weak bonds), including modification by blocking of terminal groups.

(5) Deactivation of impurities by their incorporation into stable complexes.

In all of the above reactions new chemical compounds and new groups of lower reactivity, including those containing stronger bonds, are formed which are actually responsible for the stabilisation effect.

The great variety of reactions causing the deterioration of polymer properties (ageing)[5,10,27-30] renders the classification of inhibition processes extremely difficult. The above-listed inhibition reactions may be divided into two groups, generally classified as 'chain' and 'non-chain' inhibition. Chain inhibition involves deactivation of the active sites of a chain process, i.e. their conversion to inactive products so that they can no longer take part in the chain propagation process. Non-chain inhibition implies deactivation of compounds taking part in any type of reaction bringing about polymer degradation. If polymer degradation is caused by chain processes, non-chain inhibition deactivates chain initiating or propagating compounds. Classification of polymer inhibition into chain and non-chain processes is helpful in developing general principles that permit a correct choice of stabilising agents. The basic polymer stabilisation methods are listed in Table 1.

Apart from the processes listed in Table 1, there are also known methods for improving polymer stability based on modification,[30] i.e. structural stabilisation, of polymers and the modification of the degrading agents and stabilisers.[13,26] All these methods are based on cage effects,[31] on the reduction of the diffusion rates of active compounds, and related reactions.[2,7] The stabilisation methods listed in Table 1 are considered in detail in the following section.

TABLE 1
POLYMER STABILISATION METHODS[13]

Degradation initiation type	Stabilisation method and stabiliser reactions	Stabiliser type
(1) Scission of macromolecules, cleavage of side (terminal) groups: $R_m—R_n → R'_m + R'_n$ $R_m—YX_2—R_n → R_m—YX—R_n + X'$	(a) Chain inhibition: $R' + YR' → RY + R' \rightleftarrows R'R$ $R' + \dot{Z}' → RZ'$ (where R' are active sites; $R'R$ and RZ' are inert products) (b) Non-chain inhibition, structural stabilisation	YR'—antioxidant (InH) or thermostabiliser; $\dot{Z}'$—a stable radical. Inclusion compound forming agents
(2) Reactions with degradation-initiating chemicals (X'): $R_m—YX_2—R_n + X' → R_m—YX—R_n + X'X$ (where $X' = O_2, O_3, H_2O, HCl$, etc.)	(a) Chain inhibition $R + \dot{Z}' →$ inert products (b) Non-chain inhibition $X' + Z →$ inert products	$\dot{Z}'$—radical acceptor; Z—an acceptor of X' (metal, metal oxide, sulphide, etc.)
(3) Reactions of weak bonds and reactive groups of impurities: $R_m—Y—Y—R_n → R_m\dot{Y} + R_n\dot{Y}$ $R_m—X_1—R_n + Y_2 → \dot{R}' +$ products (where $X_1 = $ C=C, —C≡C—, etc.; Y_2 is a chemical agent	(a) Chain inhibition $R' + \dot{Z}' →$ inert products (b) Non-chain inhibition $R_m—X_1—R_n + Z_1 →$ inert products $Y_2 + Z_2 →$ inert products	Z'—a stabiliser (InH, stable radical) Z_1—a stabiliser $(H_2, F_2,$ etc.) Z_2—chemical agent acceptor
(4) Initiation by intermediates: $R_m OOH → R_m O' + \dot{O}H$	(a) Chain inhibition $R_m O' + Z' →$ inert products (b) Non-chain inhibition $R_m OOH + Z_2 →$ inert products	Z_1—a stabiliser (InH) Z_2—a hydroperoxide decomposer (sulphide, etc.)

3. STABILISATION MECHANISMS: SPECIFIC FEATURES OF STABILIZATION AT HIGH TEMPERATURES

3.1. Chain Inhibition

3.1.1. Chain Inhibition Reactions in Thermal Degradation

Thermal degradation of polymers is in most cases due to radical-chain processes, involving polymer radicals (e.g. alkyl). Therefore chain inhibition based on the reaction of polymer radicals is most widespread.

Allyl halides are effective inhibitors of chain processes involving active radicals since they can react with the chain propagating radicals. The following reaction exemplifies this.[12,27]

$$R^{\cdot} + CH_2{=}CH{-}CH_2Cl \rightarrow RCl + CH_2{=}CH{-}\dot{C}H_2 \quad (1)$$

relatively 'stable' radical

Diphenylamine (a weak inhibitor) reacts with alkyl radicals at 100–200°C at a rate $k_z = 10\text{–}10^{-2}$ litre/mol s as follows:[12,27]

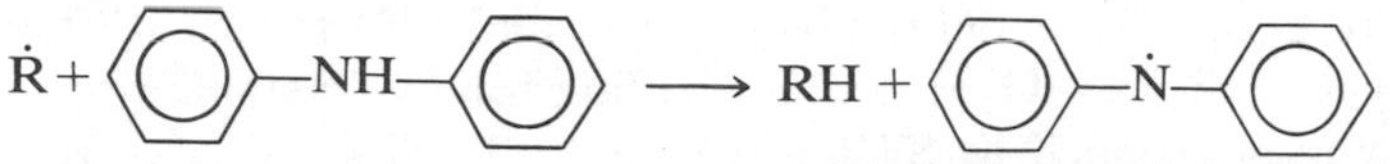

(2)

relatively 'stable' radical

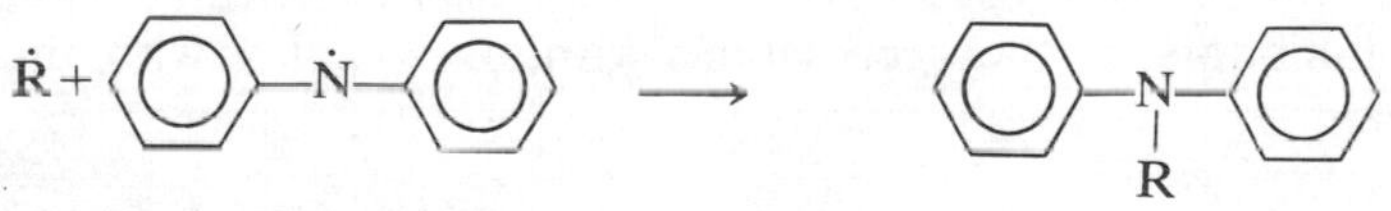

(3)

Other aromatic amines, phenols and compounds containing a labile H atom react with radicals in a similar manner.

The radicals that are formed in decomposition of tetraphenylsuccinic acid dinitrile (reaction 4) terminate the chains in the process of thermal degradation of acrylic polymers.[32]

$$N{\equiv}C{-}C(C_6H_5)_2{-}C(C_6H_5)_2{-}C{\equiv}N \longrightarrow 2\,\cdot C(C_6H_5)_2{-}C{\equiv}N \quad (4)$$

At 100–200°C, stable free radicals such as nitroxyls,[33] diphenylpicrylhydrazyl,[34] etc., can also effectively inhibit thermal degradation.

They combine with alkyl radicals at rates of 10^6–10^8 litre/mol s. However, since the resultant bonds are weak, the chain termination reactions become markedly reversible at elevated temperatures, and no inhibition effect is observed.

Inhibition of thermal degradation by the introduction of unsaturated compounds which react with the radicals taking part in chain growth, producing weakly active radicals, is also often effective. Thus thermal degradation of acrylic polymers is inhibited in the presence of itaconic acid, allyl compounds and vinyl monomers.[27,35] Itaconic acid, for instance, reacts with radicals via reaction 5:

$$\dot{R}+CH_2=C\underset{\underset{COOH}{|}}{\overset{\overset{}{|}}{}}\quad CH_2 \longrightarrow CH_3-\dot{C}R\quad \dot{C}H$$

$$\text{COOH} \quad \text{COOH} \qquad \text{COOH} \quad \text{COOH} \qquad (5)$$

relatively 'stable' radical

Compounds of metals, such as tin, can also react with radicals yielding inactive products.[2,36]

Inhibition of thermal degradation by termination of the chains at temperatures above 200–250°C may take place in the presence of various compounds of phosphorus, sulphur, halogens and finely dispersed metals. Thus finely divided iron and lead react with radicals of the type $-Si(R_2)-\dot{O}$, $R'-\dot{C}H_2$ yielding relatively stable organometallic compounds such as $Si(R_2)-O-Fe-O-Si(R_2)-O-$, $R'-CH_2-Pb-$.[14]

Various polyconjugated systems[37] may also be effective chain reaction inhibitors as they recombine with or complex with the active radicals.

3.1.2. Chain Inhibition in Thermo-oxidative Degradation

In thermo-oxidative degradation of polymers in an unlimited oxygen supply the growth of the kinetic chains may be stopped when peroxyl radicals $R\dot{O}\dot{O}$ react with an inhibitor. Substituted phenols and amines are examples of effective antioxidants reacting with peroxyl radicals with inhibition rate constants $k_7 = 10^4$–10^5 litre/mol s. Inhibition in the presence of these compounds is due to reactions of the type 6 or 7.

$$RO_2^{\cdot}+ \quad \longrightarrow \quad +ROOH \qquad (6)$$

$$RO_2^{\cdot} + \langle\bigcirc\rangle\!-\!NH\!-\!\langle\bigcirc\rangle \longrightarrow \langle\bigcirc\rangle\!\overset{\cdot}{N}\!-\!\langle\bigcirc\rangle + ROOH \qquad (7)$$

Reactions between phenols and amines and the peroxides (reaction 8) where δ is the stoichiometric coefficient, may also occur.

$$ROOH + InH \nearrow \quad R\dot{O} + I\dot{n} + H_2O$$
$$\searrow \quad RO\!-\!In + H_2O$$
$$\xrightarrow{RH,\ O_2} \quad \delta R\dot{O}_2 \qquad (8)$$

In the case of phenols secondary reactions may produce peroxy-dienones (reaction 9) (see *Developments in Polymer Stabilisation—1*, p. 1).

$$R\dot{O}_2 + I\dot{n} \rightarrow RO_2In \qquad (9)$$

At elevated temperatures these peroxides thermolyse at the O–O bonds and form radicals capable of initiating chain processes.

Inhibition of the breakdown of certain polymers by amines is characterised by their conversion to new active antioxidants which form adducts with radicals.[2,8,38,39,40] Many polyconjugated compounds, including those formed as a result of chemical conversions of antioxidants, have a high inhibiting activity.[37] Thus oxidative dehydropolymerisation of secondary aromatic amines yields polydiphenylamines that are more active inhibitors than the original amines.[31]

Data have been reported on the inhibition of the thermal degradation of polyphenyl esters at 420–440°C and the thermo-oxidative degradation at 250–300°C with organic compounds of metals (Cu, Al, Ti, Mn, Fe, Co, Ni, etc.).[41] It is shown that relatively stable free radicals of the type $RO\!-\!C_6H_4\!-\!\dot{O}$ are formed during thermolysis. On the other hand, the variable valence metal compounds react with hydroperoxides in a typical catalytic process (10, 11):

$$M^{2+} + ROOH \rightarrow \dot{M}(OH)^{2+} + R\dot{O} \qquad (10)$$
$$\dot{M}(OH)^{2+} + ROOH \rightarrow M^{2+} + HOH + ROO^{\cdot} \qquad (11)$$

It has also been noted that stabilisation of polyphenyl esters with organic cobalt salts occurs via redox reactions between the metal ions and active radicals (reaction 12).[42]

$$Co^{3+}\dot{R} \rightarrow Co^{2+} + R^+ \tag{12}$$

It may be assumed that radicals $R\dot{O}$, $RO\dot{O}$, etc., can also react in this manner. However, the efficiency of these processes is not high if inorganic salts are used.[43]

Stabilisation of high-molecular weight compounds, in the presence of bases such as KOH, appears to involve reactions of the following type.[43]

$$4KOH + 3O_2 \rightarrow 4KO\dot{O} + 2H_2O \tag{13}$$

$$KO\dot{O} + RO\dot{O} \rightarrow KOOR + O_2 \tag{14}$$

$$KOOR + R'\dot{O} \rightarrow ROR' + KO\dot{O}, \quad etc. \tag{15}$$

Peroxides of the MOOR type must be generated in the same way from the basic salts of other alkali metals. However, details of this interesting mechanism still need experimental verification. It should be noted that there is a correlation between the stabilising activity of alkali metal hydroxides and ability of these metals to form peroxides; calcium and strontium hydroxides appear to be the only exceptions.

Alkali metal hydroxides also effectively inhibit degradation of trioxane–dioxalan copolymers. It should be noted that at relatively low temperatures KOH can also inhibit the thermal oxidative degradation of polyethylene. Its effect consists in reducing the degree of branching of the polymer, that is the concentration of 'weak bonds' in macromolecules, hence preventing the formation of fragments containing unsaturated end-groups.[44]

A change in behaviour with increasing temperature has been observed for the oxidation of liquid siloxane in the presence of iron and cerium compounds.[45,46] It has been found that at temperatures below 150°C Fe^{3+} compounds promote thermal oxidative degradation, whereas at higher temperatures the same compounds show a stabilising effect, a specific concentration of Fe^{3+} assuring the minimum rate of oxidation at each temperature. The author believes[45,46] that the following processes, characteristic of degenerate-branched oxidation, accompany the degradation of liquid siloxane:

Initiation:

$$\equiv SiCH_3 + O_2 \rightarrow \equiv Si\dot{C}H_2 + \dot{O}OH \tag{16}$$

$$\equiv SiCH_3 + \dot{}OOH \rightarrow \equiv Si\dot{C}H_2 + HOOH \tag{17}$$

and/or

$$\equiv SiCH_3 + O_2 \rightarrow [\equiv SiCH_2OOH] \rightarrow \equiv S\overset{\cdot}{i} + \overset{\cdot}{O}H + CH_2O \quad (18)$$

Chain propagation:

$$\equiv Si\overset{\cdot}{C}H_2 + O_2 \rightarrow \equiv SiCH_2O\overset{\cdot}{O} \quad (19)$$

$$\equiv SiCH_2O\overset{\cdot}{O} \rightarrow \equiv Si\overset{\cdot}{O} + CH_2O \quad (20)$$

$$\equiv Si\overset{\cdot}{O} + O_2 \rightarrow \equiv SiO\overset{\cdot}{O} + \overset{\cdot}{O} \quad (21)$$

$$\equiv SiCH_3 + \equiv Si\overset{\cdot}{O} \rightarrow \equiv SiOH + \equiv Si\overset{\cdot}{C}H_2 \quad (22)$$

Branching:

$$\equiv SiCH_2OOH \rightarrow \equiv S\overset{\cdot}{i} + \overset{\cdot}{O}H + CH_2O \quad (23)$$

$$\equiv SiOOH \rightarrow \equiv Si\overset{\cdot}{O} + \overset{\cdot}{O}H \quad or \quad \equiv S\overset{\cdot}{i} + \overset{\cdot}{O}OH \quad (24)$$

$$HOOH \rightarrow 2\overset{\cdot}{O}H \quad (25)$$

$$CH_2O + O_2 \rightarrow \overset{\cdot}{C}HO + \overset{\cdot}{O}OH \quad (26)$$

Chain termination:

$$\equiv S\overset{\cdot}{i} + \overset{\cdot}{O}H \rightarrow \equiv SiOH \quad (27)$$

$$\equiv S\overset{\cdot}{i} + \overset{\cdot}{O}Si\equiv \rightarrow \equiv Si-O-Si\equiv \quad etc. \quad (28)$$

It was assumed that in the presence of iron salts the free radicals decay as a result of the reduction of the metal ion

$$M^{x+1} + \overset{\cdot}{R} \rightarrow M^x + R^+ \quad (29)$$

(where $\overset{\cdot}{R}$ is $Si\equiv$ or $\equiv Si\overset{\cdot}{C}H_2$) and that when the oxygen concentration is high enough metal ions M^{x+1} may be regenerated via:

$$M^x + O_2 \rightarrow M^{x+1} + O_2^- \quad (30)$$

Non-radical processes such as

$$\equiv Si\overset{\cdot}{C}H_2 + \{Fe^{3+}\} \xrightarrow{\text{H}_2\text{O}} \equiv SiCH_2OH + \{Fe^{2+}\} + H^+ \quad (31)$$

(where $\{Fe^{3+}\}$, $\{Fe^{2+}\}$ are complexes of Fe(III) and Fe(II) respectively) may also be assumed to take place; the product is then oxidised and the starting compound is regenerated:

$$\{Fe^{2+}\} + O_2 \rightarrow \{Fe^{3+}\}O \quad (32)$$

It is believed that the stabilisation of dimethylsiloxane liquid with iron(III) octate involves formation of a complex of the type:

$$
\begin{array}{ccc}
\text{CH}_3 & & \text{CH}_3 \\
| & & | \\
\text{—O—Si—O—Si—O—} \\
| & & | \\
\text{O:} & & \text{FeO} \\
| \\
\text{H}
\end{array}
$$

It must be noted that the nature of these complexes is not yet quite clear, and the question of whether or not non-radical reactions involving such complexes take place cannot at present be answered positively.

It has been suggested[47] that at temperatures around 350°C and upwards the stabilisation of polyorganosiloxanes in the presence of metal phthalocyanines is, perhaps, due not only to the reactions between paramagnetic ions and radicals but also to a change in co-ordination resulting in the formation of chelate compounds of complex structure.

Some investigators[48] have associated the stabilising effect of metals and their oxides at high temperatures with orientation of the polymer by the finely divided metal particles which are sites for the formation of secondary structures, or alternatively, with the formation of a new type of polymer containing metal atoms in its chains. These processes may well take place but are not necessarily the controlling ones. There is no doubt that in a number of cases the stabilising effect of many metals at high temperatures is largely due to a non-chain inhibition reaction or oxidation–reduction processes.

Published data information does not always give any unambiguous explanation for the effect of metals and many inorganic compounds in polymer ageing.[14,49–54] For instance, it is sometimes stated that salts of Fe and Sn, oxides of Al, Cr, Zn, P and B, and Pb and Cd metals promote oxidation under certain conditions. At the same time, in other studies the same compounds have been found to have an inhibiting effect.[13] The salts of Co, Cu, Ce and oxides of Fe, Mg, etc, usually stabilise temperature-resistant materials at 300–390°C, but at lower or higher temperatures they either fail to inhibit degradation or even promote the breakdown of polymer.

The use of stable free radicals or unsaturated compounds as inhibitors of oxidative degradation is not usually very effective, since they form unstable products and the rate constants of the reactions between the radical $RO_2^{\cdot}$ with the polymer PH and the unsaturated compound are often of the same order of magnitude.

The apparent contradictions in interpretation of the mechanisms of inhibitors of the high temperature ageing of polymers can be easily disentangled if one takes proper account of the experimental conditions and procedures. Unfortunately, as already noted, some investigators ignore the specifics of stabilisation of real compositions and sometimes, in trying to find new ways of interpreting the results, propose unrealistic models. A number of researchers, for example, have advanced a 'two-component polymer model'.[22] This conjecture is based on consideration of an equilibrium of the type:

$$A + D \rightleftharpoons AD \qquad \text{(i)}$$

wherein A is some chemical compound, and D is a 'defect' (a sorption centre).

Since the symbol A stands for a chemical compound, a component of the system, symbol D should also relate to a compound-component, and eqn. (i) should describe an equilibrium between three chemical entities—A, D and AD. But the authors[22] hypothesise a model of an interaction between A (a chemical component) and D (a component which is assumed to be a certain defective zone having a definite extension and being actually a phase of substance; the conglomerate AD is apparently conceived as the zone containing a molecule of compound A). This kind of treatment of a physico-chemical process is fallacious: such equilibria simply do not exist in nature.[55,56] The work of Gibbs shows that the general criterion of a chemical equilibrium is:

$$\sum \nu_i \mu_i = 0 \qquad \text{(ii)}$$

where ν_i are the stoichiometric coefficients of the components and μ_i are chemical potentials at equilibrium. This criterion of equilibrium has remained unmodified since it was originally proposed. From eqn. (ii) it necessarily follows that a chemical equilibrium can only be treated in terms of chemical potentials of components (chemical entities) and the condition of equation (i) is meaningless within the context of Gibbs' conceptions. Since neither D nor AD is a chemical compound capable of being characterised by a relevant chemical potential it is impossible to apply eqn. (ii) to eqn. (i). This example illustrates the fact that when a theory is based on an unrealistic model, the result of any calculation will be mythical.

In conclusion, it should be noted that heterolytic chain degradation of polymers has so far received little attention, important though it may be at high temperatures.[14]

3.2. Non-chain Inhibition: General

A variety of non-chain inhibition processes have been applied to stabilisation.[13] They include reactions listed in Section 2, excluding the reactions of stabilisers with free radicals and ions that can only take place in chain processes. Non-chain inhibition further includes removal or deactivation of initiating agents which are effective whatever (chain or non-chain) path is taken by the polymer degradation processs.

The data of Table 1 show that non-chain inhibition opens new possibilities towards ultimately solving the problem of polymer stabilisation. However, this approach to the problem of stabilising polymers to high temperatures has been investigated only since the mechanism of stabilisation in the presence of oxygen acceptors in polymer compositions has been established.[13,14,26,57-64] It has recently been shown[56] that the non-chain inhibition technique may be useful for stabilising heat-resistant materials whose breakdown is caused not only by oxygen but also by other chemical agents.

The techniques of scavenging active low-molecular weight degradation products have long been used in polymer chemistry. One classic example is the deactivation of hydrogen chloride—an active product of the thermal degradation of polyvinyl chloride—by addition of tribasic lead sulphate, cadmium dilaurate or diphenylurea.[2,65] The basic agents used for the removal of formic acid produced in the ageing of polyacetals include carbamide, melamine, polyamides, etc.

Blocking of active sites of macromolecules and of terminal groups ('weak bonds') which take part in initiation has been used for the stabilisation of a number of polymers. The reactions involved are usually considered as polymer modification.[2] Polyconjugated systems in polymer compositions can be blocked (chemically modified) by means of the Diels–Alder reaction with maleic anhydride (reaction 33).

$$-CH{=}CH{-}CH{=}CH{-} \ + \ \begin{array}{c} CH{-}CO \\ \| \quad\quad \diagup O \\ CH{-}CO \end{array} \longrightarrow \quad \text{(33)}$$

On the whole, the range of processes directed to the blocking of the active sites of macromolecules is very broad.[30]

Decomposition of peroxides and other intermediates plays an important role in stabilisation. Peroxide decomposers are often referred to as synergists because in combination with antioxidants and thermostabilisers they exhibit non-additive (synergistic) stabilising activity.[2,66,67]

Macromolecular cross-linking processes occur under the action of the so-called 'deactivation agents'. Their effect is the opposite of that of the degradative effects of active chemical agents and can be considered to be reactions with the breakdown intermediates. Typical deactivators are mercaptobenzimidazole, mercaptopyrimidine, mercaptoxazole, etc.

Compounds capable of forming complexes with various impurities, especially with metal compounds which catalyse the degradation process, are important stabilisers for polymers. Thus traces of copper that may be catalytically active under certain specific conditions may be neutralised by addition of ethylenediaminetetra-acetic acid, which forms complexes with copper compounds. Certain metal deactivators may act as antioxidants and thermostabilisers simultaneously.

The last group of reactions that may be listed among the non-chain inhibition processes are those between stabilisers and active degradative agents penetrating into the polymer from the outside.[13] It has not previously been the practice to discuss this group separately,[2] but in view of the current progress in the study of non-chain inhibition techniques it may be convenient to consider it as a separate group. These processes include reactions of stabilisers with oxygen, water and other agents. At appropriate concentrations, this type of stabiliser may prevent the participation of these agents in the initiation reaction, leading to inert products.

When oxygen is scavenged by an appropriate acceptor, the polymer breakdown process occurs under conditions of oxygen deficiency. As a result, the original thermal oxidative degradation becomes thermal degradation and is usually characterised by a lower rate. The use of non-chain inhibition in conjunction with the generation of highly active oxygen acceptors in the polymer allows the service life of a high temperature resistant material to be extended considerably.

Hydrothermal breakdown of polymers is often brought about by water diffusing into the polymer from the environment; this may be absorbed by dessicating agents such as CaO. An easily hydrolysable additive also helps remove the water and stabilise the polymer. For example, certain 1,3-diketones readily hydrolyse as follows:

$$R-\overset{\overset{\displaystyle O}{\|}}{C}-CH_2-\overset{\overset{\displaystyle O}{\|}}{C}-R' \xrightarrow{\ H_2O\ } R-\overset{\overset{\displaystyle O}{\|}}{C}-CH_3 + R'COOH \qquad (34)$$

Diphenylcarbodi-imide (Ph—N=C=N—Ph) and other compounds which energetically remove water, thus accelerating dehydration, are highly effective.

At high temperatures, water may be removed from the polymer by way of reaction with a finely divided metal, such as iron:

$$2Fe + 3H_2O \rightarrow Fe_2O_3 + 3H_2 \qquad (35)$$

In order to determine the optimum conditions for use of non-chain inhibitors as stabilisers, it is necessary to review the available data on their reactions. A range of potential non-chain inhibitors (oxygen acceptors) have been described in the published literature.[14,68] Some examples follow of compounds that can be useful as stabilisers at high temperatures.

It has been found[69] that cobalt sulphide, Co_9S_8, undergoes extensive oxidation at 200°C producing the complex $Co_9S_8O_8$. In a number of publications the temperatures at which the oxidation rates of the sulphides SbS_3, $AlFeS$, FeS, Cu_3FeS_3, Fe_7S_8, MoS_2, ZnS, PbS, Cu_2S_3, CdS, etc., are maximal have been determined and the reaction mechanisms established. It has been shown, in particular, that oxidation of halcopyrite is described by the following equations:

$$nCuFeS_2 + 4O_2 \rightleftharpoons 2FeO + Cu_2S(n\text{-}2)CuFeS_2 + 3SO_2 \qquad (36)$$

$$3FeO + \tfrac{1}{2}O_2 \rightleftharpoons Fe_3O_4 \qquad (37)$$

$$2Fe_3O_4 + \tfrac{1}{2}O_2 \rightleftharpoons 3Fe_2O_3 \qquad (38)$$

and the oxidation of iron and copper sulphides follows the schematic pathways 39 and 40.

$$2FeS \xrightarrow{O_2} [2FeSO \xrightarrow{O_2} 2FeSO_2 \longrightarrow Fe_2SO_3 2SO_2]$$
$$\downarrow{-2SO_2} \qquad (39)$$
$$Fe_2O_3$$

$$2CuS \xrightarrow{O_2} [Cu_2SO \xrightarrow{\frac{1}{2}O_2} Cu_2SO_2 \xrightarrow{O_2} Cu_2SO_3 \xrightarrow{-SO_2} Cu_2O$$
$$\downarrow \qquad (40)$$
$$2CuO$$

It should be noted that many metal sulphides are in addition effective synergists and decompose peroxides by a molecular mechanism. Thus it has been shown[70] that WS_2, which accelerates hydroperoxide decomposition, effectively reduces the radical yield in the decay process.

Antimony is slowly oxidised at temperatures below 350°C; at 500°C the process is violent and results in formation of Sb_2O_3 and then Sb_2O_4. Intensive oxidation of silicon and lead is observed at temperatures above 400°C.

Oxidation of a number of hydroxides of variable valence metals— $Fe(OH)_2$, $Co(OH)_2$, $Mn(OH)_2$ and the like—has been studied thoroughly. Manganese(II) hydroxide absorbs oxygen strongly even at low temperatures:

$$Mn(OH)_2 + \tfrac{1}{2}O_2 \rightarrow MnO_2 + H_2O \qquad (41)$$

Some oxides of variable valence metals are also good oxygen acceptors. For example, oxidation of cobalt oxide is accompanied by oxygen absorption:

$$3CoO + \tfrac{1}{2}O_2 \rightarrow Co_3O_4 \qquad (42)$$

Dithionites (such as sodium hyposulphite, $Na_2S_2O_4$) which are known to react with O_2 in the presence of alkalis can also be used as oxygen acceptors:

$$S_2O_4^{2-} + 2OH^- + 1\tfrac{1}{2}O_2 \rightarrow 2SO_4^{2-} + H_2O \qquad (43)$$

However, dithionites are usually unstable and decompose on heating:

$$2M_2S_2O_4 \xrightarrow{\ 100°C\ } M_2S_2O_3 + M_2SO_3 + SO_2 \qquad (44)$$

where M is a monovalent metal. Soluble alkali metal salts and zinc dithionite, ZnS_2O_4, are more stable.

In model reactions, sulphites in admixture with pyrogallol, alkalis and other compounds are used as oxygen acceptors.[14]

3.2.1. Non-chain Inhibition with Agents Generated in the Degradation Process

It is well known that very finely divided metals may be active oxygen acceptors at high enough temperatures. High surface area copper absorbs oxygen even at low temperatures. At 400–500°C the process is quite violent; however, it is not possible to remove oxygen from a system completely because of the high dissociation pressure of copper oxide. Metallic silver and silver supported on carriers (Al_2O_3, etc.) absorb O_2 at 200–300°C; palladium catalyses the reaction between H_2 and O_2 at low temperatures rapidly and completely. Fine iron powders also react vigorously with oxygen at low temperatures yielding oxides.[14,71,72]

The thermodynamics of the oxidation of metals and other solid-phase materials has been well studied. The thermodynamic aspects of corrosion protection of high temperature resistant materials by the use of metal additives has also been investigated[73] but the kinetics of these processes have received much less attention.

The high activity of metals toward oxygen, which is responsible for their high inhibiting effect, at the same time reduces their practical utility. A freshly formed metallic surface very rapidly undergoes oxidation on contact with ambient air and thus loses its activity. This gives rise to the problem of generating metallic particles in the right form in polymers during high temperature operation.

A large number of chemical processes are known to be suitable for preparing highly active oxygen acceptor compounds or free radicals. Many of these reactions may find application in the preparation of effective polymer stabilisers and may serve as model processes for the investigation of the mechanism of inhibited breakdown of high temperature resistant polymers. The variety of chemical processes involved makes it necessary to take into account the properties of the particular polymer system being considered in choosing the stabilising system.

Of all active inhibitors, finely divided metals and their oxides obtained from organic salts of the metals are of special interest.[14,48,74,75] For example, decomposition of Co, Ni, Cu and Pb formates can follow one of the following paths,

$$
\begin{array}{c}
\text{M}\Big\langle\begin{array}{c}\text{O—C(=O)—H}\\\text{O—C(=O)—H}\end{array} \longrightarrow \left(\text{M}\Big\langle\begin{array}{c}\text{O—C(=O)}\vdots\text{H}\\\text{O—C(=O)}\vdots\text{H}\end{array}\right) \longrightarrow \text{M}+2CO_2+H_2 \quad (45)
\end{array}
$$

$$
\begin{array}{c}
\text{M}\Big\langle\begin{array}{c}\text{O—C(=O)—H}\\\text{O—C(=O)—H}\end{array} \longrightarrow \left(\text{M}\Big\langle\begin{array}{c}\text{O—C(=O)}\vdots\text{H}\\\text{O—C}\vdots\text{H}\end{array}\right) \longrightarrow \text{MO}+CO+CO_2+H_2 \quad (46)
\end{array}
$$

whereas alkali metal formates decompose to give oxalates and carbonates by reaction 47(a) or 47(b) at 300–400°C. Decomposition of these

$$M_2C_2O_4 + H_2$$

$$\overset{(a)}{\nearrow}$$

$$2HCOOM \qquad\qquad\qquad (47)$$

$$\overset{(b)}{\searrow}$$

$$M_2CO_3 + H_2 + CO$$

compounds may be accompanied by secondary reactions such as 48–51 yielding new products. However, the role of side reactions in decomposition of Co, Ni, Cu and Pb formates is small and the basic products are M, MO, CO, CO_2 and H_2.

$$3H_2 + CO \rightarrow CH_4 + H_2O \qquad\qquad (48)$$

$$2H_2 + CO \rightarrow CH_3CHO \qquad\qquad (49)$$

$$H_2 + CO \rightarrow HCHO \qquad\qquad (50)$$

$$2HCHO \rightarrow HCOOCH_3 \qquad\qquad (51)$$

Metals and their oxides can also be easily produced by decomposing oxalates by reaction 52:

$$M(COO)_x \rightarrow M + CO + CO_2 + M_nO_m \qquad\qquad (52)$$

This scheme implies various secondary processes, such as 53–56.[14,76]

$$M(COO)_2 \rightarrow M + 2CO_2 \qquad\qquad (53)$$

$$M + CO_2 \rightarrow MO + CO \qquad\qquad (54)$$

$$M(COO)_2 \rightarrow MO + CO + CO_2 \qquad\qquad (55)$$

$$MO + CO \rightarrow M + CO_2, \quad \text{etc.} \qquad\qquad (56)$$

Oxalates of transition metals, such as iron and nickel, decompose producing pyrophoric products:

$$Fe + 2CO_2$$

$$\overset{(a)}{\nearrow}$$

$$FeC_2O_4 \qquad\qquad\qquad (57)$$

$$\overset{(b)}{\searrow}$$

$$FeO + CO + CO_2$$

$$NiC_2O_4 \rightarrow Ni + 2CO_2 \qquad\qquad (58)$$

whereas alkali and alkali-earth metal oxalates decompose into stable oxides, e.g. by reaction 59.

$$MgC_2O_4 \rightarrow MgO + CO + CO_2 \qquad\qquad (59)$$

Metals and their oxides may be also obtained by decomposing other metal compounds.[48,77-80] This is liable to produce various gaseous substances, apart from metals; for instance, decomposition of formates involves liberation of H_2 and CO which at high temperatures may take part in chain termination reactions during thermal degradation. Besides, hydrogen can recover the metal and also hydrogenate the 'weak points' where there are impurities, thus improving both the initial polymer properties and its heat resistance.[64,81]

Decomposition of carbonates involves formation of oxides that can play the role of active fillers. If the oxide is formed from one of the lower valence transition metals, it can serve as an oxygen acceptor. A classical example is decomposition of Co(II) carbonate yielding cobaltous oxide:

$$CoCO_3 \xrightarrow{330-340°C} CoO + CO_2 \qquad (60)$$

Thermochemical decomposition of the carbonyl, cyclopentadienyl and arene compounds of metals produces pure fine particle metals or easily oxidisable compounds. For example, the decomposition of iron pentacarbonyl yields the so-called 'carbonyl iron' by reaction 61.

$$Fe(CO)_5 \xrightarrow{200°C} Fe + 5CO \qquad (61)$$

A typical arene compound—dibenzenechromium—decays at temperatures above 300°C producing free (pyrolytic) chromium and benzene.[82]

Decomposition of certain other chemical compounds may also yield oxygen acceptors. Thus basic cobalt silicates yield cobalt oxide or divalent cobalt silicates which undergo rapid oxidation in air.[83]

$$Co_3Si_2H_4O_9 \xrightarrow{400-500°C} 3CoO + 2SiO_2 + 2H_2O \qquad (62)$$

$$Co_3Si_2H_2O_{12} \xrightarrow{500-700°C} 3CoSiO_3 + SiO_2 + H_2O \qquad (63)$$

However, water is also a product of reactions 62 and 63 which can itself promote degradation.

Acceptors of oxygen and free radicals can be easily obtained in polymers by reducing various metal compounds to pure metals or lower valence oxides. Many methods of reducing various compounds of Ag, Cu, Fe, Ni and Co with reducing agents such as H_2, CO and CH_4 have been described in the literature. Reducing agents can be obtained in polymers by initiating decomposition of carbonyls, formates, oxalates and many other compounds. Sodium formate decomposition (reaction 47) is one example of such a process.

Solid-phase reduction of metal oxides with graphite has been found to be effective; for example reaction 64 can take place in the temperature range 100–200°C, provided there is a good contact between the reagents.

$$2Ag_2O + C \rightarrow 4Ag + CO_2 \tag{64}$$

Oxygen acceptors generated in polymers may often also be active acceptors of degradation-initiating radicals. It will suffice to name the well-known reactions of Pb, Zn, Sb, As and Bi with alkyl radicals yielding organometallic compounds,[27] for example reaction 65.

$$Pb + 4\dot{R} \rightarrow PbR_4 \tag{65}$$

Radical acceptors may be selected on the basis of the strength of the bonds being formed (e.g. C—M, Si—O—M, etc.) and temperature range in which the polymer will be used. When using metals as radical acceptors one must also take into account the possibility that they may at the same time be chain transfer agents and cleave certain atoms from molecules.[84] The result may be not inhibition but initiation of polymer breakdown.

At relatively high temperatures, chain decomposition processes may be inhibited with such radical acceptors as CO, SO_2, etc. which take part in copolymerisation reactions yielding less reactive radicals (R—$\dot{C}$O, R—$\dot{S}O_2$, etc.) compared with the original alkyl radicals.

Inclusion compounds occupy a special place among stabilising agents.[12,13] In such compounds, the active stabilising material is spatially occluded in the structure of another component and is barred from reacting with oxygen or other reagents when introduced in a polymer. However, when subjected to heat during service, an inclusion compound may decompose yielding the active stabiliser. This method of chemical structure stabilisation of inhibitors and flame retardants such as I_2, H_2O, CO_2, etc., will doubtlessly find extensive application in non-combustible polymeric compositions.

4. STABILISER EFFICIENCY

The efficiency of a stabiliser (inhibitor) may most conveniently be estimated using the ratio:

$$S_\tau = \tau_Z / \tau_H \qquad \text{(iii)}$$

where τ_Z, τ_H are times in which a given specimen characteristic varies by a predetermined magnitude in the presence (Z) and absence (H) of an inhibitor.

Obviously, the higher the experimentally measured values of the ratio (iii), the higher the inhibitor efficiency. When the degradation and stabilisation mechanisms are known, the ratio (iii) may be calculated theoretically with a high degree of accuracy.

Often it is convenient to use simplified qualitative efficiency criteria based on approximate estimates of process kinetics. As an example, we will consider one of the general kinetic schemes (Scheme 1) of oxidative degradation of polymers in the presence of a non-chain inhibitor/oxygen acceptor.[13,26]

SCHEME 1.   Kinetic Scheme for Oxidation Degradation.

1.1 *Initiation*:

$$\text{weak bond} \xrightarrow{k_{in}} 2R^{\cdot}_{prim} \qquad \text{(a)}$$

$$R^{\cdot}_{prim} + RH \xrightarrow{k'_{in}} R^{\cdot} + R_{prim}H \qquad \text{(b)}$$

$$RH + O_2 \xrightarrow{k_0} R^{\cdot} + HO^{\cdot}_2 \qquad \text{(c)}$$

$$2RH + O_2 \xrightarrow{k'_0} 2R^{\cdot} + H_2O_2 \qquad \text{(d)}$$

1.2 *Chain propagation*:

$$R^{\cdot} + O_2 \xrightarrow{k_1} RO^{\cdot}_2 \qquad \text{(a)}$$

$$RO^{\cdot}_2 + RH \xrightarrow{k_2} R^{\cdot} + ROOH \xrightarrow{k_5} \text{inert products} \qquad \text{(b)}$$

$$\downarrow k_3$$

$$\text{products} + 2R''^{\cdot}$$

1.3 *Chain termination*:

$$R^{\cdot}+R^{\cdot} \xrightarrow{\;k_4\;} R{-}R$$

$$RO_2^{\cdot}+R^{\cdot} \xrightarrow{\;k_5\;} ROOR$$

$$RO_2^{\cdot}+RO_2^{\cdot} \xrightarrow{\;k_6\;} products$$

1.4 *Non-chain inhibition*:

$$Z+O_2 \xrightarrow{\;k_Z\;} inert\ products\ (ZO_2)$$

1.5 *Inhibitor recovery*:

$$ZO_2+reducer \xrightarrow{\;k_Z\;} Z+products$$

If k_Z is high enough, conditions can be adjusted so that oxygen will react practically with Z alone and will not take part in any other elementary reactions. Ignoring process 1.5 and using the following values of rate constants of the elementary reactions 1.1(c), 1.1(d), 1.2(a) and component concentrations (250–300°C):[40,85,86]

$$k_0 = 10^{-2}\text{–}1{\cdot}0\ litre/mol^2\ s; \qquad k_0' = 10^{-3}\text{–}10^{-3}\ litre/mol^2\ s;$$

$$k_1 = 10^8\ litre/mol\ s;$$

$$[RH] = 10\ mol/litre;\ [O_2] = 10^{-4}\ mol/litre;\ [R^{\cdot}] = 10^{-8}\text{–}10^{-7}\ mol/litre,$$

the rates of the corresponding reactions with oxygen are:

$$W_{RH+O_2} = k_0[RH][O_2] = 10^{-5}\text{–}10^{-3}\ mol/litre\ s \qquad (iv)$$

$$W_{2RH+O_2} = k_0'[RH]^2[O_2] = 10^{-5}\text{–}10^{-3}\ mol/litre\ s \qquad (v)$$

$$W_{R+O_2} = k_1[R^{\cdot}][O_2] = 10^{-4}\text{–}10^{-3}\ mol/litre\ s \qquad (vi)$$

We will also assume $W_{Z+O_2} = 10^{-2}$ mol/litre s. Calculated for a heterogeneous process these values correspond to the real case of stabilisation at 350°C of siloxane elastomers with pyrophoric iron generated in the polymer on decomposition of iron oxalate[60] (Fe concentration 5 wt%, surface area $S \simeq 10$ m^2/g, oxidation rate constant $k_{Fe} = 10^{-14}$ g/cm^4 s).

It is found that for a number of systems:

$$W_{Z+O_2} \gg W_{R+O_2},\ W_{RH+O_2},\ W_{2RH+O_2} \qquad (vii)$$

If we take into account the fact that the equilibrium O_2 concentration is small in polymers, it will be easy to show that the processes 1.1(c), 1.1(d) and 1.2(a) are practically unimportant until Z has totally reacted with oxygen. It should be noted that in this estimate we have used values of S and k_{Fe} which are somewhat too low. In real systems, pyrophoric metals with a surface area of $100 \, m^2/g$ and upwards can easily be obtained. In addition, the real value of the oxidation rate constant may be much higher than $10^{-14} g^2/cm^4 \, s$. For example, for copper at 350–400°C, $k \simeq 10^{-8} \, g^2/cm^4 \, s$, and the rate of the reaction

$$Cu + O_2 \xrightarrow{} Cu_2O \xrightarrow{O_2} CuO$$ must be as high as 10^2 mol/litre.

When the oxidative breakdown mechanism is known, the efficiencies of stabilisers Z can be evaluated and S_r calculated.[13,14] It is thus easy to derive the stabilisation efficiency criteria for non-chain inhibitors.

It would be of special interest to estimate the efficiency of non-chain inhibition when there are diffusion limitations. Thus in the overwhelming majority of practically important cases of oxidative degradation, diffusion is the controlling factor. During the ageing of a specimen of sufficient mass, the inhibitor InH is spent according to Scheme 1 as oxygen diffuses into the polymer, and the service lifetime of the article is determined by the oxygen diffusion rate. Calculations show that even for polymers characterised by large oxygen penetrability factors (ca 10^{-6} ml cm (cm s atm)$^{-1}$, the lifetime of 0·1–1·0 cm-thick specimens can be as long as many dozens or even hundreds of hours at temperatures between 300 and 400°C.

As an example, consider the case where all the oxygen that has diffused into a system reacts with the active stabiliser Z and the specimen lifetime is determined solely by oxygen diffusion rate.[60–63] Suppose the polymer lifetime is the time over which the specimen is 80% decomposed ($t_{80\%}$). If the specimen is in the shape of a plaque, oxygen diffuses into it from two sides and both reaction 1.4 fronts will move perpendicularly to the plate faces towards the centre (there is no outward diffusion of Z).

In this case eqn. (viii) is valid:

$$P = -D_{O_2}(\partial c/\partial x) \tag{viii}$$

where P is penetrability (the amount of material passing in a unit of time through a unit area of unit thickness for a normal concentration or pressure drop), D_{O_2} is diffusivity and $\partial c/\partial x$ is the concentration gradient. Furthermore

$$P = D_{O_2} \frac{[O_2]_0 - [O_2]_x}{x} \qquad \text{(ix)}$$

where $[O_2]_0$ is the oxygen concentration at the interface (mol/cm^3), and $[O_2]_x$ is the concentration in the specimen at depth x.

Denoting the specific rate of reaction 1.4 (equal to the number of moles of Z that have reacted in a unit of time over the unit area in the reaction zone) as W', we have:

$$W' = [Z] \frac{dx}{dt} \qquad \text{(x)}$$

where $[Z]$ is concentration of Z (mol/cm^3); x is distance (cm); t is time (s).

Since (if we assume that one mole of O_2 reacts with one mole of Z) the reaction rate is equal to the oxygen flow rate within the reaction zone, and $[O_2]_0 \gg [O_2]_x$, we may write:

$$W' = [Z] \frac{dx}{dt} = \frac{D_{O_2}[O_2]_0}{x} \qquad \text{(xi)}$$

$$x \, dx = \frac{D_{O_2}[O_2]_0}{[Z]} \, dt \qquad \text{(xii)}$$

Integration of eqn. (xii) gives eqn. (xiii).

$$\frac{x^2}{2} = \frac{D_{O_2}[O_2]_0}{[Z]} \, t \qquad \text{(xiii)}$$

Thus the movement of the reaction front in a specimen is described by a parabolic function.

We obtain further:

$$t_{80\%} \equiv \tau = \frac{x^2}{2} \cdot \frac{[Z]}{D_{O_2}[O_2]_0} = k \frac{l^2[Z]}{D_{O_2}[O_2]_0}, \qquad \text{(xiv)}$$

where l is specimen thickness, k is a coefficient approximately equal to 0·1 and determined by the stoichiometry of reaction (xii) and the assumptions used.

Considering the relationships between D_{O_2}, $[O_2]$ and temperature, we have:

$$\tau = \frac{kl^2[Z]}{D_0 e^{-E_D/RT} \sigma'_0 e^{-\Delta H/RT}} \tag{xv}$$

where E_D is the activation energy of diffusion, H is the heat of gas dissolution and D_0 and σ'_0 are the respective pre-exponentials.

However, for many polymers ΔH is small (between -2 and $+2$ kcal/mol). Therefore one may take:

$$\tau = \frac{kl^2[Z]}{D_0 e^{-E_0/RT}[O_2]_0} \tag{xvi}$$

Since in polymer compositions oxygen diffusivities may be small enough (of the order of 10^{-8} or even 10^{-9} cm^2/s), the lifetime of polymer products may be as long as several thousands of hours.

Equations (xiv) and (xvi) have been obtained assuming that in the absence of an inhibitor Z the polymer decomposes almost instantly. In the general case the following relationship is valid:

$$\tau = k\frac{l^2[Z]}{D_{O_2}[O_2]_0} + \tau' = k'l^2[Z] + \tau' \tag{xvii}$$

where τ' is a quantity commensurate with the lifetime of an unstabilised specimen.

Equation (xvii) has been derived for the case where there is no stabiliser diffusion, and oxygen concentration variation due to its reaction with the polymer may be neglected. A detailed mathematical model has been derived[58,87] without taking these limitations into account. It was possible to give a theoretical explanation of the discrepancies from eqn. (xvii) for sufficiently thick polymer specimens. It has also been shown that the value τ' may be considered equal to the lifetime of an unstabilised specimen only when dealing with highly effective stabilisers for which $\tau' \ll \tau$. When, on the other hand, the stabilisation effect is small, eqn. (xviii) holds:

$$\tau' = t_b + (2t_b t_D^{(1)} \cdot \alpha)^{\frac{1}{2}} \tag{xviii}$$

where t_b is polymer breakdown time, provided that oxygen concentration is $[O_2]_0$ at all of its points; $t_D^{(1)}$ is oxygen diffusion time equal to l^2/D_{O_2}; $\alpha = [Z]_0/\nu[O_2]_0$, wherein ν is the stoichiometric coefficient of non-chain inhibition (i.e. the number of inhibitor molecules reacting with one oxygen molecule).

If a stabiliser Z is an inhibitor of not only the oxidative but also the thermal breakdown and can react with practically all primary radicals yielding stable compounds of the RZ type (chain inhibition), the simplified kinetic Scheme 2 of inhibited degradation is valid.

SCHEME 2.   Simplified Inhibition Scheme.

(1)   Weak bond $\xrightarrow{k_{in}}$ $R_{prim}^{\cdot}$

(2) $R_{prim}^{\cdot} + Z^{\cdot} \xrightarrow{k_Z''} R_{prim}Z$

(3) $O_2 + Z \xrightarrow{k_Z}$ inert products

Schemes 1 and 2 do not allow for the possible chain transfer through the stabiliser $(R_1R_2 + Z^{\cdot} \rightarrow R_1^{\cdot} + ZR_2)$ and some other elementary reactions. Nevertheless, experimental agreement is satisfactory, indicating that the stabilisation mechanisms discussed closely represent what actually occurs.

Calculation of s_τ for non-chain inhibitors shows that in a number of practical cases this quantity may vary within broad limits, being sometimes very high. As an example, we will make the pertinent calculations for polymers characterised by high gas penetrability when $s_\tau = \tau/\tau' > 10$ and the simple equation (xiv) is valid.

In filled rubbers and other polymeric compositions the oxygen diffusivities at 100°C may be small, of the order of 10^{-8}–10^{-9} cm^2/s or lower.[88,89] The activation energy of gas diffusion into an elastomer in the range 0–100°C has an order of magnitude of 10 kcal/mol, and above 100–150°C, E_D markedly decreases. It therefore follows that in the case of oxygen diffusion into a filled rubber composition at temperatures around 300°C we may take $D_{O_2} = 10^{-3}$–10^{-5} cm^2/s. Let $[Z] = 10^{-3}$ mol/cm^3; $l = 0 \cdot 1$ cm; $D_{O_2,300°} = 10^{-4}$ cm^2/s; $[O_2] = 0 \cdot 5 \times 10^{-7}$ mol/cm^3 (if it is assumed that the article is used in the open air and $[O_2]_{300°}$ corresponds to gas solubility in the sample; $\sigma = 10^{-2}$ cm^3 (cm^3 atm)$^{-1}$. Taking the above values for the pertinent parameters and assuming that Z reacts with O_2 via the reaction: $Z + \frac{1}{2}O_2 \rightarrow ZO$, i.e. one mole of oxygen reacts with two moles of Z, and using an equation of the type (xiv) we obtain (xix):

$$\tau = \frac{10^{-3} \text{ mol/cm}^3 \times (0 \cdot 1)^2 \text{ cm}^2}{10 \times 2 \times 10^{-4} \text{ cm}^2/\text{s} \times 0 \cdot 5 \times 10^{-7} \text{ mol/cm}^3} = 10^5 \text{ s} \approx 30 \text{ h} \quad \text{(xix)}$$

Simple calculations demonstrate that it is possible to predict theoretically the polymer lifetime on stabilisation with highly active agents. However, in the general case they still remain semi-quantitative, because the formulae are derived using simplifying assumptions. It must be remembered, besides, that the calculation of diffusivity and solubility of an active agent in a polymer at high temperatures on the basis of experimental data obtained for temperatures between 0 and 100°C cannot but be qualitative. Thus it has already been noted that the temperature dependence of D does not follow the Arrhenius law over a broad range of temperatures; as the temperature increases, the value of E_D may drop abruptly.[63] Therefore, in order to check whether an equation like (xvi) is valid or not it is necessary to use D and σ values determined for a given temperature. It should be pointed out in this connection that sometimes it is stated that non-chain inhibition is effective only in polymeric materials intended for short-term usage.[25] This statement assumes that stabilisation with conventional antioxidants (chain inhibition) is more effective. It is wrong in principle.

The above typical calculation for a very high coefficient of diffusion and a relatively thin specimen refutes this totally unjustified opinion. In most other polymer compositions the values of D are two to four orders of magnitude lower than that assumed in the calculation, and article thicknesses are usually considerably greater than 0·1 cm. It must be evident from this that when there is only oxidative degradation, the lifetime of most real polymer products stabilised with oxygen acceptors both at low and high temperatures may be as long as several years. This is largely due to the conditions specific to a solid-phase material characterised by low reagent diffusivities. Figure 1 presents lifetimes of polymer articles in the form of plaques of various thicknesses; it is assumed that the specimens are subject only to oxidative degradation (Z concentration = 10^{-4} mol/cm^3; oxygen concentration in specimen is taken to be 5×10^{-8} mol/cm^3; the inhibitor regeneration rate is assumed to be slow, i.e. $f \simeq 1·5$).

In real polymer compositions with $D_{O_2} = 10^{-6}$–10^{-9} cm^2/s, the solubility of oxygen is considerably lower than that assumed in the calculations used in Fig. 1. This means that the calculated values are unduly low in these cases. In addition it is sometimes possible to introduce into or generate in the material a quantity of stabiliser in excess of 10^{-4} mol/cm^3 (nearly 0·1 mol/kg). This latter measure may also be expected to extend the lifetime of the article. Diffusivities of many other chemical agents capable of initiating the degradation of a

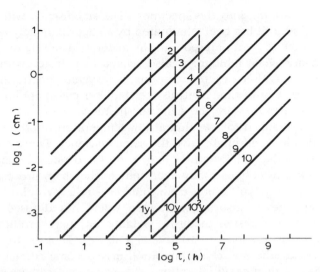

FIG. 1.   Lifetimes of polymer plaques undergoing oxidative degradation: l is plaque thickness (cm), τ is lifetime (h) and y represents years. The straight lines relate to different oxygen diffusivities in the material, D (cm^2/s): 1, 10^{-4}; 2, 10^{-5}; 3, 10^{-6}; 4, 10^{-7}; 5, 10^{-8}; 6, 10^{-9}; 7, 10^{-10}; 8, 10^{-11}; 9, 10^{-12}; 10, 10^{-13}.

The calculations were made using the equation:

$$\tau = f[Z]_0 l^2 / 8[O_2]D_{O_2}$$

for

$$f = 1 \cdot 5, \ [Z]_0 = 10^{-4} \ \text{mol/cm}^3, \ [O_2] = 5 \times 10^{-8} \ \text{mol/cm}^3.$$

polymer are as low as 10^{-12}–10^{-13} cm^2/sec at moderate temperatures,[90] which may be very beneficial for stabilisation efficiency.

In connection with what has been said above it may be interesting to compare the efficiencies of the chain and non-chain inhibition in practically important cases where the reagent (e.g. oxygen) diffusion rate is the limiting factor.

If the acceptor concentration in the material is low or if it is generated during service and its generation rate is not much higher than the rate of oxygen diffusion into the specimen, the efficiencies of both chain and non-chain reactions may be substantially the same. In practice chain inhibition may be effective when the inhibitor InH reacts with the radicals $RO_2^\bullet$; the oxidation chain length being close to unity, the number of incipient chains is commensurate with the amount of oxygen penetrating into the specimen. The stabiliser Z will be spent completely after it has been eliminated by an approximately equivalent amount of the reagent diffused into the material. The rate of consumption of a non-chain inhibitor is also related to its rate of removal by O_2. Thus in both cases the amount of oxygen penetrating into a specimen per unit of time controls the stabiliser consumption rate, and when their quantities are the same the lifetimes of the specimens will be commensurate. It follows that at low temperatures both methods of stabilisation may, all in all, have the same efficiency. However, at high temperatures the stabiliser InH (e.g. substituted phenol, amine, etc.) loses its effectiveness and may even initiate the chain process. At the same time, many typical non-chain inhibitors (e.g. pyrophoric iron) at high temperatures absorb the oxygen which has diffused into the specimen almost instantaneously yielding an inert product (which remains completely inert to 1000°C and above). Thus as the temperature increases, the non-chain inhibition may become commensurately more efficient than chain inhibition. This behaviour is characteristic of the stabilisation of solid organic materials.

There is every reason to expect that in future it will be possible to identify effective inorganic stabilisers capable of inhibiting the thermal breakdown of polymers at high temperatures. There have been a number of suggestions in the literature[14] that this is possible. Remarkable stabilisation may also be achieved by the use of agents which are chemically inactive (fillers) but which lead to changes in the physical structure of polymer. Restructurisation often reduces the diffusivity of attacking agents into a polymer which naturally prolongs its lifetime.

Recently, non-chain inhibitors are beginning to find application as stabilisers against thermal oxidative degradation at low temperatures too; the efficiency of these methods is often not inferior to that of chain stabilisation methods.[91]

5. EXPERIMENTAL VERIFICATION OF NON-CHAIN INHIBITION

Unfortunately, there are few reliable quantitative results available today that allow an unambiguous conclusion about one or another stabilisation mechanism at high temperatures. We will, therefore, mention only those experimental results which support the non-chain mechanism of inhibition of oxidative degradation of polymers. But first it should be noted that the relatively small number of quantitative experiments is no obstacle to drawing general conclusions. The fact is, the non-chain inhibition concept is general enough and when diffusion controls the lifetime of a polymer (and these are the practically important examples) the theory gives very accurate estimates.

Non-chain inhibition by metals, oxides, sulphides and other compounds permits an explanation of facts which it was impossible to explain previously. Thus the stabilising effect of acetylacetonates of cobalt and other metals[14] may be due not only to the termination of oxidation chains but also to O_2 absorption via the non-chain inhibition reaction. In a similar manner, other Co^{2+}-containing systems may convert oxygen to the ionic form and thereby transfer it to the inactive state:

$$Co^{2+} + O_2 \rightarrow Co^{3+} + O_2^- \tag{66}$$

It should be noted that a reaction of the type 66 may often be one of the elementary steps of the chain mechanism, but it may also be part of a non-chain inhibition process. However, the lack of quantitative evidence does not permit its role in stabilisation to be established.

The stabilising effect of certain sulphides or colloidal metals obtained electrolytically in a polymeric or some other matrix can, as has already been noted, be due not only to a chain mechanism but also to a non-chain inhibition reaction $Z + O_2$.[13,14]

We have been able to obtain a number of results which appear to demonstrate the possibility of inhibition of oxidative degradation of polymers via a non-chain process and explain some of the facts known from earlier experiments.

A case in point is the high temperature oxidative stabilisation of polyorganosiloxanes (300–500°C) with such active oxygen acceptors as Fe, Cu, Co, NiO, FeO, etc., generated by decomposition of their oxalates in polymer matrices.[60–62] The following data support the non-chain stabilisation mechanism.

(1) The amount of oxygen acceptor generated in the polymer matrix through oxalate decomposition is consistent with that calculated for reactions 67 and 68.

$$M(COO)_2 \rightarrow M + 2CO_2 \tag{67}$$

$$M(COO)_2 \rightarrow MO + CO + CO_2 \tag{68}$$

(2) As oxygen diffuses into the specimen, all the metal (or oxide)

transforms into the corresponding stable oxide:

$$M + O_2 \rightarrow M_n O_m \qquad (69)$$

$$MO + O_2 \rightarrow M_n O_m \qquad (70)$$

(3)   In the absence of stabiliser the polymer breaks down almost instantaneously, but in the presence of metals it begins to decompose intensively only after all the metal has been spent.

(4)   The theoretically derived formula (xvii) is satisfied with a high degree of accuracy.

(5)   The calculated value of τ agrees satisfactorily with the experimental one.

(6)   The reaction rates of O_2 with RH and $R^{\cdot}$ calculated on the basis of rate constants and component concentrations are much lower under the experimental conditions than the rates of reactions 69 and 70.

(7)   The effectiveness of stabilisation is determined by the ability of a metal or its lower valence oxide to form a stable oxide as well as by the oxygen absorption rate under the experimental conditions.

The data given below are for the stabilisation of temperature-resistant polymers in the presence of a metal or its mixture with the lower valence oxides generated in the polymer matrix due to decomposition of oxalates or other compounds. In what follows it is assumed that oxygen is accepted by the metal–metal oxide mixture which for the sake of brevity will be referred to simply as metal.

Figure 2 shows the variations of elongation at break and tensile strength of a siloxane-based rubber composition. It is seen that the specimens stabilised with finely divided metals retain their properties three to four times longer than if stabilised with one of the best stabilising agents of the system known previously—Redoxide.

Figure 3 shows the relationship between the lifetime of trimethylcyclopolysiloxane films and their thickness in air at 400°C in the presence of finely divided iron.[60] The Figure shows that eqn. (xvii) is satisfied with sufficient accuracy for the system.

Figure 4 shows the results of the determination of the time at which all the stabiliser is spent and the lifetimes of trimethylcyclopolysiloxane-based polymeric compositions stabilised with finely divided iron at 500°C.[13] As the iron was converted to the Fe^{3+} oxide, the ESR signal ($g = 2 \cdot 0$) was seen to increase until time τ_{ESR}.

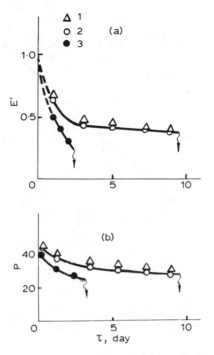

FIG. 2. Variations of elongation at break E' (a) and ultimate tensile strength P (b) of methylphenylsiloxane rubber specimens during thermal degradation at 350°C: 1, stabilised with fine Fe particles; 2, stabilised with fine Cu particles; 3, stabilised with Fe_nO_m (Redoxide). The arrows show the moment of specimen failure; initial values of E' were 200–300%. E' is in relative units and P is in kgf/cm².

Thus τ determined by the 'cracking' technique, i.e. τ_{crack}, by measuring the rate of oxygen absorption, τ_{O_2}, and by the ESR technique, τ_{ESR}, all agree well with each other (Fig. 5).

It should be noted that the degradation of thick films has been found to deviate from the parabolic law of eqn. (xvii), due apparently to the considerable absorption of oxygen by the polymer after the stabiliser has been spent[87] and to stabiliser recovery.

Analysis of published results[14,92] and the new experiments have established that the oxygen acceptor may be recovered during the service life of organosiloxane liquids and polydimethylsiloxane-based coatings in the presence of such oxygen acceptors as Fe, FeO and similar compounds of cerium, cobalt, vanadium, etc. It was found that

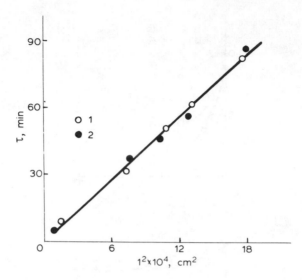

FIG. 3. The lifetime τ of cured trimethylcyclopolysiloxane films stabilised with finely divided Fe ([Fe] = 0·3 mol/kg) as a function of the square of specimen thickness, l^2, at 400°C. The lifetimes of specimens were determined according to crack formation (1) and oxygen absorption (2).

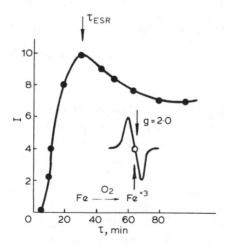

FIG. 4. Determination of the time at which all the stabiliser is spent, τ_{ESR}, in trimethylcyclopolysiloxane films ($l = 160 \, \mu$m, [Fe] = 0·3 mol/kg) by the ESR technique (by following the variation in the signal intensity I (rel. units) at 500°C).

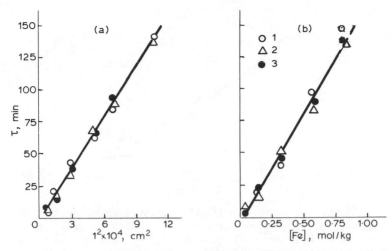

FIG. 5. Lifetimes of trimethylcyclopolysiloxane films stabilised with finely divided Fe in the presence of compounds which liberate H_2 on decomposition as a function of squared specimen thickness (a) and stabiliser concentration (b). The lifetimes were determined by film crack formation (1), oxygen absorption (2) and ESR measurements of stabiliser consumption (3).

formaldehyde formed in the process of oxidative degradation (or in the presence of reducing agents such as H_2, CO, CH_4, etc., when there is shortage of oxygen) regenerates the original stabiliser. For example, the stabilisation with iron oxide in its lowest valence state involves the following reactions:

$$FeO \xrightarrow[\text{acceptance}]{O_2} Fe_2O_3, Fe_3O_4 \xrightarrow[\text{regeneration}]{HCHO} FeO + products \quad (71)$$

The simultaneous use of thermogravimetry and thermal volatilisation analysis is known to be of special interest in accelerated tests[94,95] of stabilisers at high temperatures. We will give one example of preliminary estimation of the efficiency of finely divided metals and their lower oxides recovered from the corresponding formates. The variable valence metal formates are introduced into 10–15 μm polymer films in the dispersed form and are then subjected to a derivatographic investigation with concurrent analysis of the gaseous products.

Figure 6 shows the results of derivatographic analysis of polymethylorganosiloxane specimens; Fig. 7 shows the results of measurements of the degradation rate for this polymer obtained by the

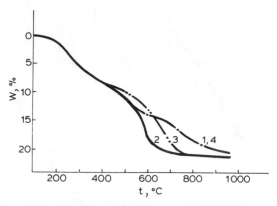

FIG. 6. Results of simultaneous TG/TVA investigation of unstabilised (1, 2) and stabilised (3, 4) polymethylorganosilicate (PMS) specimens in air and helium: 1, PMS in helium; 2, PMS in air; 3, PMS with Fe + FeO generated from iron formate, in air; 4, PMS with Fe + FeO generated from iron formate, in helium. Temperature rise rate $7·5°\,min^{-1}$; specimen weight $7 \pm 0·2$ mg; film thickness 10–15 μm; W is specimen weight variation (%); t is temperature (°C).

derivatographic and chromatographic methods. Consideration of Fig. 7 reveals that air has a substantial effect on the second and third stages of the breakdown of pure polymer (Fig. 7a,a′). Addition of a non-chain inhibitor in the form of a Fe + FeO mixture generated from iron formate in the temperature range 300–320°C has little effect on the rate and nature of degradation in helium (Fig. 7b,b′) but strongly affects the polymer degradation in air (Fig. 7c,c′). Not to overwhelm the reader with too much information which is readily evident from comparison of Figs. 6 and 7, we would note only that these results present just another proof of the high activity of oxygen acceptors as non-chain inhibitors.

Recent studies have been devoted to the mechanisms of formation and degradation of metallopolymer compositions.[95] It has been shown that these materials, like some of the polymeric and other heterogeneous compositions, exhibit useful properties. However, some authors[95] are reluctant to attribute their high thermal oxidative stability to the reactions between active metals and oxygen, i.e. to non-chain inhibition processes. As a rule, these investigations were carried out under conditions where the thermal degradation process could be limited by the diffusion of oxygen and degradation products.[48,95,96] Apparently, the high stability of metallopolymers as well as of metal-stabilised polymers[13,14,97-99] may be attributed not only to the nature of these materials, but also to their stabilisation by non-chain inhibition.

At high temperatures, the non-chain inhibition of oxidative degradation can be effective if pyrophoric oxides of the lower valences of metals such as FeO or CoO, sulphides, finely dispersed silicon and

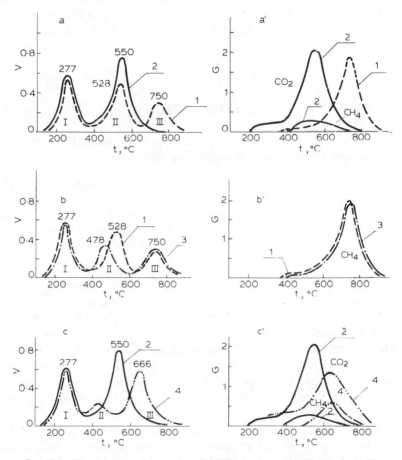

Fig. 7. Results of simultaneous TG/TVA investigations (a, b, c) and chromatographic analysis (a', b', c') of stabilised and unstabilised PMS specimens in helium and air: 1, PMS in helium; 2, PMS in air; 3, PMS with Fe + FeO generated from iron formate, in air; 4, PMS with Fe + FeO generated from iron formate, in helium. Temperature rise rate $7.5°\,min^{-1}$; specimen weight 7 ± 0.2 mg; film thickness $10-15\ \mu$m; V is relative rate of polymer mass loss $(\times 10^2, min^{-1})$; G is relative rate of gas liberation (rel. units).

other oxygen acceptors[13,14] are used. It has been shown recently that stabilisation of organosiloxane coatings with FeO or silicon makes their service lifetimes at temperatures from 500 to 1000°C an order of magnitude longer.[13]

As regards future investigations in this field, the most important results will obviously be obtained through combined kinetic and thermodynamic approaches to the problem of polymer stabilisation at high temperatures. In this way thermally stable (stabilised and filled) compositions possessing improved degradation properties should be obtained.[100]

6. CONCLUSIONS

The stabilisation of polymeric materials may be attained by inhibiting the degradation process in two different ways:

(1) By inhibiting chain processes by adding chain terminating agents; and

(2) By removing from the system the chemical agents (present in, penetrating into or formed in the polymeric composition due to ageing), weak bonds, etc.

The first method involves 'chain' inhibition and the second 'non-chain' inhibition. At elevated temperatures the second method is often more effective. Non-chain inhibition may considerably extend the service lifetimes of many polymeric materials. In solving practical problems, non-chain inhibitors should most conveniently be generated in the polymer matrix itself, whether in the process of its preparation and processing or during service.

Recent results on high temperature stabilisation mechanisms rationalise experimental facts obtained earlier and permit the selection of appropriate stabilisers on a scientific basis.

In many cases effective stabilisation may be obtained by adding fillers to a polymeric composition. This process affects the physical structure of the specimens, and hampers diffusion of attacking agents and their penetration into the materials.

Many earlier publications have been based on assumptions which are sometime erroneous and contradictory. The authors of this chapter have, therefore, restricted discussion to their own experimental work. This does not, of course, mean that all the unquoted papers are not worthy of attention: we are still unable to interpret many interesting and important facts contained in them in an unambiguous way.

REFERENCES

1. SCOTT, G., *Atmospheric Oxidation and Antioxidants* (1965). Elsevier, Amsterdam.
2. VOIGT, J., *Die Stabilisierung der Kunststoffe gegen licht und wärme* (1966). Springer-Verlag, Berlin, Heidelberg, New York.
3. THINIUS, K., *Stabilisierung und Alterung von Plastwerkstoffe*, Vol. 1 (1969). Akademie-Verlag, Berlin.
4. *Stabilization of Polymers and Stabilizer Processes, ACS Adv. Chem. Ser.*, **85** (1968). Am. Chem. Soc., Washington.
5. GRASSIE, N., *Chemistry of High Polymer Degradation Processes* (1959). Butterworths, London.
6. EMANUEL, N. M., *Vestnik Akad. Nauk SSSR*, **27**(7) (1969) 41.
7. EMANUEL, N. M., 'Problems facing fundamental investigations in the field of polymer stabilization and degradation', *4th School on Polymers*, Lecture 1 (1970), State Committee on Science and Technology, Moscow.
8. EMANUEL, N. M. and LYASKOVSKAYA, YU. N., *Inhibition of the Oxidation of Fats* (1961). Pishchepromizdat, Moscow.
9. GORDON, G. YA., *Stabilization of Synthetic Polymers* (1963). Goskhimizdat, Moscow.
10. MADORSKY, S. L., *Thermal Degradation of Organic Polymers* (1964). Wiley, New York, London, Sydney.
11. FRAZER, A. H., *High Temperature Resistant Polymers* (1969). Wiley, New York, London, Sydney.
12. GLADYSHEV, G. P. and POPOV, V. A., *Radikalische Polymerisation bis zu hohen Unsätzen* (1978). Akademie-Verlag, Berlin.
13. SHUSTOVA, O. A. and GLADYSHEV, G. P., *Uspekhi Khimii*; **45** (1976) 1965.
14. GLADYSHEV, G. P., ERSHOV, YU. A. and SHUSTOVA, O. A., *Stabilization of High Temperature Resistant Polymers'* (1979). Khimiya, Moscow.
15. EMANUEL, N. M., in *Encyclopedia of Polymers*, Vol. 3 (1977) pp. 478–86. Sovetskaya Entsiklopedia, Moscow.
16. KORSHAK, V. V., *Chemical Structure and Temperature Behavior of Polymers* (1970). Nauka Press, Moscow.
17. SCOTT, G., *Europ. Polym. J.*, **5** (1969) 189.
18. JELLINEK, H. H. G., Ed., *Aspects of Degradation and Stabilization of Polymers* (1978). Elsevier, Amsterdam.
19. ALLARA, D. L. and HAWKINS, W. L., Eds., *Symp. 173rd Meet. New Orleans, 21–25 March 1977, ACS Adv. Chem. Ser.*, **169** (1978).
20. ARNOLD, C. J., *J. Polym. Sci., Macromol. Rev.*, **14** (1979) 265.
21. SCOTT, G., *J. Appl. Polym. Sci., Appl. Polym. Symp.*, **35** (1979) 123.
22. SHLYAPNIKOV, YU. A., *Dokl. Akad. Nauk SSSR*, **202** (1972) 1377.
23. SHLYAPNIKOV, YU. A., *Uspekhi Khimii*, **50**(6) (1981) 1105.

24. NATANSON, E. M. and UL'BERG, Z. R., *Colloid Metals and Metallopolymers* (1971). Naukova Dumka, Kiev.
25. SOBOLEVSKY, M. V., SKOROKHODOV, I. I., DITSENT, V. E., SOBOLEVSKAYA, L. V. and EFIMOV, V. M., in *Preparation and Investigation of Effectiveness of Chemicals for Polymer Materials*, Vol. 3 (1969) p. 198. Tambov.
26. GLADYSHEV, G. P., *Methods of Stabilization of High Temperature Resistant Polymers* (1972). Institute of Chemical Physics, USSR Academy of Sciences, Moscow.
27. WALLING, CH., *Free Radicals in Solution* (1960). Foreign Literature Publishing House, Moscow (transl. from English).
28. BURNETT, G. M., *Mechanism of Polymer Reactions*, (1954). Interscience, New York, London.
29. FLORY, P., *Principles of Polymer Chemistry* (1953). Cornell Univ. Press, New York.
30. FETTES, E. M., Ed., *Chemical Reactions of Polymers* (1964). Ind. Publ., New York, London, Sydney.
31. DOLGOPLOSK, B. A., ERUSALIMSKY, B. L., MILOVSKAYA, E. B. and BELOMOVSKAYA, G. P., *Dokl. Akad. Nauk SSSR*, **120** (1958) 783.
32. CAWLEY, P. R. and MELVILLE, H. W., *Nat. Bur. Stand. Circ.*, No. 525 (1953) 59.
33. ROZANTSEV, E. G., *Free Iminoxy Radicals* (1970). Khimiya, Moscow.
34. BUCHAHENKO, A. L. and VASSERMAN, A. M., *Stable Radicals: Electron Structure, Reactivity and Applications* (1973). Khimiya, Moscow.
35. GLADYSHEV, G. P., *Polymerization of Vinyl Monomers* (1964). Academy of Sciences of the Kazakh Republic, Alma-Ata.
36. KENYON, A. S., *Nat. Bur. Stand. Circ.*, No. 525 (1953) 81.
37. BERLIN, A. A., *Vysokomol. Soedin.*, **A13** (1971) 276.
38. INGOLD, K. U., *Inst. Petrol. Rev.* **47** (1961) 375.
39. INGOLD, K. U., *Uspekhi Khimii*, **33** (1964) 1107.
40. EMANUEL, N. M., DENISOV, E. T. and MAIZUS, Z. K., *Liquid-Phase Chain Hydrocarbon Oxidation Reactions* (1965). Nauka, Moscow.
41. STEMNITSKI, J., *ASLE Trans.*, **7** (1964) 43.
42. ARCHER, W. and BOZER, K., *Ind. EC Product Res. and Dev.* (1966) 145.
43. RAVNER, H., MONIZ, W. B. and BLACHEY, C. H., *ASLE Trans.*, **15** (1972) 45.
44. KARAKOZVA, E. I., PAUSHKIN, Y. M., KARMILOVA, L. V. and ENIKOLOPIAN, N. S., *Izv. Akad. Nauk SSSR, Ser. Khim.*, No. 2 (1973) 325.
45. NIELSEN, J. M., in *Stabilization of Polymers and Stabilizer Processes, ACS Adv. Chem. Ser.*, **85** (1968) 95.
46. NIELSEN, J. M., *J. Polym. Sci. C.*, **40** (1973) 189.
47. RAFIKOV, C. R., PODE, V. V., SHURAVLEVA, I. V., BONDARENKO, E. M. and GRIBKOVA, P. N., *Vysokomol. Soedin.*, **A11** (1969) 2043.
48. NATANSON, E. M. and BRYK, M. T., *Uspekhi Khimii*, **41** (1972) 1465.
49. JELLINEK, H. H. G., *Kem. Ind.*, **28**(4) (1979) 155.
50. CULLIS, C. F. and LAVER, H. S., *Europ. Polym. J.*, **14**(8) (1978) 575.
51. FRANCIS, C. G. and MIMMS, P. L., *J. Chem. Soc., Dalton Trans.*, No. 8 (1980) 1401.

52. CHUBAR, T. V., KHVOROV, M. M., VYSOTSKAYA, V. N., BOVKYN, Z. G. and BOTINOV, V. A., *Ukr. Khim. Zhurn.*, **46** (1980) 403.
53. BARTON, J. M., *Brit. Polym. J.*, **10**(2) (1978) 151.
54. GRASSIE, N. and MACFARLANE, L. G., *Europ. Polym. J.*, **14**(11) (1978) 875.
55. GIBBS, J. W., *The Collected Works of J. Willard Gibbs*, Vol. 1 (1928). Longmans Green and Co., New York, London, Toronto.
56. DENBIGH, K. G., *The Principles of Chemical Equilibrium* (1971). Cambridge University Press, Cambridge.
57. EMANUEL, N. M., *Uspekhi Khimii*, **43** (1974) 811.
58. BRUN, E. B., KUCHANOV, S. I. and GLADYSHEV, G. P., *Dokl. Akad. Nauk SSSR*, **225** (1975) 1339.
59. EMANUEL, N. M., *J. Polym. Sci., Polym. Symp. Edn.*, **51** (1975) 69.
60. SHUSTOVA, O. A. and GLADYSHEV, G. P., *Dokl. Akad. Nauk SSSR*, **221** (1975) 399.
61. OVCHARENKO, E. N. and SHUSTOVA, O. A., *Vysokomol. Soedin.*, **B17** (1975) 864.
62. GLADYSHEV, G. P., *J. Polym. Sci., Chem. Edn.*, **14** (1976) 1453.
63. GLADYSHEV, G. P., *Vysokomol. Soedin.*, **A17** (1975) 1257.
64. SHUSTOVA, O. A., in *Int. Symp. Chemistry of Organosilicones*, 1975. *Report Summaries*, Vol. II, Part 2 (1975) p. 55, NIITEKHIM, Moscow.
65. MINSKER, K. S. and FEDOSEEVA, G. T., *Degradation and Stabilization of Polyvinylchloride* (1972). Khimiya, Moscow.
66. LEVIN, P. I. and MIKHAILOV, V. V., *Uspekhi Khimii*, **39** (1970) 1687.
67. LEVIN, P. I., *Inhibition of Polymer Oxidation with Stabilizer Mixtures* (1970). NIITHKHIM, Moscow.
68. DENBIGH, K. G. and BEVERIDGE, G. S. G., *Trans. Instn. Chem. Engrs*, **40** (1962) 23.
69. PENSKY, A. V., BABENKO, A. R. and KEFER, R. G., *Izv. Vuzov. Nonferrous Metallurgy Series*, (1) (1973) 37.
70. KITAYEVA, D. KH., TSEPALOV, V. F., GUMARGALIEVA, K. Z. and GLADYSHEV, G. P., *Vysokomol. Soedin.*, **B16** (1974) 501.
71. HAUFFE, K., *Reactions in Solids and on their Surfaces*, (1963). Foreign Literature Publishing House, Moscow.
72. GALWAY, A. and GRAY, P., *J. Chem. Soc., Faraday Trans.*, **68** (1) (1972) 1935.
73. HUGGINS, R., *J. Compos. Mater.*, **4** (1970) 434.
74. SHISHIDO, SH. and MASUDO, Y., *J. Chem. Soc. Japan, Chem. Ind. Chem.*, (1973) 185.
75. KHIMCHENKO, YU. I., VASILENKO, V. P., RADKEVICH, L. S., MYNKOVSKY, V. V., CHUBAR, T. V. and CHEGORYN, V. M., *Poroshkovaya Metallurgiya*, **5** (1977) 7.
76. KATAMIZU, N., *J. Japan. Chem.*, **28** (1974) 74.
77. TUMMAVUORI, J. and SUONTAMO, R., *Finn. Chem. Lett.*, (6) (1979) 176.
78. SMITH, T. W. and WYCHICK, D., *J. Phys. Chem.*, **84**(12) (1980) 1621.
79. YAMAMOTO, Y., *Coord. Chem. Revs.*, **32**(3) (1980) 193.
80. ABEL EWAND STONE, F. G. A., *Organometallic Chem.*, **8** (1980) 181.
81. SHUSTOVA, O. A. and GLADYSHEV, G. P., in *XIth Mendeleev Congr.*

General and Applied Chemistry, Alma-Ata 1975. *Report Summaries* No. 2 (1975) p. 247. Nauka, Moscow.

82. FISCHER, E. and FRITZ, H. P., *Angew. Chem.*, **73** (1961) 353.
83. DALMON, J. A., MARTIN, G. A. and IMELIK, B., *J. Chem. Phys. Phys.-chem. Biol.*, **70** (1973) 214.
84. JACKSON, R., *Essays in Free-radical Chemistry*, Ed. Norman, R. O. C., (1970). Chemical Society, London.
85. DENISOV, E. T., *Rate Constants of Liquid-phase Homolytic Reactions* (1971). Nauka, Moscow.
86. TOBOLSKY, A., *Properties and Structure of Polymers* (1964). Khimiya, Moscow (transl. from English).
87. BRUN, E. B., SHUSTOVA, O. A., KUCHANOV, S. I. and GLADYSHEV, G. P., *J. Polym. Sci., Polym. Chem. Edn.*, **18**(8) (1980) 2461.
88. REILINGER, S. A., *Uspekhi Khimii*, **20** (1951) 213.
89. BEAR, E., *Structural Properties of Plastics* (1967), p. 193. Khimiya, Moscow.
90. STANNETT, V. T., KOROS, W. J., PAUL, D. R., LONGSDALE, H. K. and BAKER, R. W., in *Advances in Polymer Sciences*, Ed. H. J. Cantow *et al.* (1979), p. 69. Springer-Verlag, Berlin, Heidelberg, New York.
91. ELSON, V. G., SEMCHIKOV, YU. D., HUATOVA, N. L., BICHKOV, V. G., TITOVA, S. N. and RAZUVUEV, G. A., *Vysokomol. Soedin.*, **B22**(7) (1980) 494.
92. PAPKOV, V. S., BULKIN, A. F., ZHDANOV, A. A., SLONIMSKY, G. L. and ANDRIANOV, K. A., *Vysokomol. Soedin.*, **A19**(4) (1977) 830.
93. SEVASTIANOV, V. I., OVCHARENKO, E. N, SHUSTOVA, O. A., GLADYSHEV, G. P., *Vysokomol. Soedin.*, **A19**(5) (1977) 1094.
94. GLADYSHEV, G. P. and TSEPALOV, V. F., *Uspekhi Khimii*, **44** (1975) 1830.
95. BAGLEY, N. N. and BRYK, M. T., *Ukr. Khim. Zhurn.*, **42** (1976) 41.
96. NATANSON, E. M. and BRYK, T. M., *Ukr. Khim. Zhurn.*, **36** (1970) 1017.
97. GOLUBKOVA, L. S., SHUSTOV, V. P. and YURKEVICH, O. R., *Dokl. Akad. Nauk Byelorussian SSR*, **20**(3) (1976) 228.
98. SKORIK, YU. I. and KUCHAEVA, S. K., *Zhurn. Prikl. Khimii*, **12** (1974) 2621.
99. ZHILINSKAITE, L. P., MISEVICHUS, P. P. and MACHULIS, A. N., in *Trans. 9th Republican Scientific Conference on Problems of Investigations and Applications of Polymer Materials*, (1968) p. 118. Vilnius,
100. CLARK-MOKS, C. and ELLIS, B., *J. Polym. Sci., Polym. Phys. Edn.*, **11** (1973) 2089.

INDEX